The Biology of Coastal Sand Dunes

The Biology of Coastal Sand Dunes

M. Anwar Maun

OXFORD
UNIVERSITY PRESS

OXFORD
UNIVERSITY PRESS

Great Clarendon Street, Oxford OX2 6DP

Oxford University Press is a department of the University of Oxford.
It furthers the University's objective of excellence in research, scholarship,
and education by publishing worldwide in

Oxford New York

Auckland Cape Town Dar es Salaam Hong Kong Karachi
Kuala Lumpur Madrid Melbourne Mexico City Nairobi
New Delhi Shanghai Taipei Toronto

With offices in

Argentina Austria Brazil Chile Czech Republic France Greece
Guatemala Hungary Italy Japan Poland Portugal Singapore
South Korea Switzerland Thailand Turkey Ukraine Vietnam

Oxford is a registered trade mark of Oxford University Press
in the UK and in certain other countries

Published in the United States
by Oxford University Press Inc., New York

British Library Cataloguing in Publication Data
Data available

Library of Congress Cataloging in Publication Data
Data available

Typeset by Newgen Imaging Systems (P) Ltd., Chennai, India
Printed in Great Britain
on acid-free paper by
CPI Antony Rowe, Chippenham, Wiltshire

ISBN 978–0–19–857035–6 (Hbk)   ISBN 978–0–19–857036–3 (Pbk)

10 9 8 7 6 5 4 3 2 1

This book is dedicated with love to my grandchildren—Kaleem, Nida, Hashir, Ali, Zenab, Bilal, and Maidah.

Preface

Coastal sand dunes around the world are unique systems. There are large variations between continents, but along coastlines many of the selective forces are similar. Because of that certain evolutionary processes converge along different coastlines. This book is a compilation of work done by me and my graduate students over more than 30 years. The painstaking work done by graduate students in spite of the odds, such as burning sun, biting deer flies, albedo from the sand surface and cold numbing sand at harvest time, all added up to create tremendous hardships; however, we persevered. I can assure you that working in sand dune systems is not 'a day at the beach'!

Coastal zones are becoming increasingly topical (and politically sensitive) as they face relentless pressures from urban expansion, recreational development and sea level rise due to climatic change. This timely book provides a comprehensive introduction to the formation, dynamics, maintenance and perpetuation of coastal sand dune systems. It describes interactions between living organisms and the physical processes of geomorphology, and due to the increasingly endangered status of dunes particularly emphasizes conservation and management. A global range of examples enhance the book's international appeal. Included is coverage of the latest methods/techniques and experimental approaches with suggestions for student-based field studies and projects.

This is a basic text for students in ecology and biology of coastal sand dunes. It is intended only to introduce the vast subject of sand dunes along coasts, which we depend on for energy and resources.

This text is also suitable for both senior undergraduate and graduate students enrolled in courses in coastal zone management, marine biology, plant ecology, restoration ecology and conservation biology, as well as for professional ecologists and conservation biologists requiring a concise but authoritative overview of the topic. The book also will be relevant and useful to coastal managers, planners and naturalists.

Publisher's Note

Anwar Maun sadly passed away in September 2007, shortly before completing the manuscript for this book. Dianne Fahselt wrote the final chapter on rising sea levels, and Irene Krajnyk took on the considerable job of checking the scientific nomenclature and collating and proofreading the manuscript. The publisher would like to thank both of them for their hard work in making this a fitting tribute to Anwar's life and work.

Acknowledgements

I am indebted to many who helped over the years and provided valuable discussions and criticism. First I thank my 27 graduate students, both MSc and PhD, who are now working in North American, Europe and Australia. I would like to mention several in particular who completed substantially more work than was required for their degrees: Irene Krajnyk, Anita Payne, Irene Westelaken, Josée Lapierre, Felicite Stairs, Tao Yuan, John Zhang and Peter Baye. I'm also thankful to Mary Leck and Dianne Fahselt for critically evaluating the text before submission (Mary Chapters 3 and 4, Dianne all chapters), Tom Poulson for correcting my thoughts on the theory of succession, Brooke Stephens for preparing the figures using ORIGIN and Ian Craig for converting them to EPS and TIFF. I would also like to express my gratitude to the Natural Sciences and Engineering Research Council of Canada for their support of my research programme over the past 30 years.

Anwar Maun

Richard Gardiner kindly verified the scientific nomenclature of fungi, Anthony Davy reviewed Chapter 13 and Brooke Stephens generously donated her time to help verify the references in Chapter 15. Michelle Turner, Terrence Bell, Jessica Hurley, Amanda Adams, Marc Possmayer and Hollydawn Murray cooperated to free Irene Krajnyk for editing and proof-reading, while Helen Eaton, Ian Sherman and Carol Bestley at OUP greatly facilitated production. Anwar's three daughters, Fakhra, Farzana and Zohra, encouraged him throughout the time he was writing this book.

Irene Krajnyk and Dianne Fahselt

Anwar Maun, sand dune ecologist (1935–2007)

Professor Emeritus in the Department of Biology at the University of Western Ontario, London, Ontario, Canada

Anwar was born in Pakistan on January 1, 1935. He attended Punjab University in Lahore, where he received a Merit Scholarship as well as both a BSc and MSc (agricultural botany) in 1953 and 1958, respectively. In 1961 Anwar won an open scholarship to the American University of Beirut in Lebanon. He earned his second MSc (in crop ecology) in 1963. Winning the Graduate Exchange Scholarship to Washington State University brought Anwar to Pullman, Washington in 1964. Upon completion of his PhD (also in crop ecology) in 1968 Anwar came to the University of Western Ontario as a postdoctoral fellow. In 1971 he was appointed to Assistant Professor rising to the rank of Professor in 1986. Anwar's first introduction to sand dunes was in 1972 during a field trip to Pinery Provincial Park and in 1973 he started his sand dune research programme, which would last for more than 30 years. The aim of Anwar's research was to understand how plants survive and interact in a sand dune environment. He approached the problem from several different fields and points of view, including population dynamics, ecophysiology, and soil–plant interactions. This enabled Anwar to contribute not only to the knowledge of particular processes and species responses but also to a more comprehensive view of the dynamics of the sand dune ecosystem and thus to its conservation and management. A letter from the president of World Wildlife Fund and from several pro-

vincial park officials confirmed that Anwar's research had a strong bearing on management policies for Ontario's sand dune ecosystems. In 1993, Anwar was awarded a Certificate of

Recognition by the Ontario Government for 20 years of dedicated assistance in the development of resource management programmes in the Pinery and Ipperwash Provincial Parks. His work on the vegetation of coastal sand dunes along the Great Lakes earned Anwar a world-wide reputation leading to the award of the prestigious George Lawson Medal by the Canadian Botanical Association in 1997. That same year a student of Anwar's was awarded the J.S. Rowe Award for the best student paper published in 1996 in the Journal of Ecology. In 2003 Anwar was awarded Recognition of 25 years of Excellence in Research by the president of the Natural Sciences and Engineering Research Council of Canada.

During his research career at UWO Anwar supervised the work of 27 graduate students (MSc and PhD) and two postdoctoral fellows. He was the author of more than 100 scientific papers and articles, with the majority focused on sand dune ecology, a field in which he became recognized as a world expert. Anwar travelled widely to present his work, consult, and examine theses for 31 MSc and PhD students from Canada, USA, India, Mexico, and the Netherlands. He spent one sabbatical at the University of California, Davis (1981) and another at the Institute for Ecological Research in Oostvoorne, Holland (1988). For many years Anwar was an Associate Editor of the *Canadian Journal of Botany*, in addition to being on the editorial board of the *Journal of Coastal Research*. He reviewed papers submitted for publication to more than ten other international journals. Anwar was a member of many academic and professional societies: the British Ecological Society, Botanical Society of America, Canadian Botanical Association, Environmental Reviews, Ecoscience, Federation of Ontario Naturalists, McIlwraith Field Naturalists Club, Canadian Water Resources Association Scholarship Committee, and Lake Huron Centre for Coastal Conservation. He served for many years on the Board of Directors for Friends of the Pinery.

When Anwar was appointed to the faculty he was asked to develop, coordinate, and run a new second-year level course in ecology with laboratories and tutorials. The course started in 1972 with an unprecedented enrolment of more than 800 students. At UWO Anwar taught more than 12 different courses and received an Award of Teaching Excellence in Plant Ecology by the UWO University Students Council 1995–1996. Probably Anwar's most important contribution to teaching at UWO was the founding and development of an innovative and successful interdisciplinary Environmental Science Graduate Program (MSc and PhD), bringing together faculty from six departments in five faculties—science, social science, engineering, law, and medicine—of which he was the Director for seven years until his retirement in 2000.

Among his many other contributions, Anwar served as acting chair of the Department of Plant Sciences, spent many years as chair of both the undergraduate and graduate education committees, served on many other major committees both in the department and Faculty of Science and was an enthusiastic supporter of the Honours Ecology and Evolution Program at UWO.

Anwar's interests ranged from photography, woodworking and gardening to the collection of Inuit lithographs, prints of Bikkers, and Pakistani carpets. He also wrote more than ten popular articles about sand dunes. Anwar enjoyed going to the theatre in Stratford (Ontario), observing nature, cooking, and travelling—especially to places where there were sand dunes.

Anwar's love and passion for the dunes was unwavering. During the last few years, in spite of failing health, Anwar visited the dunes several times each year and this gave him tremendous pleasure and happiness. He left many projects unfinished and many pleasures still anticipated. He is missed greatly by his family, friends, and colleagues in North America, Pakistan, and in numerous other parts of the world where his insight, abilities, and spectacular smile earned him enthusiastic followings.

Paul Cavers, Dianne Fahselt, Professors Emeriti and
Irene Krajnyk, Lecturer, Department of Biology

Contents

CHAPTER 1

Geomorphology

1.1 Introduction

Geomorphology is the study of form and structure of sand dunes. Dunes are found in three types of landscapes: sea coasts and lakeshores, river valleys, and arid regions. Coastal dunes are formed along coasts in areas above the high water mark of sandy beaches. They occur in both the northern and southern hemisphere from the Arctic and Antarctic to the equator, and in arid and semi-arid regions (Fig. 1.1). They are very common in temperate climates but are less frequent in tropical and subtropical coasts. Dunes are also common around river mouths where the sand carried in water is deposited (Carter *et al.* 1990b). During floods rivers overflow their banks and deposit sand in river valleys that is subsequently dried by wind and shaped into dunes. In dry regions with less than 200 mm of precipitation per year, the weathering of sandstone and other rocks produce sand that is subject to mass movement by wind because of sparsity of vegetation. There are many similarities in processes and patterns of dune form and structure among these three systems, however each location has its own unique features. In this chapter the emphasis will be on the geomorphology of dune systems along the coasts of oceans and lakes.

Coastal geomorphologists have been attempting to classify the coastal land forms but they defy a simple classification because of tremendous variability in plant taxa, sand texture, wind velocity, climate, sand supply, coastal wave energy and biotic influences including human impact. According to Carter *et al.* (1990b) the great variety of coastal land

forms around the world is primarily related to sediment availability, climate, wave energy, wind regime and types of vegetation. Classification based on these criteria would be more useful in distinguishing between shoreline dune forms than the use of subjective terms—for example white, grey or yellow dunes—sometimes employed by plant ecologists (Tansley 1953). Cowles (1899) said 'a dune complex is a restless maze' because the great topographic diversity depends on changes in the dune terrain from day to day, month to month, season to season and year to year.

There is tremendous geographic variation in coastal development along the sea coasts of the world (Davies 1972). Dune systems include stationary dunes, large hollows carved out by wind erosion, conical dune forms around the base of trees or shrubs and fossil beaches with pebbles on bare sand surfaces. Even though the general advance of a dune is in the direction of prevailing winds, forms advancing in other directions may also be found in the dune system. The embryo dunes close to the sea coast are subjected to the full force of wind, those behind the dune ridges are somewhat protected. The supply of sand varies from place to place and plant species composition and distribution changes at different locations within dune systems. When the sea level rises, embryo dunes may be destroyed, and populations of perennials on the high beach may be lost. However, when the seashore recedes the beach is enlarged and a new bare area becomes available for colonization by plants. The population of one plant species may be completely buried, whereas another one close by may be eroded out of existence. Nevertheless, according to

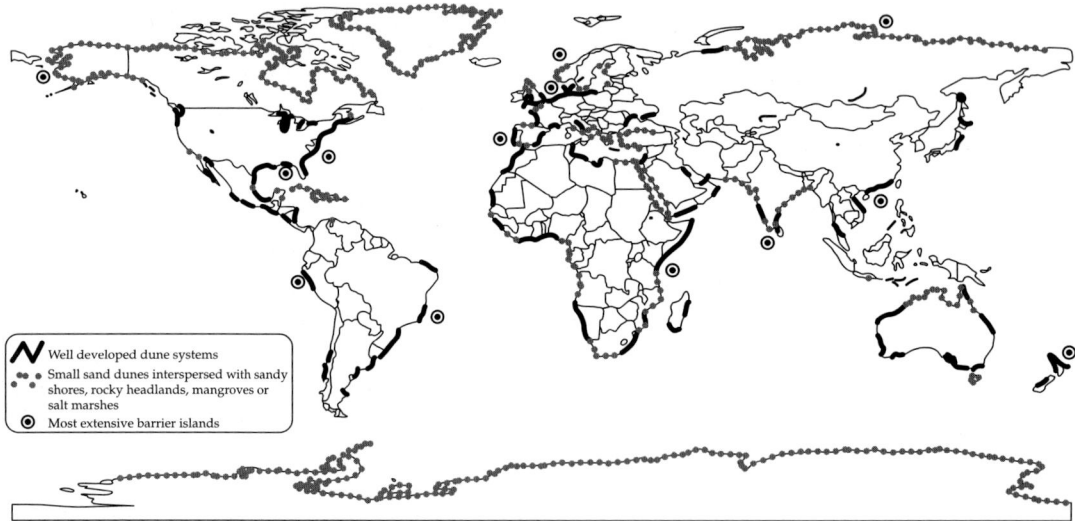

Figure 1.1 Main locations of sand dune systems along sea coasts of the world (Martínez *et al.* 2004) and along the Great Lakes.

Cowles (1899) there is 'simplicity in the complexity', because on the whole a dune complex largely reflects the direction of the prevailing dominant wind, with other environmental stresses of the region playing a secondary role in determining dune geomorphology. A dune complex refers to the open dune systems colonized primarily by grasses, forbs and shrubs and extending from the beach to the forested dune ridges. The light intensity penetrates to the soil level and the area is under the influence of winds. Doing (1985) defined it as a geographical, geomorphological and ecological functional unit with a complete range of dune ridges from the beach to the forest.

1.2 Sediment supply

Sediments available for coastal dune formation originate from two major sources: continental shelves and coastal dune fields such as beaches, dunes and estuaries (Swift 1976). Large-scale changes in coastal supply occur because of fluctuations in sea level. Decline in ocean volume and sea levels occur during glacial periods because of loss of water to

snow and ice and during lowering, sediment is fed from the shelf to expanding beach areas. On low-angle high-energy coasts with abundant sediment, this may result in a sustained sand supply to the shore over several thousand years and create extensive beach and dune ridge complexes. During interglacial periods, ice caps and glaciers melt, ocean volume increases, sea levels rise and large amounts of sediments are released from ice and the erosion of coastal dunes. Following the last major glaciation sea level stabilized at its current level about 6000 years ago. Thus, the dunes store sediment during glacial periods and release it to the oceans during the interglacials, it is eventually reused for the formation of new dune systems. Major sources of sediment for coastal dune systems at present include cliff and coastal erosion, river discharge, and input from tides and washovers.

1.2.1 Cliff and coastal erosion

Fluctuations in sea level redistribute material over wide areas and ultimately determine local levels of wave attack. The erosion of cliffs and bluffs by waves increases the amount of sediments in the oceans, and this sediment is then transported by longshore currents to bays and estuaries where it settles and creates coastal sand dunes. Generally, coastal cliff erosion contributes only about 0.1–0.5% of materials to dune development (Inman 2002; Emery and Milliman 1978), but local situations may differ. For example, in the Great Lakes the most important source for the Pinery sand dunes of Lake Huron in Canada is from the erosion of coastal bluffs (5–30 m in height) and near-shore zone from about 80 km of shoreline from Clark Point to Grand Bend (Fisher *et al.* 1987). The estimated sediment supply from bluffs is about 3800 m^3 year^{-1}, and the recession of shoreline is about 0.4–0.8 m per year. However, only about 14% of this sediment is sand and gravel that will be available for beach development. The cyclic rise in the levels of the Great Lakes every 10–15 years erodes the shoreline dunes, but the subsequent fall in lake levels because of corresponding variations in evaporation, precipitation and seasonal changes in discharge reverse this trend and eventually reduce levels which leads to the rebuilding of sand dunes (Maun 2004).

1.2.2 River discharge

Rivers erode embankments and transport run-off from their watersheds into lakes and oceans, a process increased by deforestation of temperate and tropical forests for timber and agriculture. Rivers were probably a major source of sediment during the early part of the present interglacial period with glaciers carrying huge loads of unsorted materials into the oceans. The river discharge plays a major role in foredune development close to the river mouths and inlets which serve as depositional zones or sinks for the transported sediment. Numerous examples of dune systems are found at river mouths along the sea and lake-shores around the world. For instance, at the Tentsmuir Forest in Scotland, the River Tay nourishes the beach with large amounts of sediments (Doody 1989) and the dune system has been advancing by about 0.5 m per year towards the North Sea.

1.2.3 Tides, hurricanes, and tsunamis

Tidal waves are generated primarily by the gravitational pull of the moon—because of its proximity to the earth—and a smaller pull (0.46 of the moon) by the sun because of its greater distance. Along sea coasts the high and low tides are really the crest and trough of this tidal wave. Even though this pull is exerted uniformly over the earth the land does not show a visible response, but water in the oceans is drawn upwards and forms a bulge. When both the sun and moon are aligned in a straight line there is maximum impact and big upwellings called spring tides are produced. Lower tides called neap tides are created when the moon and sun are aligned at 90° angles to the earth. Large amounts of sediments may be deposited by tidal waves, especially in tidal inlets, or by hurricanes and tropical storms which force immense quantities of sediments from offshore deposits as discrete washover fans or wide terraces over primary dune ridges. In the washover process the tides or storm waves carry sediment from the seaward face and deposit it on the top or leeward side of the dune ridges of the mainland or barrier island as the waves wash over the ridges. Washovers perform a useful function in that they maintain barrier islands by depositing large loads of sand during storms of hurricane intensity and have a serious impact on the survival of plant species. Barrier islands found along coasts at low latitudes are thought to have been formed in three ways: (i) they originate from underwater shoals that emerge above the water surface, (ii) spits may form on headlands that are breached later and (iii) rising sea level encroaches into the foredune and isolates parts of the dune (Schwartz 1971).

Tsunamis (harbour waves) are shock waves generated by submarine earthquakes that permanently uplift the sea floor, thus displacing the water column upwards and increasing the mean sea level. The tsunami potential of an earthquake is determined by many factors such as magnitude, depth, location and the mode of rupture (dip-slip or strike-slip), but the most important parameter is magnitude. The potential energy generated by an earthquake is converted into kinetic energy and transferred horizontally into a tsunami. Once generated a tsunami is split into two with each half moving in the opposite direction, one into the deep open ocean while the other travels in the other direction at speeds equal to the square root of water depth. As the waves travel closer to a land mass and encounter the continental slope their amplitude increases and they become steeper in their run up onshore. Tsunamis travel much farther inland than normal waves and tides because of their immense power and can completely destroy everything in their path when they hit land. The strong earthquake (9.0 on the Richter scale) on 26 December 2004 with an epicentre close to Sumatra generated a very powerful tsunami that caused huge loss of life and property in South Asia. Do tsunamis contribute material for dune building along shorelines? Because they are very infrequent occurrences their role in providing sand to the beaches or removal of beach deposits is sporadic. However, they do bring in large amounts of unsorted material that may kill herbaceous coastal vegetation by burial and woody vegetation by physically uprooting and snapping of tree trunks.

1.3 Properties of sand grains

Solid non-cohesive particles larger than dust whose terminal velocity is greater than the upward currents of air, and smaller than sand grains that can not be moved from the surface by the direct pressure of wind or the impact of other moving grains, are considered as sand (Bagnold 1960). (Terminal velocity is defined as the steady velocity of fall by a sand particle through the air when the forces of gravity and air resistance are equal.) Bagnold said that for a parent material to qualify as sand, it '(i) must be available in large quantities, (ii) must be resistant to chemical weathering, dissolution in liquids and abrasion and (iii) must be strong enough to resist fracture on impact by other sand grains'. However, not all sand conforms to this general definition. For instance, at White Sands National Monument in New Mexico, USA, sand is made of gypsum ($CaSO_4 \cdot 2H_2O$) which is moderately soluble in water and hence can persist only in this arid region. Similarly, beaches of some volcanic islands in Hawaii are made of black sand olivine [$(Fe, Mg)2SIO_4$] which is also readily soluble in water, however, supply is maintained by continuous weathering of more basaltic rocks. Of all natural substances, quartz (crystalline silica) best satisfies the Bagnold (1960) criteria and in many dune systems sand grains are in fact mostly composed of quartz. However, many other substances such as hematite (iron ore, Fe_2O_3), rutile (TiO_2), flint, or sea shell fragments (calcite, aragonite: crystalline forms of $CaCO_3$) may also qualify as sand (Siever 1988). Sand minerals may differ along a given coastline, for instance, along the Florida coast the beaches primarily consist of quartz sand in the north and calcium carbonate and quartz in the south.

Sand texture is determined by passing samples of sand through a series of sieves that have different sizes of holes with class intervals according to the Wentworth scale and recording the weight of each fraction. Most sand grains consist of more round particles because their edges and corners have been gradually rounded by running water or smoothed by abrasion. The generally accepted ranges of particle size classes are: clay < 0.002 mm, silt 0.002–0.020 mm, sand 0.020–0.200 mm, fine gravel, 0.200–2.000 mm and coarse gravel 2.000–5.000 mm (Daubenmire 1974). Shingle is defined as a mixture of pebbles and stones ranging in diameter from 5.0 to 250 mm. In dune systems, the size of sand grains usually ranges from 0.02 mm to 1.00 mm. Because of the very large range in grain sizes, a sand sample is skewed more towards coarse particles. To correct for this Bagnold (1960) suggested transforming the data into a logarithmic scale of $\log_{10}$, however because $\log_{10}$ transformation is not compatible with the Wentworth scale, the data are transformed into negative $\log_2$ of grain sizes which gives finer grain size limits, positive logs and produces positive integers called phi (φ) where:

$$\varphi = -\log_2 \text{ mm}$$

The transformed data are then plotted against phi sizes as a histogram, frequency curve or cumulative curve. For further elaboration of the technique see Pethick (1984). At present φ is the universally accepted method of presentation of grain size data.

The freshly deposited sand on a beach contains both fine and coarse material. Wind sorts this material and removes finer grains of sand, leaving behind coarse gravel and pebbles that may act as a veneer to prevent further deflation. Thus the texture of sand on sand dune

ridges is typically finer than the beach and slack. A $\log_{10}$ diagram of sand from the Pinery sand dunes (Maun 1985) showed that 17% of the sand particles were fine, 72% were medium, and 11% were coarse (Fig. 1.2). The peak particle diameter was 0.28 mm (medium in texture) and the rates of change on either side fell off rapidly from the peak, even though the rates of change were different. These values indicate similarity of texture of Pinery sand dune ridges with those of other sand dunes in Europe (Ranwell 1972) and North America (Cooper 1958). Salisbury (1952) presented data on sand textures of three coasts. The Portuguese coast has relatively coarse textured material probably, because of gale force winds with a long fetch (the distance over which wind has blown in the same direction over water) of the Atlantic coast. Wind velocity is lower on the Lancashire coast of the UK and coarse sand fraction is considerably less than Portugal but the proportion in the 0.3–1 mm fraction is almost identical on both coasts (Table 1.1). At Blakeney Point in Norfolk, England, the peak diameter was between 0.25 to 0.3 mm. Along the Pacific coast the peak diameter ranged between 0.25 and 0.5 mm at the northern site and between

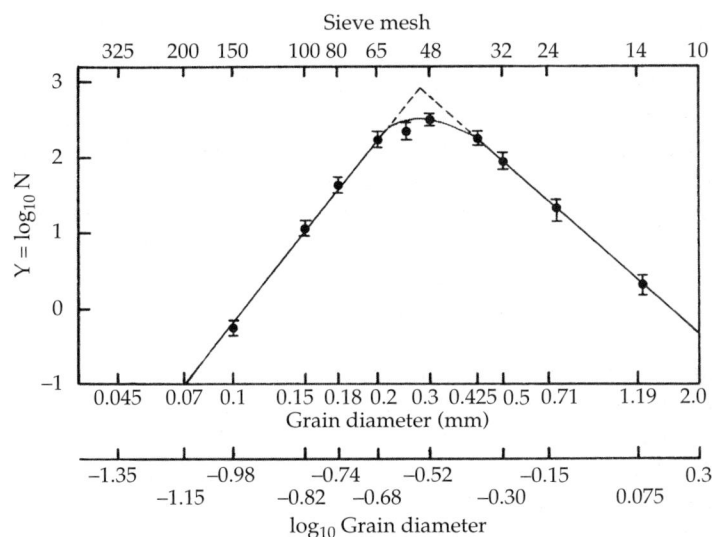

Figure 1.2 $\log_{10}$ diagram of sand texture of sand samples from the upper beach of Lake Huron sand dunes (after Maun 1985).

Table 1.1 Particle size distribution (%) of sand samples collected from three sites along the Atlantic coast in Europe and two sites along the Pacific coast of California

Grain diameter (mm)	Atlantic coast*			Pacific coast[†]	
	Composta, Portugal	Southport, Lancashire	Blakeney, Norfolk	Bodega Head	Trancas Beach
>1	8.1	1.2	0	0.7	1.1
Coarse					
0.5–1	—	—	—	5.6	25.4
0.3–1	89.8	90.5	17.9	—	—
Medium					
0.25–0.5	—	—	—	79.8	62.0
0.25–0.3	2.0	3.0	68.9	—	—
0.2–0.25	—	3.0	8.9	—	—
0.10–0.25	—	—	—	13.5	11.2
Fine					
<0.2	—	2.0	4.1	—	—
<0.1	—	—	—	0.3	0.3

* Salisbury (1952); [†] De Jong (1979).

0.56 and 1 mm at the southern site with a high-energy sea coast (De Jong 1979).

Compression of dry sand by walking on certain seashores produces a soft sound, a squeak or whistle usually called singing sand. The cause of this sound is not clearly understood (Bagnold 1960) but it may be caused by rubbing together of round sand particles against each other.

1.4 Onshore deposition

Sediments discharged from rivers, cliffs and elsewhere are sorted by grain size and reworked by waves. The fine particles of silt and clay remain in suspension and are carried further into the lake or ocean than sand and gravel that have greater mass: these are deposited nearer to shore. The longshore currents generally approach a shoreline at an oblique angle and impart a net flow of water and sediment in the same direction as the longshore drift. The deposited material is carried along the coast and then cast on the beach in the swash (the rush of water up the beach).

Waves are created by the energy contained in the winds, which in reality is solar energy that has been transformed into mechanical energy of wind. The still water of a lake or ocean is deformed by wind and transformed into the potential energy of waves. These waves are the principal source of energy input into the littoral zone. Water movement within waves consists of rotational movement of each water molecule in an almost closed circular path (Pethick 1984). At the wave crest, particles are moving in the same direction as the waves but in the troughs they are moving in the opposite direction. Thus waves transfer energy across the water surface with little or no significant net transport of water with the waves. Energy in the waves is proportional to the wave height, which is the vertical distance from the wave crest to the wave trough. Height in turn is directly related to wind velocity, wind duration and fetch.

1.5 Sediment movement

1.5.1 Sand grain movement by waves

As the waves enter shallow water, their height increases and the water depth decreases. The orbital movements of water molecules become ellipses. In even shallower water, the ellipses become flattened into merely one-dimensional horizontal trajectories and the orbital velocity (time taken by a water molecule to move from crest to trough and back to crest) of water molecules starts to exceed the critical threshold value of sediment movement. Sand grains are held on the sea bed by the force of gravity and the friction between grains and water molecules. In order to move a grain of sand the wave must exert a certain threshold force to lift it (the buoyancy component), and also overcome the friction. In deep water the frictional force of waves on sand particles is very low but as water depth decreases the frictional force increases. This exerts a shearing force or shear stress designated as τ (tau) on the seabed which is directly related to the wave energy and must be above a certain threshold value to move the sand grains. As soon as that shear stress is reached the grains are dislodged and are carried forward in the swash to the beach. The onshore velocity of swash is reduced by beach slope until all its wave energy is dissipated, at which point water starts to move backwards in backwash under the force of gravity. During their progress the waves change wave length (horizontal distance between successive crests) and amplitude and at some critical threshold of water depth relative to wave height (vertical distance between crest and trough of a wave) it breaks, thus spilling the water and forming a bore which runs up the beach face. The relationship for breaking of a wave was shown by Galvin (1972) as:

$$\text{Gamma}\ (\gamma) = \frac{H}{d}$$

where H = wave height and d = depth of water at breaking.

The ratio between wave height and water depth (γ) ranges from 0.6 to 1.2 with a mean of 0.78 and is related to beach slope (Pethick 1984). If a beach is steep, the γ values will be high, while on flatter beaches, the values will be lower. In other words, low waves will break in shallower water than waves of higher amplitude. For instance, on an average beach slope, a wave of 1.2 m height may break in

water depth of about 1.5 m. In contrast, on a steep beach, waves will travel right up to the seashore before breaking because water is still quite deep even a few metres from shore. On gently sloping beach profiles along sandy beaches the waves break at considerable distances from seashore. Waves break in different ways depending on wave height and wave period (the interval of time taken by a succession of waves to pass through a stationary position). Four basic types of breakers: spilling, plunging, collapsing and surging (Open University 1989) have been proposed by Galvin (1968) (Fig. 1.3). Spilling waves usually occur on gently sloping flat beaches and break at a considerable distance from shore. They produce foam and turbulence at the leading edge. Plunging breakers have a concave front and a convex back. The crest of the wave curls over and plunges downwards dissipating its energy within a very short distance on shallow to intermediate beach slopes. Collapsing breakers are similar to plunging types but instead of curling the front face of the crest collapses. This happens on moderately steep slopes. Surging breakers are found on very steep

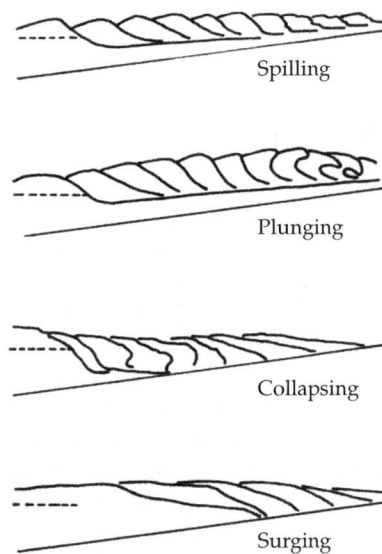

Figure 1.3 Four types of wave breakers arriving on a sea coast (after Galvin 1972).

beaches and are formed by flat low waves with relatively unbroken front faces and crests that gently slide up and down the beach.

Steeper waves (spilling and plunging) dissipate their energy over a relatively restricted area, their swash is weak and the backwash takes away most of the sediment back to the sea. In contrast, with less steep surging and collapsing waves swash is stronger than backwash, and thus deposits large amounts of sediments on the beach (Open University 1989). In other words, beaches are built up by low gentle waves and torn down by storm waves. At the land–water interface, orbital movement becomes a significant force in moving the grains of sediment both onshore and offshore. There is net movement of sand grains on a beach when the onshore velocity of water movement is greater in magnitude than the offshore movement. Shingle beaches consisting of a mixture of pebbles and stones are formed in a similar manner on exposed shorelines by storm waves. The strong swash casts heavier pebbles and stones on shore and drags lighter finer material seaward in the weaker backrush of water. The beach slope is usually greater on shingle beaches because shingle is more porous than sand and absorbs most of the water, thus reducing the backwash.

1.5.2 Sand grain movement by wind

Once sandy sediments are on a beach, wind is the predominant force in moving them inland and initiating dunes. Material deposited by the waves is dried by solar radiation, wind and drainage and then transported by wind from the mid- and upper beaches to the foredunes and first dune ridge. Wind velocity is the most important factor controlling the rate of sand transport, however the variation in sediment deposition in time and space is primarily dependent on beach width (Davidson-Arnott and Law 1990, 1996). In the process it causes sandblasting of objects it encounters in its path and increases the transpiration of plants by removing the boundary layer of air over leaf surfaces. The physical movement of sand grains by wind is a complex process that involves suspension, saltation and surface creep. Each of these components is described below.

1.5.3 Suspension

The size of grains in the deposited material whose terminal velocity of fall is less than the upward eddy currents of air within the average velocity of wind are carried up into the air as suspended particles and scattered as dust (Bagnold 1960).

1.5.4 Saltation

The major force moving sand on coastal sand dunes is saltation. In the sand–wind interaction, the air stream has a velocity profile and applies a certain force U_* (shear velocity) on the sand grains lying on the surface. Larger values of U_* are created by an increase in wind velocity and/or an increase in surface roughness. When the shear velocity (U_*) exceeds a certain threshold value called $U_{*critical}$, sand grains are ejected into the wind stream. The $U_{*critical}$ of wind depends on the square root of grain diameter. When an ejected grain rises into the air, it is moved forward by wind until it reaches the same velocity as the wind. At the same time it is acted upon by gravity and starts to lose height while being propelled forward by wind, so that its trajectory (Fig. 1.4)

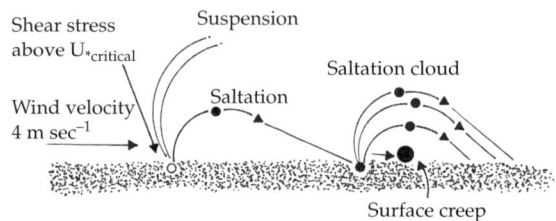

Figure 1.4 The processes of suspension, saltation and surface creep initiated by wind velocity of ≥ 5 m sec^{-1} that increases the shear stress on sand particles above the threshold ($U_{*critical}$) and ejects sand particles into the air (after Bagnold 1960).

becomes elongated and curved and strikes the ground at an angle of about 10–16° depending on the speed of wind, grain diameter and the height of its rise (Bagnold 1960). As a saltating grain impacts the surface it starts a chain reaction in which each grain causes the ejection of several other grains into the wind stream, thus accentuating the sand-moving process. For any given sand texture, the greater the wind velocity the greater the height to which a grain rises. A grain rising to a greater height will strike the sand surface with a greater force when it falls. Once initiated the process of saltation continues until the entire surface of exposed sand has been mobilized.

1.5.5 Surface creep

Some sand particles are too large to be ejected by saltation. However, if they are bombarded with sufficient force by saltating grains they are pushed forward along the sand surface in a process known as surface creep (Fig. 1.4). A high-speed grain of average diameter ejected by saltation can move a grain six times its diameter or more than 200 times its own weight (Bagnold 1960). Wind moving over the sand surface experiences a frictional drag which decreases the velocity of wind. This decrease in velocity near the surface is transmitted up through the flow as turbulent eddies exchange the slow-moving air below thus

producing a velocity profile. Bagnold (1960) showed that the mean velocity profile (v) as a function of height (z) is logarithmic rather than linear as in laminar flow. Plots of velocity profiles on the arithmetic (Fig. 1.5b) and logarithmic scales (Fig. 1.5a) reveal that the velocity of wind when plotted on the log scale decreases to zero at a certain height above the sand surface. This height is greater than 0 and was called the surface roughness constant, k by Bagnold (1960) and z_0 by Olson (1958b). It has been found that z_0 is approximately equal to 1/30th the diameter of the grain particles on a flat beach with no sand movement.

When steady winds blow over the sand surface, the drag velocity V_* is directly proportional to the rate of increase in wind velocity with log of height. Thus the mean wind velocity v as a function of height was expressed by Bagnold (1960) as:

$$v = 5.75 \ V_* \log z/z_0$$

where 5.75 is proportionality constant for $\log_{10}$, and z is height above the sand surface. A knowledge of V_* and z_0 defines the state of the wind, because z_0 denotes the point on the y axis at which the wind velocity is 0 and V_* determines the slope of the velocity. The slope will increase with an increase in V_* but z_0 remains the same. However, z_0 decreases with increasing wind velocity (Deacon 1949). His measurements showed that at wind velocities

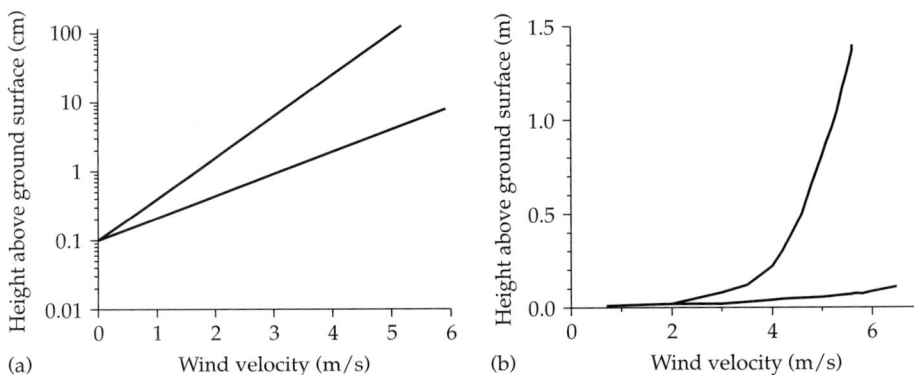

Figure 1.5 Velocity profiles over a bare sand surface plotted on (a) log and (b) arithmetic scales (after Bagnold 1960).

of 1.5, 2.5, 3.3, 4.8 and 6.2 m/sec at 2 m above long grass the z_0 values were 9, 8.2, 6.1, 4.4 and 3.7 cm, respectively. This decrease in z_0 was caused by the bending of grass blades until they became parallel to the flow of wind with a consequent decline in plant height and resistance.

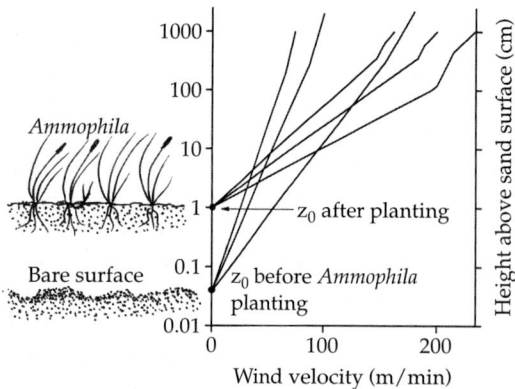

Figure 1.6 Velocity profiles over a bare sand surface before and after the planting of *Ammophila breviligulata* on Indiana Dunes, Lake Michigan. The z_0 over the bare surface is 1/30 of grain diameter. Planting of *Ammophila* raised the z_0 to 1 cm above the sand surface (after Olson 1958b).

1.5.6 Effects of vegetation

Plants increase roughness of sand surface and absorb the energy of saltating grains. Olson (1958b) compared the surface roughness constants z_0 of grain particles on a flat beach with no sand movement with that of a planting of *Ammophila breviligulata*, and found that z_0 was more than 30-fold in the vegetation (Fig. 1.6). The increase in z_0 means that the saltating sand grains will have to rise above the minimum threshold values of z_0 of a species to continue the saltation process. The proportion of grains that can succeed decreases and the sand movement declines. An increase in wind velocity increased the slope of the line but did not change the surface roughness constant. In other words, the drag velocities converged to the same z_0. According to Bressolier and Thomas (1977) the three factors controlling z_0, in order of importance, are plant density, plant height, and wind velocity. Several authors determined the surface roughness constants of some dune species and grasses of differing heights (Table 1.2). The greater the density, within or between species, the greater was the z_0. Thick grass of the same height as thin grass

Table 1.2 Surface roughness constants (z_0) of different surfaces and plant species under experimental or field conditions

Plant species	Plant height (cm)	z_0 (cm)	Authority
Euphorbia paralias	?	2.5	Bressolier and Thomas (1977)
Elymus farctus	?	16.7	Bressolier and Thomas (1977)
Ammophila breviligulata	?	1.0	Olson (1958b)
Sand surface	0	0.05	Bagnold (1960)
Grass (light wind)	60	9.0	Deacon (1949)
Grass (strong wind)	60	4.0	Deacon (1949)
Grass (mown)	3	0.7	Deacon (1949)
Grass (mown)	1.5	0.2	Deacon (1949)
Mudflat and ice	0	0.001	Sutton (1953)
Lawn grass	1	0.1	Sutton (1953)
Thin grass	10	0.7	Sutton (1953)
Thick grass	10	2.3	Sutton (1953)
Thin grass	50	5.0	Sutton (1953)
Thick grass	50	9.0	Sutton (1953)

had greater z_0. *Elymus farctus* had a z_0 of 16.7 cm compared to 2.5 cm for *Euphorbia* species and 1 cm for *A. breviligulata*, mainly because of differences in morphology and density. Tall thick or thin grasses had significantly greater z_0 than short thick or thin grasses (Sutton 1953). Grasses of the same height exposed to strong winds had lower z_0 than those exposed to light winds (Table 1.2). Vegetation also absorbs energy of saltating grains when they impact the leaves and stems of plants, thus reducing the energy output of the saltation cloud to less than the energy input by wind. As the rate of sand transport decreases there is an increase in deposition around the plants.

1.6 Sand ripples

On a beach sand ripples are formed roughly perpendicular to the wind direction by the repeated impact of grains in saltation and surface creep. Since the sand grains are not uniform in texture, they are sifted by saltation and surface creep according to size gradations and thus regularly spaced asymmetrical ridges are formed (Fig. 1.7). The profile of a ripple has four elements, stoss slope, crest, lee slope and trough (Pye and Tsoar 1990). Usually the stoss and lee slopes have an inclination of about

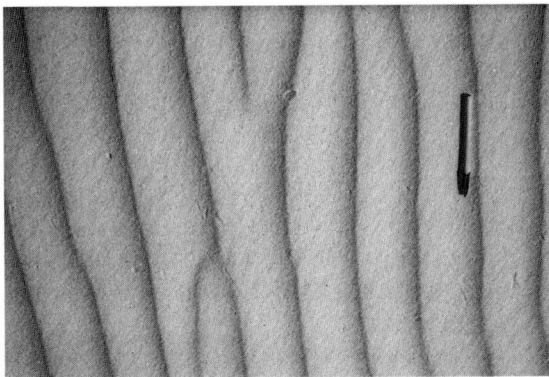

Figure 1.7 Ripples in sand at White Sand National Monument, New Mexico, formed by the repeated impact of grains in saltation and surface creep (photograph by A Maun).

8–10° and 20–30°, respectively. The coarsest material collects at the crests and the finest in the troughs. The ripple length measured as distance from crest to crest ranges between 2 and 12 cm depending on the velocity of wind. The ripple length increases with an increase in wind velocity until the velocity is so high that the ripples flatten out and are erased, leaving a smooth flat surface.

The sea bed in the intertidal zone has wave-generated ripples of various sizes and forms. The shear stress of wave energy dislodges and lifts sand grains into the water column when the shear stress by waves for grain movement exceeds a certain critical value. The ripples develop by the alternating back and forth flow of water over the sea bed. As a wave crest passes towards the seashore sand grains are moved forward to the seaward side of the ripple crest, when the wave retreats the grains are moved to the landward side of the crest, thus producing ripples. When waves travel at high speeds the shear stress on the sea bed is high, the sediment moves as a suspended sheet backward and forward over the sea floor and ripples disappear.

In the air ripples are created by the repeated impacts of grains in saltation and surface creep. In water saltation is of no consequence because the settling velocity of grains in water is 60–80 times lower than in air (Pye and Tsoar 1990) and does not dislodge other grains.

1.7 Dune morphology

For dune formation on a beach three essential requirements must be fulfilled: (i) a prevailing onshore wind above the threshold wind velocity, (ii) a continuous supply of sand, and (iii) an obstacle to reduce the velocity of wind to capture the sand load carried by the saltating cloud. All three conditions vary on different beaches around the world and the sand dune formation varies according to the prevailing conditions. The most important factor in dune formation is beach width in relation to beach morphodynamics and lake levels

(Davidson-Arnott and Law 1990). A sand
dune is a mound or a hill of sand that rises
to a single summit (Bagnold 1960). The fundamental dune shape is dictated by the life form
of colonizing plant species and their inherent
ability to grow vertically and horizontally in
response to burial by sand. Dunes are wind-
blown deposits of sand that have been fixed
by vegetation beyond the farthest inland reach
of waves. They may range from convex dune
ridges to relatively flat terraces. Several types
of sand dunes have been recognized.

1.7.1 Shadow dunes

Shadow dunes are incipient foredunes formed
by plants on dunes. On a beach when the saltating cloud of sand encounters obstacles of
living (plants) or non-living material (driftwood, rocks, flotsum and jetsum) the wind is
deflected to the sides of the obstacle whereby
the direction of travel by wind and sand load
carried by it do not coincide (Bagnold 1960).
On the windward sides of the obstacle and laterally wind velocity increases and erodes sand
while on the lee side the velocity decreases,
thus creating a wind shadow. In a wind tun-
nel experiment using *Ammophila* plants, Hesp
(1981) showed that for a 7 m sec^{-1} wind at the
centre line in front of the plant, the velocity
decreased to 0 m sec^{-1} at 2–4 cm, 0.27 m sec^{-1}
at 40 cm, 1.9 m sec^{-1} at 60 cm and 4 m sec^{-1} at
90 cm, to the lee of the plant. While the wind
shadow outside the air stream is smooth, it
consists of swirls and vortices inside (Bagnold
1960). Initially the sand load is dumped in the
relatively stagnant air in front of the plant but
as the plant gets buried, more and more of
this sand is carried to the centre line on the
lee of this plant where the two opposing vortices of air meet (Hesp 1981). Gradually, a small
dune develops around the obstacle with a
steep windward side and gently sloping long
streamlined ridge on the leeward side called a
shadow dune (Fig. 1.8). As mentioned earlier,
shadow dunes may be formed around non-
living objects such as driftwood. However,

such dunes are short-lived because as soon
as the driftwood is buried normal flow of the
wind profile resumes and sand deposition
ceases. Annual plant species such as *Cakile
edentula, C. maritima, Salsola kali, Corispermum
hyssopifolium* and biennials, namely *Artemisia
caudata, Xanthium strumarium* and other species on other shorelines also form shadow
dunes, however, they too are ephemeral and
sand is dissipated when the plants reach
maturity and die.

1.7.2 Embryo dunes

When sand is deposited around perennial
plant species, the dunes are long-lived. The
effect on sand transport and deposition is the
same as for annuals and biennials but the perennials have the ability to grow horizontally
and vertically in response to burial. They may
produce shadow dunes or low unconnected
mounds or hummocks of sand—embryo dunes
about 1 to 2 m in height. The morphology of a
dune is dependent on the plant species, plant
architecture, density, height, sediment texture,
wind velocity and the rate of sand transport
(Hesp 1989). The growth form of a large number of grass species that colonize shorelines
of North America, Europe, Africa, Australia,

Figure 1.8 Shadow dunes formed around plants of *Cakile
edentula* var. *edentula* along the Northumberland Strait of
Prince Edward Island, Canada (photograph by A Maun).

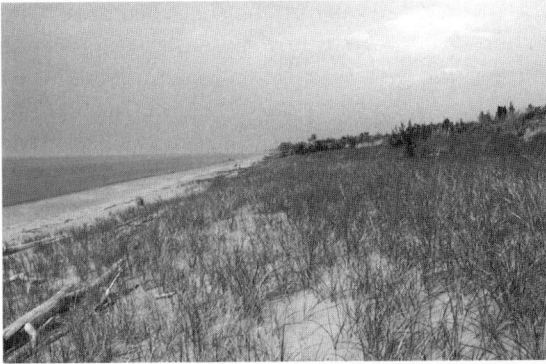

Figure 1.9 Population of *Ammophila breviligulata* along the shoreline of Lake Huron showing the formation of a more porous deposition of sand and creating a more even structure of the dune (photograph by A Maun).

Figure 1.10 Clumps of *Calamovilfa longifolia* forming convex dome-shaped dunes individually. They grow in concentric circles and eventually coalesce to form a dune ridge (photograph by A Maun).

Asia and tropical and subtropical coasts of the world is determined by the production of rhizomes or stolons. At intervals of 5–15 cm along rhizomes or stolons the plants produce discrete shoots, thus enabling the expansion of a dune. Low linear sinuous ridges result with wide empty spaces in between (Hesp 1989). The density of culms is low initially but it increases over time and small embryo dunes eventually coalesce into continuous ridges. According to Hesp (2002), high-density culms of *Ammophila breviligulata* and *A. arenaria* reduce air flow and produce dunes with a gentle seaward slope (Fig. 1.9). Caespitose growth forms such as those of *Calamovilfa longifolia, Uniola paniculata, Andropogon scoparius* (= *Schizachyrium scoparium*), *Panicum virgatum* and *P. amarum* expand in concentric circles by forming new culms at the periphery each year. The density of their culms may be quite high which leads to the formation of discrete convex domes following sand deposition (Fig. 1.10). However, if wind velocity is high for an extended period and lower leaves become covered by sand a shadow dune may be created.

Seneca *et al.* (1976) compared three species for their ability to accumulate sand along the North Carolina coast and showed that

A. breviligulata was more efficient in trapping sand than *Uniola paniculata*, and that after a 24-month period both of these species were more efficient than *P. amarum*. The amount of accumulated sand after a 27-month period was 9.5 m^3 m^{-2}, 8.2 m^3 m^{-2} and 4.3 m^3 m^{-2} of dune surface for *A. breviligulata, Uniola paniculata* and *P. amarum*, respectively (Seneca *et al.* 1976; Woodhouse *et al.* 1977). Similarly, with Great Lakes dune builders, *A. breviligulata* was more efficient in trapping sand than *C. longifolia* and both species were more efficient than *Andropogon scoparius* (Maun 1998). Hesp (2002) showed that plant species with low canopies such as *Spinifex, Ipomoea* and *Canavalia* produce dunes with a short slope on the leeward side.

Some shrub and tree species are also important dune builders. For example, *Salix interior, S. cordata, S. glaucophylloides, S. repens, Prunus pumila* and *Cornus stolonifera* possess traits similar to those of grasses. When these shrubs are buried they produce new roots along the buried shoots and the dune starts to grow in area and height. *Populus balsamifera* and *P. alba* produce root suckers which grow laterally below-ground and produce new stems some distance from the parent plants which when buried by sand also have the ability to produce new roots higher up on the buried stem, thus increasing dune height over time. The areal extent of an embryo dune is dependent on the lateral spreading ability of dune-forming plants, while the relative dune height depends on the capacity for upward growth. However, the growth pattern and ecological response of plants to burial in sand dictate topographic dune forms. For example, *Agropyron psammophilum* dunes are low, seldom more than 1 m in height, *Ammophila* dunes are broad and may be many metres tall and *Populus* dunes are tall. Dunes formed by *Salix* spp. are broader than *Prunus pumila* dunes because of its ability to expand laterally, however dunes formed by both species are tall and steep owing to their pronounced capacity to grow upwards in response to burial. In cold temperate climate regions above-ground shoots of dune-building grasses (geophytes) die in late autumn but they are still efficient in trapping sand. Nevertheless, substantial quantities of sand may be transported from incipient embryo dunes to the main dune ridges in winter.

1.7.3 Established foredunes

As incipient foredunes are colonized and stabilized by vegetation soil development is initiated. Dunes increase in height, expand horizontally and eventually coalesce, to form an established dune ridge usually parallel to the shoreline. At Magilligan Point, Northern Ireland, the transition from an incipient to an established foredune took 5 to 8 years (Carter and Wilson 1990). However, the time for transition will vary according to the sediment supply, wind velocity, climatic conditions, composition of vegetation, plant height and density, and human impact. Dune ridges are initially stabilized by a complete cover of vegetation usually consisting of monospecific stands of *A. breviligulata* in North America or *A. arenaria* in Europe. However, with the passage of time many other species invade these communities. Thus, the embryo dunes and

the dune ridge may be distinct vegetationally. When an embryo dune becomes an established dune ridge, additional incipient embryo dunes are formed close to the shoreline because of migration of the dune ridge inland or sea-level fall and expansion of shoreline seaward. Assuming an equal sediment supply, dissipative shorelines with wide beaches tend to produce large foredunes while reflective shorelines with narrow beaches have smaller foredunes (Hesp 2002).

1.7.4 Subsequent dune ridges

The youngest dune ridge parallel to the shoreline is usually called the first dune ridge. The second, third and subsequent ridges represent later stages in the evolution of the whole sand dune system. The embryo dunes, first dune ridge and slack are still actively being influenced by the action of prevailing dominant winds. The sand supply here is regular and wind is moving sand inland. Vegetation and saltation of sand are still playing a strong role in the formation and dynamics of the open dune system in this habitat. The height of the first dune ridge varies from about 5 to 50 m depending on sand supply, wind velocity, vegetation and beach width. It is usually taller than the subsequent dune ridges because it is receiving the maximum amount of sand deposition and the dominant plant community, usually *A. breviligulata*, *A. arenaria* or other species, is keeping pace with the amount of sand deposition. As one moves inland to the older dune ridges wind velocity decreases and may have little or no influence on the movement of sand, surface is stabilized, wind-borne propagules establish seedlings, humus accumulates that gives rise to the development of soil structure. Thus, the morphology of subsequent dune ridges is directly under the dominant influence of vegetation rather than the wind. Morphologically, the characteristic shape of the first dune ridge is a convex seaward slope and a precipitous leeward slope. The highest wind velocity is on the seaward side and the top surface of the dune ridge. However, as the wind passes over the crest there is a zone of still air where the sediment is dumped and saltation is no longer a factor in sand movement. Thus, the leeward side becomes the zone of deposition of sand cloud and the whole first dune may move landward. The first dune ridge usually acts as a barrier to the landward transport of sediment from the beach. Few winds are strong enough to carry sand from the beach to older ridges. Older dune ridges usually lose dune height and sharpness of contours as the centuries pass because of animal activity, drainage of water, slope wash, substrate settlement and no new sand supply. The continuous absence of major disturbances leads to the formation of an undulating topography.

1.7.5 Transgressive dune fields

On high-energy coasts in temperate zones, semi-arid coasts and in humid tropics strong winds move sand sheets inland overwhelming or transgressing terrestrial, aquatic and other plant communities. If the sand supply is too great to permit survival, dune morphology is dependent on aerodynamics of sand transport with no obstruction from plant species. Transgressive dune fields have been referred to as free dunes, mobile sands, sand sheets and sand drifts (Hesp and Thom 1990) with striking similarities to those of desert dunes. Pye and Tsoar (1990) classified them into five main types: transgressive ridges, oblique dunes, sand sheets, parabolic dunes and barchans. Asymmetric ridges shaped by the wind with gentle windward slopes (10–15°) and steep leeward slopes (30–34°) are called transgressive ridges. They undergo regular seasonal construction or destruction and the general shape is altered by wind velocity and direction. They may be oriented at right angles or oblique angles (oblique dunes) to the coast when the wind velocity blows parallel to the axis of the dune crest. They may also form U-shaped isolated dunes called barchans.

However, a distinction must be made between U-shaped dunes (barchans) in transgressive dune fields and U-shaped (parabolic) dunes along sea coasts. In transgressive dunes the horns of barchans are facing away from the wind while coastal parabolics have their faces into the wind. Arms of coastal dunes are fixed by vegetation while in transgressive dunes they are completely mobile and move freely downwind.

1.7.6 Dune slack

Dune slacks are hollows between dune ridges (Fig. 1.11) where the water table is at or just below the sand surface (Tansley 1953). They are created in two ways. First, the sea retreats and shoreline expands seaward so that space becomes available for the formation of new embryo dunes closer to the seashore thus leaving a hollow between that and the first dune ridge. Second, wind erosion creates deflation basins down to the capillary fringes between existing dune ridges. The environmental conditions in the slack differ from those of the dunes because of the coarse texture of the substrate, proximity to the water table and frequently higher soil moisture levels. (For more details see Chapter 11.)

Figure 1.11 Dry dune slack at the Pinery between the first and second sand dune ridges. Note the coarse texture of the soil and sparse vegetation (photograph by A Maun).

1.8 Creation of scarps

Scarps are steep dune faces created by the removal of beach and foredune by waves during storms. The most important determining factor for scarp formation is surge height followed by duration of surge and cohesiveness of dune profile (Carter *et al.* 1990a). Other factors are sand texture, soil moisture, cementing by mycorrhizal fungi and cohesive action of plant roots. On cohesive dunes and in temperate regions with plenty of rainfall, waves gradually undercut the foot of the foredune and create cracks on the sand surface of a dune which leads to an eventual collapse and slumping of the dune ridge at scarp angles of almost 90° (Fig. 1.12). On non-cohesive dunes with little or no vegetation and in low rainfall regions, undercutting is usually gradual and the whole surface may slide down with slope angles of about 32–34° (Carter *et al.* 1990a). As the storm subsides and normal conditions return, there is a gradual readjustment of slope due to gravity, drying of surface, reshaping by wind, and animal activity.

Rehabilitation of the foredune is initiated by slumped blocks of vegetation, rhizomes and stolons still attached to their parents, and arrival of fragments of rhizomes and stolons, ramets of herbaceous plants, branches and regenerative branches of woody vegetation and plant propagules in the swash (Maun 1989). Following the creation of a scarp along Lake Huron, Maun (2004) documented the history of recolonization of the foredune over a 10-year period and showed that within a month sandy sediment removed by wave attack began returning to the beach and dune base gradually started to build up, mainly by the re-establishment of two main grasses, *A. breviligulata* and *C. longifolia*. The slumped blocks of vegetation served as nuclei for plant growth and accumulation of sand. Although dispersed seeds of most species germinated and cuttings of trees and shrubs began to sprout, a large majority of herbaceous seedlings and woody recruits died of desiccation, excessive burial, sand blast and insect damage. After two years only the original complement of species adapted to live in this habitat had reclaimed the shoreline and generalist aliens not specifically adapted to the dune conditions had been extirpated. Within about 10 years the plant community had recovered its former species composition levels and little evidence of storm damage remained (Fig. 1.13).

Figure 1.12 The creation of a scarp by high waves in the first dune ridge during the storm in autumn/spring 1986–1987 along Lake Huron at the Pinery. The lake levels were approximately 1.5 m above normal levels (photograph by A Maun).

Figure 1.13 The recovery of plant community to approximately its former levels. Note the gradual foredune slope of the first dune ridge formed by *Ammophila breviligulata* and *Calamovilfa longifolia*. Photograph taken after 8 years of recovery in June 1995 (photograph by A Maun).

1.9 Blowouts

An erosional hollow, depression, basin, trough or swale within a dune complex created by wind removal of sand from a localized part of the dune ridge is called a blowout (Carter *et al.* 1990a). They are wind-scoured gaps in otherwise continuous transverse dune ridges (Bagnold 1960). They form readily where the crest of the ridge has been weakened by physical or biotic forces such as fire, shoreline erosion, vegetation damage by insects, rodents or other animals, trampling and formation of paths by animals and humans, and any other cause that kills the vegetation of a dune ridge. Once a weak spot has been created in a dune ridge, wind becomes the dominant erosive force and a blowout is initiated.

Blowout characteristics are controlled by two factors: vegetative cover and dune geomorphology (Jungerius and Schoonderbeek 1992). In general, species composition and percentage cover modify the erosive capacity of wind but once the vegetation is destroyed, the size and orientation of blowouts are dependent on the interaction between geomorphology and characteristics of wind. In addition, seasonal changes in weather, geology and hydrological characteristics of soil have a direct influence on blowout genesis and development. Once initiated, wind enlarges blowouts laterally and vertically. The loss of sand causes deflation of the blowout floor and steepening of its side walls. As the blowout enlarges, the wind flow is topographically accelerated and its speed is altered as it moves through the trough. These increased flows maximize sand transport and erosion within the trough. The wind velocity is generally strongest up the centre line axis of the blowout and decreases on either side, thus creating a shallow basin. As erosion of the deflation basin continues the support from the sides of blowout is weakened because of constant erosion of sand from the side walls leading to slumping of the side wall. Oblique winds may undercut only one side of the blowout wall thus creating slumping on that side only and deposition on the other wall, creating an asymmetrical blowout. All blowouts have a limit to their vertical depth because wind erodes all loose sand by saltation until the coarse sand, gravel and pebbles of a former beach or the capillary fringe that lies above the water table are exposed. Two basic types of blowouts, trough- and saucer-shaped, have been identified (Carter *et al.* 1990a).

1.9.1 Trough blowouts

Trough blowouts are elongate depressions in a sand dune ridge which have been shaped by the wind flowing through the trough at high speeds and eroding sand from the depression and pushing it over the steep back slope. The sides of blowouts remain vegetated and steep, narrow and relatively deep with more pronounced downwind depositional lobes and marked deflation basins (Carter *et al.* 1990a). Trough blowouts are particularly well developed in relatively high dune ridges and eventually evolve into U-shaped or parabolic dunes.

1.9.2 Saucer blowouts

Saucer blowouts develop on low gradient slopes on the windward side of large foredunes or on low rolling topography from which the vegetation has been denuded. According to Carter *et al.* (1990a) 'they are shallow, ovoid, dish-shaped hollows with a steep marginal rim and commonly a flat-to-convex downwind depositional lobe'. The lateral margins of these blowouts are steep and sand is deposited immediately downwind, giving it the shape of a saucer (Fig. 1.14).

1.9.3 Anatomy of a trough blowout

Byrne (1997) examined the seasonal change in sand movement in a trough blowout (Fig. 1.15) (250 m long × 75 m wide) on the south-eastern

Figure 1.14 Photo of a saucer blowout in the second dune ridge at the Pinery along Lake Huron (photograph by A Maun).

Figure 1.15 Photo of a trough blowout on a sand dune at the Pinery along Lake Huron (photograph by A Maun).

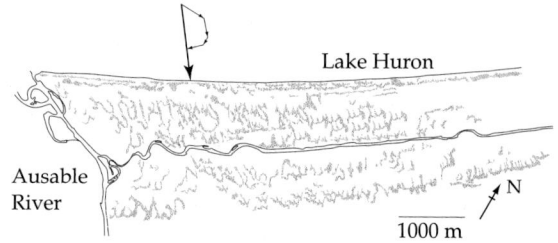

Figure 1.16 Parabolic sand dunes along Lake Huron at Pinery Provincial Park (after Bowles 1980).

shoreline of Lake Huron at the Pinery Provincial Park, Canada. The winds blow primarily from the north, west and south. The south and south-southwest winds blow at an average velocity of 15 km h^{-1} and account for about 25% of the total winds. However, the northwest, west-northwest and north-northwest winds although less frequent (15% of total winds), blow directly onshore and are stronger with an average velocity of about 19 km h^{-1}. She found that the bulk of sand movement (+ or −) occurs during the autumn and winter months when plant species are dormant with relatively small volumes moved during the growing season. This pattern of sand movement was explained by the seasonal shift in wind direction and topographic steering of wind up the windward slope. In summer, the oblique winds erode the eastern wall of the blowout and deposit the sand on the western side, thus shifting the long-term axis of the blowout to the east.

Carter *et al.* (1990a) presented three stages in the development, growth and stabilization of blowouts: (i) initiation of deflation, (ii) continued deflation which leads to slumping of side walls and sand movement to the leeward side and (iii) eventual re-vegetation, stabilization and infilling of blowout. All these stages can be seen in the old parabolic dune ridges (Bowles 1980) at the Pinery dune system (Fig. 1.16). They were probably trough blowouts at one time and then stabilized by vegetation. Dech *et al.* (2005) quantified the net cover changes of bare and vegetated areas of ten blowouts of the Pinery dune system using a geographic imaging system of colour photos taken in 1973 and 1998. They found that individual blowouts were at different stages of development in 1973 and that after 25 years some experienced a significant increase in bare areas while others showed significant colonization. However, taking all sampled blowouts as a whole the bare areas created by the destruction of existing vegetation and regained by colonization were similar. Both destruction and colonization occurred simultaneously in these blowouts.

The windward slope of the blowout generally has a gentle gradient of about 5° with a precipitous leeward slope of about 30°. The over-steepening of lee slope is corrected from time to time, primarily by spillage or sheering under the force of gravity. In topography the windward slope is uneven and the texture of sand is coarse while the leeward slope is smooth and even with a fine sandy texture. A few annual plant species are able to establish in the trough of a blowout. For example, *Corispermum hyssopifolium*, an annual, germinates and produces dense populations of seedlings in spring. However, most of these plants remain rather small in size (< 10 cm), produce a few seeds by the end of summer and die.

1.9.4 Stabilization of blowouts

When a trough blowout is eventually stabilized by vegetation, a parabolic dune consisting of a long deflation basin with coarse sand, an eroding windward crest, elongated vegetated side ridges and an actively depositing lee slope results. As a dune retreats farther and farther from the coast, the effective velocity of wind decreases and its energy is largely dissipated before it reaches the crest of the dune ridge because of surface roughness of the long trough on the windward side or on the formation of another ridge between the coast and the crest. This reduces the burial of plants on the leeward side and permits gradual re-colonization of bare areas. Stabilization progresses both from the windward and the leeward sides. On the lee slopes of blowouts the surviving trees and grasses begin to recover and increase in biomass and coverage. Additional woody species also begin to re-establish. The windward side takes longer to stabilize because wind is still a factor, soil texture is coarse and the terrain is unstable. This habitat is morphologically similar to the mid and high beach areas and is colonized first by annual and biennial species, and later by perennial grasses, shrubs and trees.

1.10 Coastal dune formation in the tropics

Jennings (1964, 1965) suggested that coastal sand dunes are virtually absent in humid tropical and subtropical climates where rainfall is very high. For example, on the west coast of Malaysia and Gulf of Papua New Guinea, wave-constructed coastal land forms are small sand masses in comparison to temperate dune systems. As the swash of each wave moves in, it rides the small sand mass leaving behind small amounts of sand. A detailed review (Hesp 2004) of dune fields in the tropics showed that substantial coastal dunes were missing in some coastal locations in South East Asia, tropical West Africa and the Asian Red sea coast. At other tropical, subtropical, arid and semi-arid locations between tropic of Cancer (23.5°S) and Capricorn (23.5°N) dune systems are massive (Swan 1979; Pye 1983; Hesp 2004). The major reasons for these differences are high variability in precipitation, length of dry season and moisture availability.

Why would the dunes be absent from some locations in the humid tropics? Jennings (1964, 1965) suggested two reasons. First, the beach does not dry out long enough for winds to carry the sand inland and second, vegetation establishes very readily and stabilizes the sand. Both of these hypotheses have been challenged. The beach does not have to be dry for winds to carry the sand inland because high velocity winds are able to dry the top millimetres of sand surface and move sand grains when the sand is wet or even when it is raining. Secondly, even if vegetation covers a sand formation, wind can still deposit sand over it and a sand dune will be formed. A third reason applies more to seasonally dry humid tropics where intense insolation produces a thin crust of salt and sand over the beach surface which prevents wind from dislodging sand grains and moving them inland (Morton 1957). Similar cementing action, however, has not been observed in all coastal locations (Jennings 1964). Even on a cemented dune the new sand can be deposited on top of the salt and sand crust, thus increasing dune height. However, Morton (1957) described an exception on a tropical West African coast where cementing action of salt was not a factor and a wind-blown sand dune formation was found primarily because of coarse light-weight sand made of broken shell fragments.

The major factors for the absence of dune systems in humid tropical coasts are weak variable winds or offshore winds, a lack of dry season, dense growth of tall trees and an inadequate supply of sand (Swan 1979). Jennings (1964) showed that the percentage frequency of gale-force winds and average velocity on humid tropical coasts along Australian coastline was almost a magnitude lower than extratropical coasts. Because of a general lack of sand mobility low terrace-type incipient and sometimes established dunes are formed (Hesp 2004). However, in certain locations with marked dry seasons along the tropical coast of Sri Lanka, within 10° of the equator, transverse, transgressive and hill type dunes are found (Swan 1979). Strong persistent onshore winds, a long dry season and ample supply of sand are required for best dune development.

1.11 Differences between coastal and desert dunes

Based on his studies in desert dunes of Libya, Bagnold (1960) showed that the principles governing the physics of sand movement and aerodynamic processes of wind action are the same in desert and coastal tropical and temperate dune systems. The major differences are in the amount of precipitation, seasonal distribution, length of dry season, temperatures, wind velocity, composition of plant communities and seasonal variability in wind. Along the tropical coast of Africa near the equator occasional violent storms occur during the

major growing period, the wet season, while in temperate regions, these storms are common at the equinoxes (Lee 1993). In deserts precipitation as rain is infrequent and unpredictable while in coastal dunes salt spray is a regular recurrent phenomenon. All these environmental factors and their variability have an impact on the evolution of traits of plant species occupying each dune system. For instance, there is a general absence of therophytes in the tropics (Lee 1993), while they are common in temperate dunes and become very abundant after rains in deserts. Grasses of desert dunes have rhizosheaths while they are absent in most grasses of coastal dunes and probably the tropics as well (Danin 1996). The rate of colonization by plants in the tropics is very rapid and the most characteristic feature of this zone is the flattened or prostate rosette forms that limit interception of salt spray. Dominant pioneers and plant elements in the beach vegetation are stoloniferous chamaephytes or rhizomatous geophytes. In deserts the species composition depends on total annual rainfall and its predictability. In extreme deserts no plant species may be able to survive in the dunes except in somewhat moister valleys or oases where the majority of plants are xerophytes.

1.12 Summary

Coastal sand dunes occur along sea coasts and large lakes throughout the world, however they are more common in temperate climates. For dune formation, a steady supply of sand is required which fluctuates during glacial and interglacial periods. During glacial periods, expanding beach areas receive sediments from the continental shelf and large dune systems are formed. During interglacial periods, these dunes are destroyed and sediments become available which are reused to form new dune systems at the new sea–land interface. At present, three major sources of sand originate from river discharge, cliff and coastal erosion and input from sediments from the sea floor brought in by tides and washovers. Sand may be defined as grains ranging in diameter from 0.02 to 1.00 mm whose terminal velocity of fall is greater than upward currents of air and lower than sand grains that can not be moved by wind or by saltation. The sand on the sea floor is held there by the force of gravity and friction between water molecules and sand grains. Strong wave action exerts a shearing force that overcomes friction, lifts the grains, propels them towards the shore and eventually deposits them on the beach.

Once on the beach the inland movement is mediated by a complex physical process that includes suspension of very fine particles, saltation and surface creep. Saltation of sand particles is caused by wind speeds above a certain value that lifts sand particles into the air and moves them forward while they lose height due to the force of gravity, hit the sand surface and eject more sand grains into the wind stream, thus accentuating the process. Particles that are too large to be lifted into the air stream by saltation may be moved forward due to bombardment by saltating grains—a process called surface creep. As sand grains are not uniform in size, saltation and surface creep sifts them according to their size and may form sand ripples on bare sand surfaces. Sand dunes are formed by plant species that increase surface roughness, absorb the energy of saltating grains, decrease the rate of sand transport and lower the wind speed so that the sand is deposited on top and to the lee of the plants. Continuous movement of sand by saltation over time deposits more sand on plants until sand dunes (mounds or hills of sand rising to a single summit) are formed. The inherent capacity of plant species such as *Ammophila* and other pioneers to emerge through burial and subsequent burial episodes increases dune height.

Dune geomorphologists have classified sand dunes into shadow dunes, embryo dunes, dune ridges, transgressive dune fields and several other types depending on wind velocities, energy levels along coasts, vegetation types, rainfall patterns and climate. Violent wave storms can erode dune systems and create scarps by variable wave surges, duration of surges and cohesive properties of the dune system. Once created, these dunes may take years before they are rehabilitated by vegetation to their former levels. Dune ridges are held together by vegetation and any physical or biotic force that kills vegetation exposes them to wind erosion and blowouts are created. Once initiated wind erodes sand laterally and vertically, deflates the floor, and steepens the side walls. Its size and orientation depend on characteristics of wind, weather conditions, geomorphology and soil hydrology.

The two basic types, saucer and trough blowouts, have been recognized. Saucer blowouts are shallow, ovoid, dish-shaped hollows on the windward sides of large foredunes. Trough blowouts are elongate depressions in a dune ridge with steep vegetated ridges, marked deflation on the windward side and deposition on the leeward side. The high deposition on the lee of the blowout kills vegetation in its path and creates bare areas. Eventually when a dune blowout retreats farther from the coast, the effective velocity of wind declines because its energy is dissipated before it reaches the crest of the dune ridge. A blowout is then gradually stabilized from both the leeward and windward sides by woody and herbaceous plant species. In humid tropical regions with high persistent rainfall sand dunes are missing, however, in many tropical and subtropical regions with abundant rainfall, a dry season and ample supply of sand, large dune fields are found. The principles governing the physics of sand movement and aerodynamic processes of wind are the same in desert and coastal dune systems. Nevertheless, there are variations mediated by seasonal distribution of rainfall, wind velocities and characteristics of vegetation.

Further reading

Students of sand dune ecology are encouraged to develop an appreciation for dune geomorphology. The following references provide an excellent discussion of the subject.

Bagnold RA (1960). *The physics of blown sand and desert dunes.* Methuen and Co. Ltd., London.

Pethick J (1984). *An introduction to coastal geomorphology.* Edward Arnold Ltd., London.

Open University (1989). *Waves, tides and shallow-water processes.* Pergamon Press, Oxford, UK.

CHAPTER 2

The sand dune environment

2.1 Introduction

The micro-environmental conditions of different soil habitats are influenced by prevailing vegetation, aspect, soil texture, soil colour and other variables that influence the incoming and outgoing solar energy. The variability is especially pronounced in sand dunes because of shifting substrate, burial by sand, bare areas among plants, porous nature of sand and little or no organic matter, especially during the early stages of dune development. Even within a dune system there is disparity in radiative heating of different habitats that is manifested as variation in micro-environmental factors such as relative humidity, temperature, light, moisture content and wind turbulence. The major factor affecting these changes is the establishment of vegetation that stabilizes the surface, adds humus, develops shade, aids in the development of soil structure and reduces the severity of drought on the soil surface. The system changes from an open desert-like sandy substrate on the beach to a mature, well-developed soil system with luxuriant plant communities. The principal topics discussed in this chapter include accounts of micro-environmental factors of coastal sand dunes that influence the growth and reproduction of colonizing species.

2.2 Soil moisture

The water content of the substratum in sandy soils is one of the most important limiting factors in plant growth. Sandy soils have high porosity and after a rain most of the water is drained away from the habitat because of

the large interstitial spaces between soil particles and the low capacity of sand to retain water. Evaporation in open dune systems also removes substantial quantities of water. Lichter (1998) showed that evaporation was greater on non-forested dune ridges than on forested areas and the rate of soil drying was influenced by soil depth and dune location. After 3 days of a heavy rainfall there was a drastic decrease in the percentage of moisture (67–80%) at 0–5 cm levels in open habitats compared to only 30–36% in the forested dune ridges (Table 2.1). The same measurements at 10–15 cm depths showed much lower reduction in the percentage of moisture. In the swale (slack) even though the evaporative demand was the same, there was actually an increase in moisture because of seepage from the dune ridges.

Gooding (1947) measured soil moisture along a transect on sand dunes of Barbados at 20 m intervals from 50 m above high tide to 150 m inland and found an increase in soil moisture content from 0.2% in the pioneer zone to 2% in the *Coccoloba* zone, probably due to the accumulation of humus (Fig. 2.1). Along the Pacific coast of California at Bodega Head, soil moisture values ranged between 1–6% (De Jong 1979) while along the Atlantic coast of North Carolina (Oosting and Billings 1942) the values at 10 and 20 cm depths ranged between 2.5–5% on the basis of oven dry weight of soil from the end of July to early August (Table 2.2). Although the moisture equivalent of these soils was about 2%, the actual soil moisture was consistently higher. There was very large variability in soil moisture content depending on the season,

Table 2.1 Percentage of moisture at 0–5 cm and 10–15 cm soil depths on dunes, swale and forested dune of different ages on a sand dune chronosequence along Lake Michigan following a heavy rainfall and 3 days after the heavy rainfall (after Lichter 1998)

	Soil depth			
	0–5 cm		10–15 cm	
	Initial (%)	3 days (%)	Initial (%)	3 days (%)
Open dune (25 years)	6.4 ± 0.6	1.3 ± 0.5	5.7 ± 1.4	3.5 ± 0.3
Open dune (145 years)	4.5 ± 0.4	1.5 ± 0.6	4.3 ± 0.3	4.0 ± 0.8
Forested dune (285 years)	10.1 ± 1.1	6.6 ± 0.8	6.2 ± 1.4	4.0 ± 1.1
Swale (95 years)	16.0 ± 2.1	18.5 ± 4.3	16.3 ± 1.9	17.7 ± 1.7

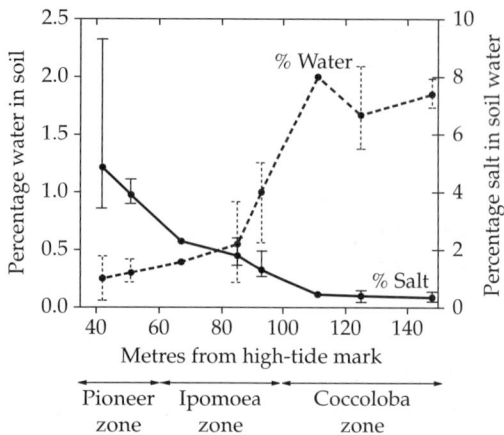

Figure 2.1 Percentage of moisture on a dry weight basis in soil from the Pioneer zone through the *Ipomoea* zone and to the *Coccoloba* zone on the sand dunes of Barbados (after Gooding 1947).

precipitation level, location and microhabitat (Oosting 1954).

In the Pinery sand dune system the moisture content of the soil was rather high even in shallow layers of soil at the beginning of summer (April and May) (Fig. 2.2). The levels decreased in the months of June and July at all depths, but decrease was more pronounced at the 5 and 10 cm layers of soil (Baldwin and Maun 1983). The system started to recharge again in August and by May of next year the soil moisture was again high because of spring rains and melting snow. Soil moisture also varied between habitats. For example, significantly higher soil moisture was recorded at 5 and 10 cm depths in the transition zone (800 years old) compared to the first dune ridge (150 years old) and slack habitats (200 years old). However, these differences disappeared at the 25, 50 and 75 cm depths of soil (Fig. 2.2). Means calculated over the entire sampling season showed that the moisture content in the dune and slack habitats ranged between 1.6–3.0% at 5–10 cm depths compared to 4–5% in the transition zone. The surface layer of sand had the lowest moisture ($< 1\%$) in the beginning of summer. Even this moisture was lost readily in the dry month of June and the top 5–10 cm of soil became completely dry. The air in the pore spaces of this dry layer contributes to insulation of soil from thermal conduction of the surface layer. Thus, the moisture beneath this dry layer is conserved. Even microhabitats within the same sand dune ridge have differences in moisture content. For example, the top of the dune ridge is drier because of greater evaporation under the influence of higher wind velocities (Olsson-Seffer 1909). Similarly, the south slopes in the northern hemisphere and north slopes in the southern hemisphere are warmer and drier than those facing other directions. Some plant species such as *Sorghum bicolor* and *Ipomoea pes-caprae* growing along sea coasts are able to utilize dew as a source of moisture (Akhtar and Shaukat 1979).

Table 2.2 Mean moisture (± SE) content (% moisture by dry weight of soil), moisture equivalent and percentage moisture of sand at wilting, collected from 10 and 20 cm depths at 5 stations (n = 5) on each of two transects along the coast at Bogue Bank, North Carolina (calculated from Oosting and Billings 1942)

Dates of collection	Transect	(%) Moisture	
		10 cm	20 cm
27 July	1	3.36 ± 0.16	4.70 ± 0.95
	2	3.12 ± 0.63	3.40 ± 0.46
4 August	1	3.92 ± 0.48	4.78 ± 0.34
	2	3.68 ± 0.38	4.22 ± 0.47
9 August	1	2.68 ± 0.34	5.10 ± 0.94
	2	2.66 ± 0.69	4.04 ± 0.54
Moisture equivalent	1	1.94 ± 0.11	2.06 ± 0.04
	2	1.98 ± 0.08	1.66 ± 0.03
Wilting percentage	1	0.138 ± 0.020	0.143 ± 0.031
	2	0.136 ± 0.022	0.154 ± 0.020

Percentage of moisture at one station in the moist depression (slack) between dune ridges averaged over the three dates was 9.5 ± 3.7% at 10 cm and 14.5 ± 3.7% at 20 cm depth of soil.

2.2.1 Moisture level changes during succession

Along a chronosequence at Wilderness State Park, Lake Michigan, Lichter (1998) found that the percentage of soil moisture of the upper soil horizon increased with dune age from about 2% in the foredunes to 12% in the 1000–2500 year old dune ridges (Fig. 2.3) because of complete vegetation cover on older dune ridges. The moisture-holding capacity of the soil also increased with an increase in dune age. The field capacity (moisture retained in the soil after the drainage of excess water by the force of gravity) of samples from the first dune ridge was 10.65 ± 0.95% (n = 42) and increased to 12.75 ± 1.75% (n = 42) for samples from the transition zone (Baldwin and Maun 1983). Sandy soils of coastal dunes may have low field capacity but all this moisture is readily available to the plant. In contrast, soils with larger silt and clay fractions release a relatively smaller proportion of water.

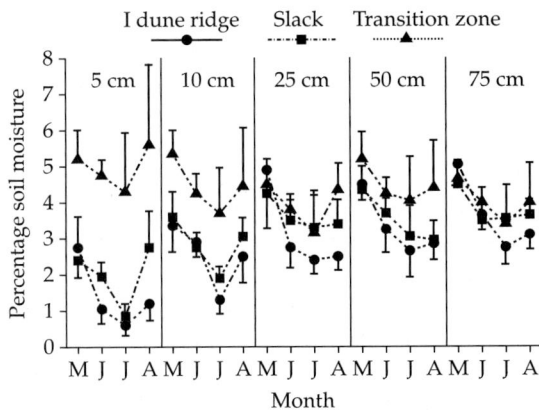

Figure 2.2 Percentage of soil moisture contained in soil samples collected at different depths from three habitats (first dune ridge ●, slack ■, and the transition zone ▲) on Lake Huron sand dunes. Vertical bars indicate ±1 SE (after Baldwin and Maun 1983).

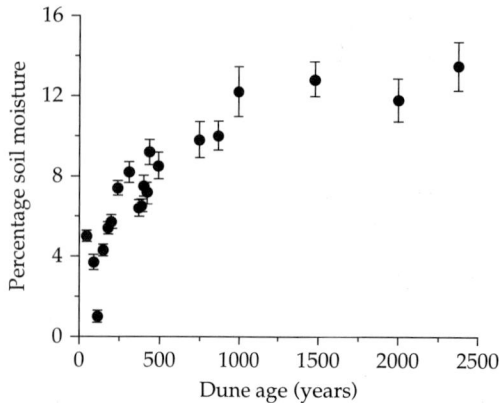

Figure 2.3 Changes in percentage of moisture in the upper 15 cm of mineral soil on dune ridges ranging in age from 25 to 2375 years along Lake Michigan (after Lichter 1998).

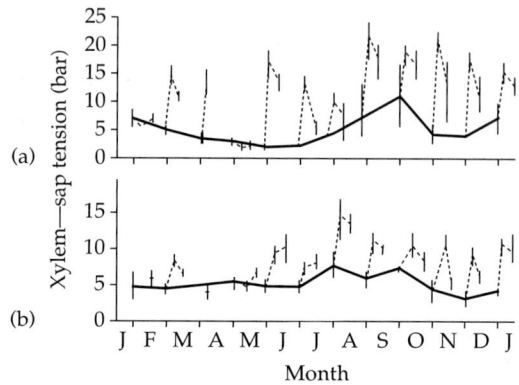

Figure 2.4 The seasonal (——) and daily (-----) course of xylem sap tension for (a) *Atriplex leucophylla* and (b) *Cakile maritima* at the northern site during 1975. The daily measurements were taken at dawn, midday and dusk. The vertical bars present the range (after De Jong 1979).

2.2.2 Plant–water relations

In plant–water relations, the moisture content of soil and plant is expressed in water potentials. Water potential is a term used to express the energy status of water in the soil, the plant or atmosphere. Pure water has the highest water potential and has been assigned a value of 0: thus a soil saturated with water has a water potential of 0. Water potential (ψ) can be expressed in terms of pressure or energy and the unit megapascal (MPa) is used.

$-1\text{MPa} = -10$ bar
while a bar $= 10^3$ dynes cm^{-2}
or 10^2 joules kg^{-1} or 0.987 atm.

One dyne equals the amount of force required to alter the speed of a mass of 1 g by 1 cm per second for each second the force acts.

Several authors (Oosting 1954; Kearney 1904; Chapman 1976) have suggested that coastal dune plants may be classified as xerophytes. De Jong (1979) questioned this designation and showed that even though dune species are drought resistant they do not quite fit the definition of xerophytes, especially in temperate regions. He compared the water relations of beach taxa on a moist cooler site (Bodega Head, 38°20′ N lat., 123°04′ W long., annual

precipitation $= 83.2$ cm) and a dry warmer site (Trancas Beach, 34°02′ N lat., 118°51′ W long., annual precipitation $= 39.2$ cm). The xylem sap tensions of *Atriplex leucophylla* and *Cakile maritima* showed great seasonal variation while *Abronia maritima* and *Ambrosia chamissonis* varied within a narrower range (Fig. 2.4). Although the xylem sap tensions in the warmer drier site of southern California were greater than the moister cooler site of northern California, they did not approach those of xerophytes. The top layers (0 to 30 cm) dried out to very low water potentials (–3.0 to –8.5 MPa) depending on temperature and precipitation of the site, but the soil at both locations remained relatively moist at about 100 cm depth. The surface of southern California site dried out to greater depths. However, the soil water table remained the same at both sites throughout the year. Nevertheless below the dried upper layer at both sites, the sand contained relatively greater amounts of moisture and soil water potential was seldom below –1.5 MPa.

De Jong (1979) also showed evidence that in coastal dune soils water moves up into the upper layers from below. He dug two pits ($100 \times 100 \times 60$ cm) in the pioneer zone of vegetation at the northern California site.

One of the two pits was lined with plastic (0.13 mm thick) to prevent lateral and vertical movement of moisture while the second served as control. The water potential of sand above the plastic in the pit was substantially lower (−3.6 MPA), than control, −1.0 MPa. The sand above the plastic in the pits dried out very quickly. However, below the plastic the water potential was −0.3 MPa which was even higher than the control. The tests clearly showed that the water moved up from below under normal conditions, possibly by capillary movement, or along a water potential gradient, because water potential in upper layers of sand was more negative than the lower layers of sand and only a very small quantity of water will be needed to raise the water potential to between −1.0 MPa and −1.5 MPa.

Nevertheless, Olsson-Seffer (1909) suggested that the rise of water by capillarity in dune sand was extremely low and the most probable mechanism may be internal dew formation, whereby water moves up as vapour and then condenses in upper cooler layers of sand, especially at night. According to Salisbury (1952) there was no evidence for the hypothesis that dew formation at night increased the moisture a few cm below the sand surface. On a South African foredune, Ripley and Pammenter (2004) reported that leaf water potentials of three semi-succulent species, *Arctotheca populifolia*, *Ipomoea pescaprae* and *Scaevola plumieri* did not drop below the turgor point, transpiration was related to atmospheric demand, stomatal conductance was normal and in a normal year the water utilized by transpiration was less than the annual rainfall. However, in a dry year with below average rainfall, plants were utilizing more than actual rainfall by tapping into other water sources such as internal dew, stored water or ground water. Among the three species, *I. pes-caprae* showed stomatal control of water loss under dry conditions. Soil water remained above the lower limit of about 0.5% at which plants can extract water from sand dune soil.

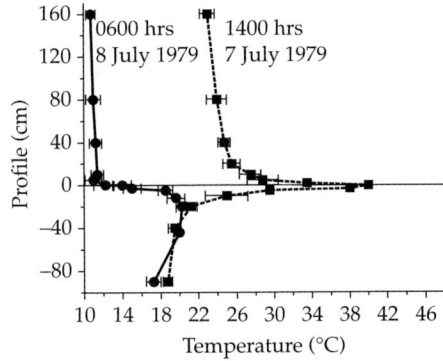

Figure 2.5 Temperature profiles in and above the sand on a lacustrine sand dune system at 0600 and 1400 hrs in July. Horizontal bars represent ±1 SD (after Baldwin and Maun 1983).

2.3 Soil and air temperatures

On clear summer days the temperature of the surface layer of sand may reach up to 60°C, mainly because sand is a poor conductor of heat owing to large pore spaces between particles. At the Pinery, the soil surface temperature reached 40 ± 6°C at 1400 hrs on 7 July 1979 at the slack site (Fig. 2.5). However, immediately below the sand surface, temperature declined rapidly producing very steep lapse rates (Baldwin and Maun 1983). For example, at 5 cm below the sand surface, the temperature had decreased to about 30 ± 4°C, a drop of almost 10°C. The soil temperature continued to drop even further until it had reached 19°C at 90 cm below the sand surface. In the evening, however, the situation was reversed. The thin hot surface layer of sand cooled rapidly because the low thermal conductivity of sandy soil did not allow the replenishment of heat from lower layers of soil. Thus, there was a large diurnal temperature range at the sand surface. Olsson-Seffer (1909) recorded a diurnal temperature range of 25.6°C between day and night at the Hangö dunes in Finland. However, the diurnal range within the soil decreased with an increase in soil depth. At about 40 cm depth the soil temperature was the same both during the day and night. In summer, immediately above the sand surface, the air temperature was about 2°C warmer than at 5 cm above the sand surface, primarily due to convection from the soil surface that may be several degrees higher than the air above. Then the temperatures decreased rapidly, producing a steep lapse rate (Fig. 2.5). At night, the air temperature immediately above the sand was lower than at 5 cm above the soil layer. The lapse rate was negligible up to a height of 160 cm above the soil surface at 0600 hrs on 8 July 1979.

The daily march of soil and air temperatures showed that the soil surface layer attained the highest temperatures at 1400 hrs, remained hot until 1800 hrs and dipped down to its lowest value at 0600 hrs (Baldwin and Maun 1983). Decline in soil temperature was lower at 5 and 20 cm depths because of time required for the conduction of heat through the soil profile, but the lowest temperatures were again recorded at 0600 hrs. The curves for changes in daily air temperatures followed very similar trends. Highest temperatures of 48–53°C during the day were recorded at 0 cm followed by 5, 40 and 160 cm above the sand surface. Similar observations on air and soil temperatures have also been well documented in the dune system at Blakeney Point in England (Salisbury 1952) and the Pacific coast of California (De Jong 1979). High temperatures increase energy absorption by leaves and strand plants usually avoid overheating by rolling of leaves, waxy surfaces, vertical orientation and hairy leaves, thereby also avoiding moisture stress.

2.4 Vapour pressure deficit

Vapour pressure deficit (VPD) is the difference between the actual vapour pressure of the atmosphere minus the saturation vapour pressure at any given temperature. It is assumed that a fully turgid plant has a vapour pressure equal to that of pure water at the same temperature. Thus, the vapour pressure of leaf intercellular spaces would essentially maintain a saturation vapour pressure

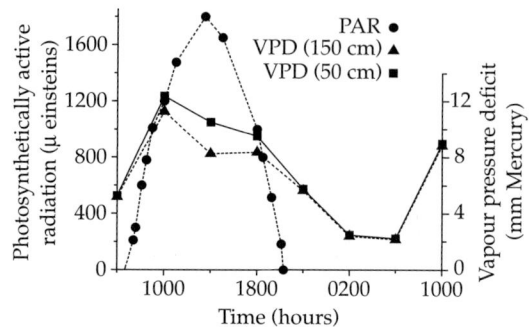

Figure 2.6 Daily march of photosynthetically active radiation (PAR) and vapour pressure deficit (VPD) between 7 July 1979 and 1000 hrs, 8 July 1979 (after Baldwin and Maun 1983).

at the prevailing leaf temperature. The VPD (mm Hg) of the Pinery sand dunes (Fig. 2.6) reached high values during the day at about 1000 hrs and then decreased (Baldwin and Maun 1983). The lowest values were recorded from 0200 to 0600 hrs. The values were lower at 150 cm than at 50 cm above the sand surface, probably due to higher temperatures at 50 cm. However, VPD decreased at both heights to the same values at night because of cooling and exchange of dry air with moist air from the adjacent lake surface.

2.5 Photosynthetically active radiation

The range of light between the 400 nm and 700 nm wavelengths is defined as photosynthetically active radiation (PAR). Sunny days on sand dunes are bright and hot during June and July. The values for PAR reach their highest values (1800 μ einsteins) at 1400 hrs and then decline throughout the afternoon (Baldwin and Maun 1983) (Fig. 2.6). Albedo (reflecting) from the sand surface also increases the light intensity on the lower parts of the leaf surfaces of plants.

The proportion of sunlight reaching the soil surface decreases with the establishment of a forest over-storey. Lichter (1998) measured the proportion of full sunlight received in seral communities at 1 m above soil surface in foredunes

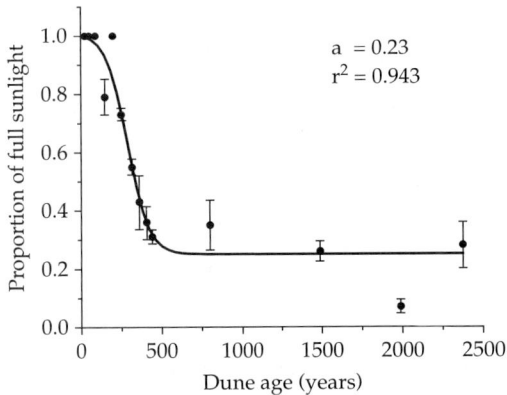

Figure 2.7 Proportion of full sunlight reaching 1 m above the soil surface in plant communities ranging in age from 15 to 2375 years in a sand dune chronosequence along Lake Michigan. Vertical bars indicate ± 1 standard error (SE) (after Lichter 1998).

to 2500-year-old dune ridges in a sand dune chronosequence along Lake Michigan. He found a sharp decline in light intensity with an increase in diversity and establishment of trees and shrubs. The values decreased from full sunlight in the open dunes to about 23% of full sunlight in 500-year-old dune ridges and remained at that level in plant communities up to the 2500-year-old dune system (Fig. 2.7). Most of the plant species of open sand dunes are heliophytes and well adapted to withstand and utilize maximum PAR values of the dune systems on bright sunny days (Perumal and Maun 2006). Measurements made on a South African coast showed that peak CO_2 assimilation rates of *A. populifolia*, *I. pes-caprae* and *S. plumieri* plants on a foredune were 25.9 ± 6.9, 9.7 ± 5.2 and 20.1 ± 4.5, respectively (Ripley and Pammenter 2004). The rather low values of *I. pes-caprae* were primarily due to low chlorophyll concentrations per unit leaf area. The plant also exhibited marked photoinhibition and stomatal control of water loss.

2.6 Precipitation

Coastal sand dune formation and vegetation development are strongly influenced by the amount of precipitation per year and its distribution. Precipitation has several significant impacts on the growth and development of coastal sand dunes. Sand movement is controlled by the quantity of water in soil. Wet soil increases coherence of sand particles and increases resistance of sand grains to wind action. High amounts of precipitation in the tropics increases the colonization of habitat by plant species, significantly reduces the movement of sand and does not allow an increase in height of a dune system. In regions of high rainfall, the impact of salt spray is minimal because salt is washed off the plants and leached out of the rhizosphere. Conversely, in dry regions salt spray may form a salt crust over the sand surface that binds sand particles and does not allow sand movement. In low rainfall areas, sand movement is high (depending on wind velocity) which curtails the establishment of plant species: large dune fields with no vegetation may be very common. Development of the soil profile is accelerated in high rainfall regions because of faster decomposition of organic matter.

Another important factor is the distribution of rainfall. Well distributed rainfall allows much better growth and establishment of vegetation than seasonal rainfall. According to Olsson-Seffer (1909), even though the total amount of yearly rainfall on the Brisbane coast of Australia is more than twice (129.5 cm) that of the Baltic coast (59.3 cm), the growth of vegetation is much better in the latter region because of distribution over a much longer period of the growing season.

2.7 Wind velocity

Wind velocity is an important force in coastal sand dunes because it influences vegetation by increasing evaporative demand, burial, erosion, sandblasting and wind training. The constant evaporative loss of moisture from the soil and increased transpiration reduce the stature of shrubs and trees along sea coasts. High wind velocities carry sand grains in saltation which strike windward sides of trunks and lower leaves of trees and shrubs on the seashore, eroding the bark and shredding green leaves.

The wind velocity at the Pinery during summer 1979 and 1980 ranged from 2–8 km h^{-1} in 1979 and 1980 (Baldwin and Maun 1983). The velocity was highest on the beach and the first dune ridge and declined in the slack and transition zone. It was also higher at 150 cm compared with 50 cm above the sand surface. Thus the velocities were not high enough during summer for sand movement, and little or no burial took place during the summer months. However, during late autumn, winter and early spring wind velocities frequently exceeded 5 m sec^{-1} with frequent burial of plant species.

2.8 Soil genesis

In coastal sand dune systems the sand deposited by the waves on the beach is relatively unaltered parent material consisting of weathered rocks of different types. For example, at the Pinery along Lake Huron, the parent material consisted of limestone, dolomite, chert, sandstone, siltstone, Precambrian felsics, mafics and metamorphic rocks (Cooper 1979). Although the parent material in other coastal dunes of the world may be quite different, the processes of soil genesis are similar. With the passage of time the soil profiles show progressive development from unsorted, unaltered parent material to well-developed horizons through the process of podsolization. The most obvious evidence is the release of carbonates of Ca^{++} and Mg^{++} and oxides of iron and aluminium from the A horizon or the zone of leaching and litter accumulation, to the zone of deposition in the B horizon where they are precipitated. The primary reason for the release of these cations is a decrease in soil pH. The production of organic acids by the decomposition of litter in the A horizon dissolves bases which are leached to the lower layers. Similarly, organic matter is

also transported downwards in the soil profile. The rate of leaching is variable in different parts of the world depending on the amount of annual precipitation, mean annual temperature and length of the frost-free season. For example, the rate of leaching on sand dune systems along Hudson Bay (56°N lat., 86°W long.; precipitation 51 cm; mean temperature –5.5°C) is only one third (Protz et al. 1984) that of a Great Lakes sand dune (43°15′ N lat., 81°45′ W long.; precipitation 81 cm; mean temperature 8°C). The rate of loss of carbonates from the top 10 cm of mineral soil is greater in the earlier stages of succession than later, probably because of differences in the solubility of minerals. Calcite is more soluble and hence is lost from the system first while dolomite dissolves more slowly and takes longer to leach out (Olson 1958a).

The depth to which carbonates will leach is also related to the age of the system and pH of the leaf litter: the greater the age, the greater is the depth to which carbonates are moved and precipitated. Similarly, plants such as conifers that produce more acids on decomposition are more efficient in the release of cations and carbonates from the A horizon. Thus, in the same region, the depth of carbonate movement will be greater in habitats with plants producing acid-rich litter. VandenBygaart (1993) studied soil genesis and gamma radioactivity of the Pinery dunes by examining soil profiles of different ages from the present to 5000 years before the present (BP) and showed the development of a podsol profile. Sand dunes of 100 years of age did not exhibit any horizons (VandenBygaart and Protz 1995) but by about 1000 years (Fig. 2.8) the A and B horizons were clearly distinguishable. These layers of leaching and deposition became more pronounced with time, the depth to which carbonates leached increased and by 4700 years and 2270 m from the lake shore, a clear podsol profile became evident. It became even more pronounced in plant communities on sandy substrate by 12 000 years BP (Sparling 1965).

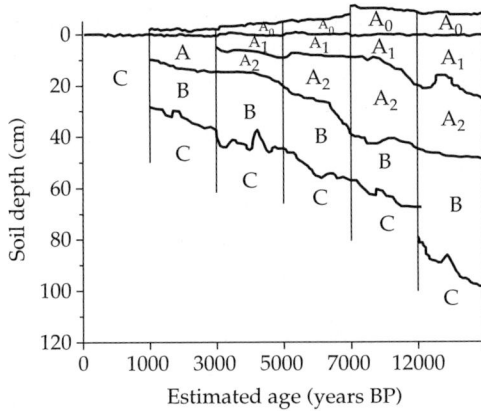

Figure 2.8 Soil profile development on a lacustrine sand dune system at Pinery Provincial Park along Lake Huron. A_0, A_1, A_2 are the zones of leaching, B, the zone of deposition and C, the unaltered parent material (adapted from VandenBygaart 1993; Sparling 1965).

2.9 Bulk density

Bulk density (weight per unit volume of soil) is a useful measure to estimate soil structure and the extent to which the root penetration may be restricted by the soil. In the open sand dunes of about 100 years of age, the bulk density was about 1.55 g cm^{-3} (VandenBygaart and Protz 1995). It decreased with an increase in age of the sand dune system to about 1.3 g cm^{-3} at 5000 years. Similar results were reported by Lichter (1998) who showed that the bulk density of the top 15 cm of soil decreased from 1.6 g cm^{-3} in foredune soils to 1.15 g cm^{-3} in 1000–2500-year-old soils with a forest canopy (Fig. 2.9). The main reasons for a decrease in bulk density are an increase in soil organic matter and humus, root growth and their decay, and burrowing activity of soil animals.

2.10 Weathering in sand dunes

There was a general trend of an increase in silt and clay fractions with an increase in age of the dune systems along Lake Michigan (Olson 1958a), Pinery dunes (Fig. 2.10) along Lake Huron (VandenBygaart and Protz 1995)

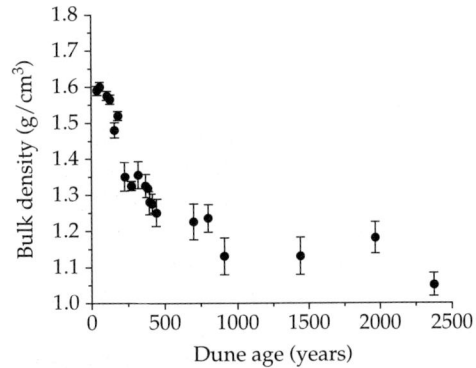

Figure 2.9 Changes in bulk density of soil on dune ridges ranging in age from 15 to 2375 years in a sand dune chronosequence along Lake Michigan. Vertical bars indicate $\pm$ 1 SE (after Lichter 1998).

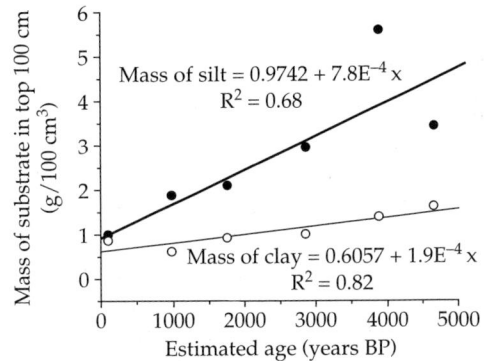

Figure 2.10 Mass of silt and clay in the top 100 cm of soil in the sand dunes at Pinery Provincial Park (after VandenBygaart and Protz 1995).

and in aeolian soils in many areas of the world (Thompson 1981). VandenBygaart and Protz (1995) suggested that because of a positive linear relationship (Fig. 2.10) between time and mass of silt and clay, it was most probably caused by aeolian deposition of fine particles from active dunes, weathering of feldspar and ferromagnesium into clay, and larger particles to smaller size particles. Nevertheless, another possibility could be that the sand deposited by waves at an earlier time period contained more silt and clay.

2.11 Organic matter

The single sand grains deposited on the beach by wave action do not have any biotic component in them. The first addition of organic matter in this habitat occurs by the deposition of detritus thrown onshore by the waves. The detritus consists of dead remains of plants and animals, flotsam and jetsam cast into the sea by human activities or live aquatic fauna that get beached and killed. By far the largest quantity of material consists of seaweeds (along the Baltic), *Fucus vesiculosus* (Olsson-Seffer 1909) on seashores and very large banks may be created depending on the location, wave action, nutrient content of seawater and its productivity. The length of time taken for the decomposition of this material depends on the climate of the region; for example, in warm tropical regions with high rainfall, the material decomposes rapidly while in temperate regions it takes much longer. Burial of this material by sand hastens decomposition.

The foredunes soon start to accumulate organic matter following the establishment of plants and accumulation of litter. In a 5-year-old foredune along Lake Michigan, the organic carbon averaged less than 0.05% (Olson 1958a); however, within about 50 years the organic matter content in the foredunes of the Pinery ranged from 0.1–0.5% in the surface 5–10 cm of sand (Baldwin and Maun 1983; Morrison 1973). In a sand dune system along Lake Michigan the above-ground litter production increased to a maximum of 350 g m^{-2} year^{-1} at 440 years of dune age (Lichter 1998).

The quantity of carbon accumulation in the total ecosystem through 440 years of succession was 102 mg ha^{-1} while the rates of accumulation of carbon in biomass, soil and total ecosystem were 21.3, 7.9 and 23.2 g m^{-2} year^{-1}, respectively (Table 2.3).

Measurements in a stabilized habitat of the dune system in the transition zone (800 years old) showed organic matter values of 4.5, 2.5 and 1% at soil depths of 5, 10 and 25 cm, respectively (Baldwin and Maun 1983), indicating that organic matter is high in the surface layers but decreases with soil depth. These values are considerably lower than those of a garden loam soil which may have 10–30% organic material. The mass of organic matter in the chronosequence of a lacustrine sand dune system at the Pinery dune along Lake Huron showed a logarithmic relationship with age in the top 100 cm of sand during the first 2000 years, which then levelled off and reached a steady state (Fig. 2.11), indicating that organic matter addition was equal to the loss from the system (VandenBygaart and Protz 1995). The carbon to nitrogen ratio of pioneer species (usually C_3 and C_4 grasses) on foredunes may be higher than 80:1 and it takes greater length of time for the decay of grass litter to humus compared to tree litter at later stages of succession.

2.12 Nutrient status

Coastal sand dune soils in general lack three macronutrients: nitrogen (N), phosphorous (P)

Table 2.3 Average rate of carbon and nitrogen accumulation in living above- and below-ground biomass, in soil (O horizon plus upper 15 cm of mineral soil) and total ecosystem in a sand dune along northern Lake Michigan (after Lichter 1998)

Pool	Total quantity mg ha^{-1}		Rate of accumulation g m^{-2} year^{-1}	
	Carbon	Nitrogen	Carbon	Nitrogen
Biomass	67.4	0.47	21.3	0.11
Soil	34.7	1.20	7.9	0.27
Total ecosystem	102.1	1.67	23.2	0.38

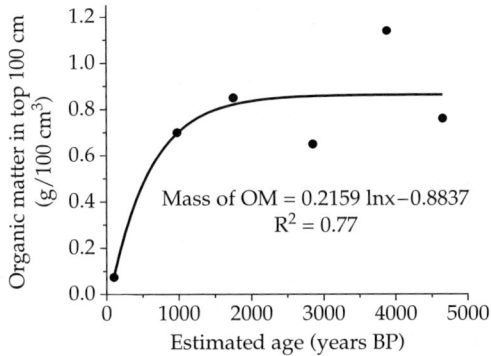

Figure 2.11 Change in organic matter (OM) content of the top 100 cm of soil with an increase in soil age (after VandenBygaart and Protz 1995).

and potassium (K). The seminal papers by Willis and Yemm (1961), and Willis (1963) on mineral nutrient status of Braunton Burrows sand dunes showed that the sparse growth of vegetation was primarily caused by the deficiency of N and P and to a lesser extent K, rather than other environmental stresses such as exposure to the elements or the water table. Similarly, Tilman (1986) showed that N was the most important limiting nutrient in sandy soils of Cedar Creek Natural History Area, Minnesota. Addition of nitrogen to the habitat brought about a change in species diversity and relative composition because of differential response of species to the total soil nitrogen concentration. Houle (1998) also showed marked improvement in productivity of *Elymus mollis* by the addition of nutrients on a dune system along Hudson Bay primarily because of luxury consumption by plants. However, as soon as the nutrient source was exhausted the plant communities reverted to the original state. Addition of micronutrients did not show an improvement in growth or a change in relative coverage of plants.

2.12.1 Nitrogen cycle

Although nitrogen is the most abundant gas in the atmosphere, it can not be directly utilized by plants because they are unable to break the

stable bond between the two nitrogen atoms. However, atmospheric nitrogen is intimately associated with soil nitrogen through symbiotic nitrogen fixation and atmospheric deposition. In an ecosystem, nitrogen is found in three pools; nitrogen contained in plants growing in the community, retained in the soil as organic matter, adsorbed on soil particles or in soil solution, and atmospheric dry and wet nitrogen inputs including biological nitrogen fixation. Loss of nitrogen from the system occurs by soil leaching and denitrification. Thus, all of these components interact with each other and during primary succession in sand dunes, the amount of nitrogen increases in the early stages because gains exceed losses, but after the elapse of a certain period of time, a stable equilibrium may be reached. Olson (1958a) showed that 100-year-old sand dunes contained only 0.005 to 0.002% N, however it increased following stabilization and increase in age of the dunes for approximately the first 1000 years and then levelled off. In another sand dune system along Lake Michigan, Lichter (1998) showed that soil nitrogen content increased from 0 in freshly deposited sand in the foredunes and stabilized at about 1.36 mg ha^{-1} after about 440 years of age. The relative availability of both NH_4^-N and NO_3^-N in older forested dunes was approximately three times that of open sand dune ridges.

The atmospheric inputs of organic and inorganic nitrogen in coastal sand dunes consist of material contained in precipitation and lightning, dry fallout as atmospheric dust and gases and deposition of material suspended in fog and clouds. Dry deposition is approximately equal to wet deposition. On a sand dune at Point Reyes, California, Holton (1980) estimated an annual input of 0.9 Kg ha^{-1} in the form of ammonia and 0.7 Kg ha^{-1} in the form of nitrate. Sea spray and fog contributed an additional 1.4 Kg ha^{-1} as nitrate and 1.1 Kg ha^{-1} as ammonia. Estimates by Lichter (1998) along Lake Michigan showed that atmospheric nitrogen deposition was about

0.51 g m^{-2} year^{-1}, however, a large part of the deposited nitrogen was lost in drainage and absorbed by plants while only a small proportion was retained in the soil. The total soil nitrogen pool also increased with an increase in age of a dune system. The amount of soil nitrogen in the 600-year-old sand dune ridge along Lake Huron was significantly higher than the 120-year-old dune ridge, probably because of accumulation of litter, atmospheric inputs and decay of organic material (Maun and Sun 2002). Similarly, along Lake Michigan, total nitrogen pool in soil continued to increase exponentially until it stabilized at 1.2 mg ha^{-1} after 440 years (Table 2.3). Over this period the annual rate of nitrogen acquisition by the total ecosystem was 0.38 g m^{-2} year^{-1}. The total nitrogen pool stored in vegetation increased with time because of an increase in species diversity, soil organic matter and stabilization of plant communities. Maun and Sun (2002) estimated the amounts of nitrogen inputs from the atmosphere, nitrogen pools in vegetation and soil, and leaching loss from the soil on the open sand dune ridges of the Pinery sand dune system. The atmospheric wet and dry inputs contributed approximately 1.6 g m^{-2} year^{-1}. In some heavily polluted areas of Europe the amount of wet and dry deposition of nitrogen from pollution sources may be up to 40 kg N ha^{-1} year^{-1} (Killham 1994). According to Sun (2000) the dry inputs of nitrogen were primarily in the form of nitrous oxides and ammonia, while wet inputs were in the form of nitrates and ammonium. The above- and below-ground biomass contained about 0.29–0.84 g m^{-2} of nitrogen, while the mineral soil contained about 10.5–24.7 g m^{-2} (Table 2.4). A very small amount was lost to leaching each year. Sun (2000) did not take into consideration the input made by biological nitrogen fixation because there were no leguminous plant species in this sand dune system.

How do plants of coastal sand dunes cope with such low levels of nitrogen in the soil? They rely on association with symbiotic micro-organisms in the rhizosphere, nutrients contained in salt spray, precipitation and dry fallout, and accumulation of organic matter (Holton 1980). Even when levels of N in the soil were extremely low, the average tissue concentrations of nitrogen in *Cakile edentula* and *Corispermum hyssopifolium* were within the normal range of 1.5–3.0% along Lake Huron (Hawke and Maun 1988). Nitrogen in live plant tissue of *Elymus mollis*, *Ammophila arenaria* and *Cakile maritima* was 2.8, 1.4 and 2.7%, respectively along the Pacific Coast (Holton 1980). Similarly, even though soil nitrogen pools were low, the leaf nitrogen content of *A. populifolia*, *I. pes-caprae* and *S. plumieri* plants on a South African foredune were within normal limits (Ripley and Pammenter 2004). According to Tilman (1986), early successional species were better adapted to grow in N-poor soils and ranked nine species in the following order

Table 2.4 An estimate of nitrogen and phosphorous inputs (g m^{-2} year^{-1}) from the atmosphere, nitrogen and phosphorous pools in grassy vegetation and soil (g m^{-2}), and leaching loss from the soil (g m^{-2} year^{-1}) on the open sand dune ridges of a sand dune system along Lake Huron (after Maun and Sun 2002)

	Nitrogen	Phosphorous
Atmospheric input (g m^{-2} year^{-1})	1.622	0.016
Above-ground plant biomass (g m^{-2})	0.195–0.398	0.007–0.042
Roots and litter biomass (g m^{-2})	0.093–0.437	0.007–0.038
Soil pool (gm^{-2}) (top 15 cm)	10.50–24.70	19.10–24.50
Leaching loss (g m^{-2} year^{-1}) 50 cm depth	0.010–0.023	0.001–0.003

on the basis of increasing biomass attainment under nitrogen limiting conditions: *Ambrosia artemisiifolia, Achillea millefolium, Chenopodium album, Agropyron repens, Agrostis scabra, Poa pratensis, Sorghastrum nutans, Schizachyrium scoparium* (= *Andropogon scoparius*) and *Liatrus aspera*. From these data he concluded that early successional species grew more rapidly and acquired more nitrogen from nitrogen-poor soils than late successional species.

Some of the perennial pioneer species such as *Ammophila breviligulata* may receive some of the nitrogen by their association with diazotrophic bacteria (Azotobacter) in the rhizosphere (Ralph 1978; Hassouna and Wareing 1964). Apparently the majority of nitrogen required by *A. arenaria* may be acquired by this means (Wahab 1975). However, the rhizosphere nitrogen fixation in sand dunes has not been independently confirmed. For example, Holton (1980) in his studies on *Ammophila arenaria* and *Elymus mollis* could not detect any nitrogen fixation in sand and root samples and was unable to culture nitrogen-fixing bacteria from root and rhizosphere collections. How important is nitrogen fixation to primary succession in sand dune systems? Several studies show that although occasional nitrogen-fixing plant species may occur in early, mid and late stages of succession, their contribution to the nutrition of other co-occurring plant species is not significant (Barbour *et al.* 1985; Lichter 1998), because the number of nitrogen-fixing species occurring along a coast is very small and distributed irregularly. The main reasons for their absence are high energy costs by the plant in fixing a molecule of nitrogen and a very high requirement of other nutrients, especially phosphorous (Sprent 1993), which are normally in short supply during early stages of succession in sand dunes.

Nevertheless, several nitrogen-fixing species, *Myrica gale, M. pennsylvanica, Shepherdia canadensis, Alnus incana* and *A. rugosa*, do occur at later stages of succession. According to Lichter (1998) and Olson (1958a) even the contribution of these species to nitrogen

accumulation is negligible and thus the gradual build up of nitrogen in the soil depends on the life history processes of a species (Walker *et al.* 1986), acquisition of atmospheric inputs and build up of organic matter over time.

The acquisition of nutrients by dune plants is also facilitated by their association with mycorrhizal fungi. The primary reason for the increased uptake of macro- and micronutrients is the exploitation of increased soil volume by extra-matrical mycelium of the arbuscular mycorrhizal (AM) fungi. The hyphae are more efficient than plant roots in the absorption of soil nutrients. More details are presented in Chapter 9.

2.12.2 Phosphorous cycle

The second most important nutrient in sand dunes is phosphorous. Unlike nitrogen, phosphorous does not occur in the atmosphere as a gas. The primary source of phosphorous in sandy substrates is the weathering of minerals such as apatite ($Ca_5(PO_4)_3X$ where X may be hydroxide [OH], chlorine [Cl] or flourine [F]). Since the amount of apatite in sand is very low, coastal sand dunes are deficient in P (Fig. 2.12). Nevertheless, atmosphere does make a small contribution to the overall input of phosphorous from aerosol spray from oceans, lakes and rivers, fine dust particles from volcanic ash, rocks and soil, material of organic origin such as smoke, ash, fine litter, humus, pollen grains and combustion of fossil fuels. Estimates of total phosphorous input from the atmosphere range from 0.07 to 1.7 kg ha^{-1} year^{-1} (Newman 1995). This input is very small, but it is extremely important for plants in sand dune systems with little or no phosphorous at the start of primary succession.

In a sand dune chronosequence along Lake Michigan, the concentration of phosphorous in the upper mineral soil increased to a maximum until about 345 years but a drastic decline occurred from 345- to 485-year-old dunes (Lichter 1998). At 2375 years of age the P levels were even lower than at 25 years,

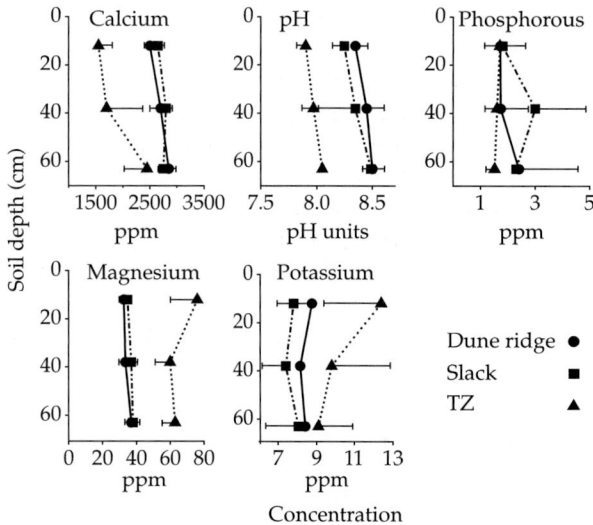

Figure 2.12 Concentrations of calcium, magnesium, potassium, phosphorous and soil pH at different depths in soil in three habitats of the dune complex at the Pinery along Lake Huron. Horizontal lines indicate $\pm$ 1 SD (after Baldwin and Maun 1983).

probably because of a decline in pH and storage of phosphorous in plants, especially tree trunks. Estimates showed that approximately 50% of phosphorous absorbed by plants is returned later as organic matter to the soil in the form of litter. However, this organic phosphorous must first be converted to inorganic form by the activity of micro-organisms before it can be absorbed by plants. In addition, plants must compete with soil microbes for available phosphorous.

Soil pH has a direct influence on weathering of minerals and availability of phosphorous in sand dunes. At high pH, phosphorous released from mineral soil precipitates as calcium and magnesium phosphate. As the soil pH decreases, the availability increases until it reaches a maximum between pH 6.8 and 7.2. A further decrease in pH results in decline of phosphorous availability again because it forms insoluble precipitates with iron and aluminium. Since insoluble phosphates are released very slowly into the soil solution, phosphorous, in spite of its abundance as a precipitate, is always a limiting nutrient in sand dunes. In sand dunes with an acidic reaction (pH 4.5–5.5) such as those prevailing at Greenwich and Basin Head, Prince Edward

Island, the amount of available phosphorous was slightly higher—6.2 parts per million (ppm)—in foredunes than inland dunes— 5.2 ppm. Sun (2000) estimated the amounts of phosphorous inputs from the atmosphere, phosphorous pools in vegetation and soil, and leaching loss from the soil on the open sand dune ridges of a sand dune system along Lake Huron. He found that atmospheric inputs contributed about 0.016 g m^{-2} year^{-1}. The above- and below-ground biomass contained about 0.014–0.080 g m^{-2}, while the soil pool contained about 19.10–24.50 g m^{-2} (Table 2.4). A very small amount was lost to leaching each year.

2.12.3 Cations

The cation retention capacity of sandy soils is very low. As explained earlier in the section on soil genesis, low pH releases cations from the upper horizons of the soil and moves them down to lower layers by leaching. According to Boyce (1954) coastal sand dune soils have abundance of cations in the upper 60 cm of soil. Van der Valk (1974) summarized the inputs, outputs and distribution of cations in a coastal grassland ecosystem. Salt spray contributes a substantially larger proportion of sodium (Na), K, calcium (Ca) and magnesium (Mg) to the system than the rainfall. The total annual inputs to the dune systems are always larger than the total outputs in leaching from the system. Some of the cations are retained in the standing crop of vegetation. Thus, there is no large reservoir of cations in the soil and plants depend on regular inputs of nutrients from the atmosphere on sea coasts. On lacustrine sand dunes, however, since there is no salt spray inputs the plants depend heavily on contributions made by wet and dry deposition following precipitation events and on the release of cations from the sandy soil. According to the estimates by Environment Canada, the amount in wet deposition was equal to about 0.4–0.9, 4–5, 0.4–1.0 and 0.2–0.5 Kg ha^{-1} year^{-1} for Na$^+$, Ca^{++}, Mg^{++} and K$^+$, respectively. There is a significant change in cation exchange capacity of soil during primary succession in sand dunes. The capacity of soil to adsorb cations increases because of an increase in the percentage of humus, silt and clay.

Lichter (1998) showed that a significant decrease in Ca^{++} and Mg^{++} occurred during succession in the upper 15 cm of mineral soil, with a corresponding increase in hydrogen (H)$^+$ ions on the soil exchange complex. The increase in cation exchange capacity is rapid in the early stages of succession and then declines to a slow rate. The amount of Ca^{++} and Mg^{++} was extremely low in the surface soil after about 500 years. Similarly, iron, aluminium and organic matter are carried downwards into lower layers in the soil. In contrast a slow decrease occurred in the concentration of K$^+$ with an increase in dune age because of depletion and then accumulation in the O horizon. Apparently, K$^+$ is absorbed by plants from the mineral soil and later returned to the O horizon in the form of litter. The upper 15 cm of O horizon and the mineral soil still contained substantial quantities of K$^+$ even after 2000 years of dune age (Lichter 1998). The soil at the Pinery contained approximately 1500 to 3000 ppm calcium, 30 to 70 ppm of magnesium and 8 to 12 ppm of potassium (Fig. 2.12). Among the dune complex habitats, dune ridge and slack habitats were not significantly different in concentration of these elements according to the t test. However, higher concentrations of these elements were found in the zone of transition from grassy to forest communities especially in surface layers, suggesting that organic matter in the surface layers has higher retention capacity than the soil (Baldwin and Maun 1983).

2.13 Soil pH

The pH of the upper soil horizons declines rapidly during succession, primarily because of acid rain and the production of organic acids by vegetation (Fig. 2.12). Studies by Lichter (1998) showed a decline in pH from 8.5 to 4.3 after about 400 years of stabilization of a sand dune. A major change in pH occurred following the establishment of coniferous vegetation which produced litter with an acidic reaction. As the litter decomposes it produces organic acids such as malic and tartaric acids. The changes in pH lead to major changes in the release of other elements from the soil that move to lower layers thus initiating the development of a podsol profile.

2.14 Summary

Coastal sand dune systems provide a unique opportunity to examine the transformation of inert sand grains to a well-developed soil profile. In the process tremendous changes occur in the micro-environmental factors essential for plant growth. Sand possesses the highest porosity and after a rain storm most of the water is drained away from the habitat. The average moisture content of coastal sandy soil ranges from 2–5% at 10–20 cm depths. During early spring the amount of moisture in the soil is relatively high because of spring rains and snow melt. The moisture level decreases to the lowest levels in June and July but the system starts to recharge again in August and September. The moisture content of the soil is lower in the early stages of dune stabilization on the first dune ridge and slack, but it increases in the transition zone. In the sand dune chronosequence the moisture content increased from 2% in the foredunes to 12% in the 1000–2500-year-old dune ridges. The field capacity of sandy soil in the first dune ridge was about 10.5% and it increased to 12.5% in the transition zone soil, mainly because of greater organic matter content.

Plant species growing in coastal sand dunes are not xerophytes because water content of sandy soil maintains water potential of −1.5 MPa because of internal dew formation, capillary movement and a water potential gradient. Surface temperatures of sandy soils may rise to 40–60°C on bright sunny days but decline rapidly with soil depth to about 20°C at 90 cm depths. At night the situation is reversed with lowest temperatures at the surface, and diurnal temperature range may be up to 25°C. The vapour pressure deficit or the difference between actual vapour pressure and the saturation vapour pressure was highest at about 1000 hours and then decreased. Photosynthetically active radiation (PAR) reaches its maximum values of 1800 μ einsteins at 1400 hrs and then decreases. As the plant community develops from open sand dunes to closed forest communities, the PAR values at the forest floor decline to <50% of full sunlight.

The amount of precipitation and its yearly distribution have a significant impact on the growth and establishment of vegetation because of its influence on cohesion of sand, plant colonization, decrease in salinity, and the development of the soil profile. Wind velocity of a region has a strong impact on plant growth and community development because of its influence on burial, erosion, sandblast, evapotranspiration and dune building. During the sand dune chronosequence, there is soil development mediated by the activity of plants, climate, rainfall and micro-organisms. Along the Great Lakes, a podsol profile developed after about 5000 years because of the release of carbonates of Ca^{++} and Mg^{++} and oxides of iron and aluminium from the A horizon and their movement to lower horizons. The soil bulk density decreases with soil development due to an increase in organic matter, root growth and activity of biotic organisms. During sand dune chronosequence there was in general an increase in weathering of soil and an increase in silt and clay. The source of this silt and clay may be weathering, aeolian deposition, or higher silt and clay content of sand deposited at the beginning of sand dune formation. The amount of organic matter builds up over time. On the beach the first addition of organic matter occurs by the deposition of detritus. As plants establish the organic matter builds up and humus content increases with the age of the dune system. The organic matter of a dune system along the Great Lakes increased from 0.1% in foredunes to 4.5% in the transition zone (800 years old). Development of forest increases the organic matter content to even higher values because of regular addition of leaf litter, root decay and deposition of dead plant material.

The essential nutrients required for plant growth in coastal dunes increased over time from very low in foredunes to acceptable levels in later stages in the chronosequence. In general, three essential nutrients, N, P, and K, are in short supply. Pioneer plants generally require very low quantities of nitrogen and phosphorous and depend primarily on detritus washed up on the beach, association with symbiotic organisms, and nutrients deposited by rainfall, salt spray and dry fallout. The primary source of P in sandy soils is from the weathering of apatite, salt spray and atmospheric input. Phosphorous content gradually builds up over time. Estimates showed that atmospheric inputs were about 0.016 g m^{-2} $year^{-1}$, above- and below-ground biomass contained about 0.026–0.054 g m^{-2}, and mineral soil had about 19 g m^{-2} of soil on the beach and 20–25 g m^{-2} on the dunes. Plants depend on release of cations from sandy substrate input from salt spray and wet deposition by rainfall. Soil pH declines rapidly during succession from about 8 to 4.5 within a period of about 500 years.

In spite of the low concentrations of essential nutrients in sand dune systems, the plant species growing there are very well adapted to the prevailing nutrient levels. Deficiency symptoms are not manifested by plant species, cell division and plant growth are normal and there is evidence of internal cycling of nutrients.

CHAPTER 3

Seed dispersal

3.1 Introduction

Dispersal is a term used for the dissemination of detached reproductive structures from parent plants to a new site. Disseminules include spores, seeds, fruits, whole inflorescences, whole plants, fragments of the parent plant, bulbs and bulbils. Fruit attributes related to a particular dispersal agent or dispersal syndromes are complex and have resulted from millions of years of evolution. In practice, dispersal is mainly local, although some species of sea coasts are well adapted for long-distance dispersal. Knowledge of the modes of plant dispersal is vital to the study of coastal dune ecology because of the clear correlation between diversity and dispersal mechanisms. From the evolutionary point of view, dispersal improves fitness of species: the progeny is able to colonize a new site and extend the range of the species. The fitness here will be defined as getting to a coastal site by using any vector for dispersal, colonization of the new site (germination, establishment and reproduction) and dispersal of the propagules of the immigrant from the new site.

Dispersal confers many benefits to the populations of plant species. It reduces competition for limited space and resources in the parental location and the more widely dispersed the propagules, the greater are the chances for the offspring to colonize elsewhere. Dispersal increases the chances of survival and evolution of more fit strains of a species by occupying more diverse habitats than the parents, and speciation may eventually occur in response to new selective pressures. For species adapted to live along sea coasts, dispersal by sea is primarily directed for dissemination to another site by the sea coast. During dispersal several physiological changes may occur in the disseminules that facilitate colonization of the species at the new habitat. For example, Barbour (1972) reported that immersion of upper fruits of *Cakile maritima* in seawater stimulated their subsequent germination under controlled conditions. Seed coat dormancy may also be broken by abrasion of seeds in sand while being rolled along the sand surface. Considering the large number of species along coasts and on islands, only a very few species may be successfully disseminated in seawater. The propagules of most species are either unable to survive immersion in seawater or are destroyed by predators at sea. However, a few highly adapted species are able to travel thousands of kilometres in sea currents (Guppy 1917).

The treatise on dispersal published by Ridley (1930) is still the most comprehensive compilation of observations and records by botanists, ecologists and naturalists. Subsequently many accounts on specific plant species of coastal sand dunes have appeared in the *Journal of Ecology* under the heading 'Biological flora of British Isles'. A student of dispersal is also referred to several reviews of the topic by Harper (1977), Van der Pijl (1982), Estrada and Fleming (1986) and Murray (1986a). In this chapter, I will confine my remarks to particular dispersal adaptations of plants on coastal sand dunes. The primary agents of dispersal of seeds, fruits and vegetative fragments of plant species on coastal sand dune systems are water, wind, gravity and animals.

3.2 Dispersal in water

Because of the large areas of sea coasts, dune plant species have tremendous possibilities of dispersal to distant coasts. Water dissemination is normally confined to hydrophytes because reproductive structures of most land plants become waterlogged soon after immersion. However, some plant species of coastal beaches are exceptions because their propagules are adapted to travel in seawater. In order to travel in seawater propagules must possess three traits (Ridley 1930).

1. Seeds should remain inactive during transport, i.e. not imbibe water and germinate and buds on fragments of rhizomes or stolons should not sprout.
2. They should be able to float in seawater for extended periods of time without waterlogging and sinking to the bottom or being injured by salt water.
3. Seeds, seedlings or buds on fragments must resist damage by salt water and maintain viability during transport in seawater. Since violent storms are frequent during sea dispersal the flotation apparati of fruits must be able to withstand turbulence (Van der Pijl 1982). Propagules along bodies of freshwater dunes must meet the same requirements except they need not withstand salt in the flotation medium.

According to Dirzo and Domínguez (1986) long-distance dispersal in water is an accidental rather than an adaptive event because there is very high mortality during dispersal and most propagules are unsuccessful in colonization. However, it can be argued that many coastal species remain buoyant in water for long periods and establish on distant coasts. Adaptations that facilitate floatation are not as elaborate as those for wind dispersal.

3.2.1 Fruits and seeds

Fruits and seeds of most coastal beach species meet the three criteria of dispersal in seawater and have a world wide distribution. For example, species of the genus *Cakile*, including *C. edentula*, *C. maritima* and *C. lanceolata*, are annuals that grow along lacustrine and oceanic beaches and have a dimorphic fruit which is a two-segmented silique (Fig. 3.1). The fruit is constricted in the middle and at maturity an abscission layer is formed between upper and lower segments. The upper segment then breaks off while the lower segment remains firmly attached to the parent (Barbour 1972; Rodman 1974). The fruits are buoyant in water because of the hard, corky and indehiscent pericarp surrounding each rather large seed. Fruit dimorphism confers *Cakile* species with adaptive advantages for dispersal, emergence of seedlings and physiological dormancy. The upper fruits that break off the parent have higher shell mass and are longer and wider than the lower segments, thus they are better adapted for long-distance water dispersal (Barbour 1972; Keddy 1980; Maun and Payne 1989). Secondly, the upper fruit segments have higher seed mass than the lower fruit segments and following germination their seedlings are

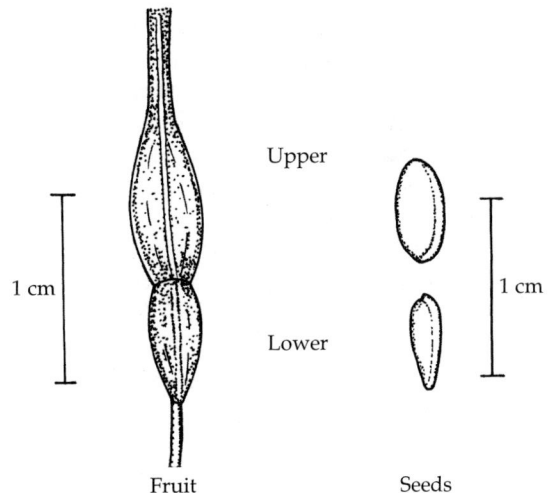

Figure 3.1 A schematic drawing of the whole fruit of *Cakile edentula* var. *lacustris* with upper and lower fruit segments and upper and lower seeds enclosed in them (after Payne and Maun 1981).

able to emerge from greater burial depths. Thirdly, the upper fruit segments are more exacting in their germination requirements than lower segments (Maun and Payne 1989). Rodman (1974) and Guppy (1917) reported no germination of upper fruits while they were floating in seawater.

Payne and Maun (1981) spray-painted (fluorescent orange) several mature plants of *Cakile edentula* var. *lacustris* growing on the beach at the Pinery and then recorded the distance of travel of upper fruits from the parent location after two, four and six weeks. Wave action

removed the upper fruits and transported them upwards to the driftline and laterally in either direction along the shoreline, but the lower fruits can also be moved when the plants are uprooted by waves or by the occasional severing of fruits at the pedicel. Keddy (1980) also showed the same differential dispersal of upper and lower fruits by demonstrating that seedlings growing singly away from dense clusters (parental location), originated exclusively from upper fruits.

Payne and Maun (1981) determined the floating ability of fresh and over-wintered upper and lower fruit segments by placing them in glass jars two-thirds full of water and then subjecting them to agitation that simulated wave action and control treatments. The data showed that fruits floated for a longer period of time in the agitation treatment compared to the still water controls (Fig. 3.2), probably because of aeration of the fruits. The time taken for 50% of the upper fruits to sink was also significantly greater than lower fruits in the agitated treatment (Table 3.1). The sinking index was significantly greater in the agitated lower fruit segments than upper fruit segments. The buoyancy of fruits is maintained only by the outer covering of fruits: shelled seeds floated for only a few minutes by surface tension. The floating ability of both upper and lower fruit segments was also improved

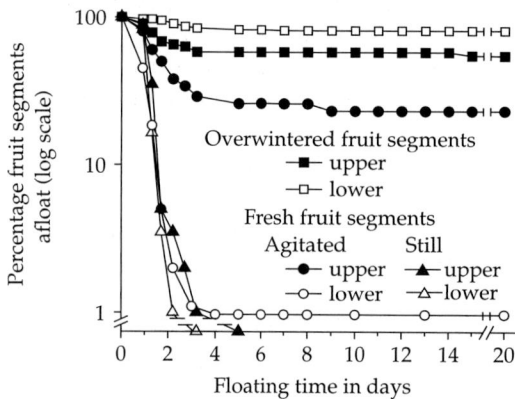

Figure 3.2 The percentage of upper and lower fruit segments of *Cakile edentula* var. *lacustris* afloat (log scale) in agitated and still water treatments after varying lengths of time in days (after Payne and Maun 1981).

Table 3.1 Length of time (hrs) taken for 50% of fresh upper and lower fruit segments of *Cakile edentula* to sink in the still and agitated treatments. The fruits were floated in glass jars filled with water from Lake Huron (adapted from Payne and Maun 1981)

Treatment	Fruit morph	Time (hrs) for 50% to sink	Sinking index*
Agitated	Upper	42.3c	43.1c
	Lower	17.8a	80.4a
Still	Upper	29.1b	61.9b
	Lower	24.8ab	68.1ab

Means in each column followed by the same letter are not significantly different at $P < 0.05$ according to Tukey's test.

*The sinking index was calculated by the summation of quotients of the number of sinking fruits divided by the number of the day on which the count was made.

by over-wintering with 65% of the over-wintered upper fruit segments and 88% of the lower segments remaining afloat after agitation for 20 days. The lower fruits were probably more buoyant because many of the pedicels had fallen off. Fruits may sink to the bottom near the shore and will later be washed up onto the beach along with sand.

Of the three taxa of *Cakile* the upper and lower fruit segments of *C. maritima* floated the longest: one hundred days after the start of the trial, 50% of the upper and lower fruit segments were still afloat (Maun and Payne 1989). The main reason for better buoyancy was larger shell mass enclosing the seeds.

A partial list of plant species of sea coasts, the buoyancy mechanism of their disseminules and flotation periods of the some species are presented in Table 3.2. The primary anatomical features conferring buoyancy in water are air cavities, space between cotyledons, spongy mesophyll, hollow pericarp, lightweight cotyledons, corky structures on fruits and a large surface to volume ratio.

The coconut palm (*Cocos nucifera*) which grows very close to the seashore and often overhangs sandy beaches of tropical and subtropical regions is well known for long-distance dispersal by ocean currents (Ridley 1930). The smooth and polished pericarp resists decay in seawater. For establishment on the shoreline the milk in the endosperm provides the much needed water and energy for germination. The fruit of *Mertensia maritima*, a fleshy herbaceous perennial of shingle beaches in the north temperate and arctic regions, also

Table 3.2 Seed or fruit buoyancy mechanisms and length of floating period of dispersal units of plant species adapted to live on foredunes of coastal sand dunes

Species name	Buoyancy mechanism	Flotation period (days)	Authority
Crambe maritima	Thick pericarp with loose spongy structure, space between cotyledons	37	Scott and Randall (1976) Darwin (1857)
Ipomoea pes-caprae	Space between cotyledons	—	Guppy (1917)
Euphorbia paralias	Buoyant tissue in testa	—	Ridley (1930)
Honkenya peploides	Spongy albumin in c otyledons	> 365 days	Ridley (1930)
Cocos nucifera	Air trapped in mesocarp	Long periods	Ridley (1930)
Xanthium strumarium	Air cavities		
Mertensia maritima	Thick hollow pericarp	18	Scott (1963)
Scaevola plumieri	?	140	Guppy (1917)
Calystegia soldanella	?	180	Ridley (1930)
Lathyrus japonicus	Lightweight cotyledons	—	Ridley (1930)
Strophostyles helvola	Lightweight cotyledons	50	Yanful (1988)
Canavalia rosea	Lightweight cotyledons	—	Ridley (1930)
Juniperus communis	Lightweight cotyledons	—	Ridley (1930)
Carex arenaria	Corky utricle	—	Noble (1982)
Rumex crispus	Corky tubercles on perianth	3	Cavers and Harper (1967)
Cakile species	Hard corky covering on seeds	3–20	Payne and Maun (1981)
Ammophila breviligulata *A. arenaria*	Large surface to volume ratio of lemma and palea enclosing the achenes	—	Maun and Baye (1989)
Pinus sylvestris	Large surface to volume ratio	7–28	Ridley (1930)
Sesuvium portulacastrum	In pumice, driftwood	—	Ridley (1930)

has a thick pericarp and four nutlets that float in seawater (Scott 1963). *Honkenya peploides*, a succulent herb of arctic shores that extends as far south as Portugal, produces few-seeded buoyant capsules with spongy albumin in cotyledons. Seeds buoyancy of coastal *Lathyrus japonicus*, *L. maritimus* and *S. helvola* in seawater is attributed to very light cotyledons. Both seeds and fragments of plant species may be dispersed in seawater in pumice (spongy, light, porous, volcanic rock) and on drifting logs (Darwin 1857).

Maritime forms of *Rumex crispus* and *R. maritimus* growing on shingle beaches along the Atlantic coasts of Europe have accrescent sepals that enlarge after flowering and enclose the small one-seeded fruit. A tubercle at the base of each of the three perianth segments enclosing the seed gives buoyancy to the fruits. In the autumn the whole panicle matures and the fruits are dispersed by wind. They may be later picked up by waves and dispersed to other shores. Cavers and Harper (1967) showed that all propagules of the maritime form remained afloat for three days while those of the inland form sank to the bottom. *Carex arenaria* is a native species that occurs primarily along the seashore on mobile and semi-mobile sand dunes, fixed dunes and in slacks of Britain and Europe (Noble 1982). Its seeds are enclosed in a buoyant utricle and can float for months. Tests of buoyancy and germination following immersion in salt water revealed that 17 maritime species occurring on the islands of southern Japan have attributes that qualify them for dispersal by sea currents (Nakanishi 1988). In many grass species, the modification of bracts improves the buoyancy of seeds. For example, in *Ammophila breviligulata* the dispersal unit consists of lemma and palea that encloses a small caryopsis and forms a boat-like structure with a large surface to volume ratio (Maun 1985). Following dispersal in water, most seeds are deposited by waves on the driftline of beaches. The detached head with globose ball of spikes of the grass *Spinifex squarrosus* may be transported long distances by

floating in water with the peduncle acting as a sail (Ridley 1930).

3.2.2 Fragments of plants

The rhizomes, stolons and roots of perennial beach species are fragmented by storm waves in autumn and early spring and transported laterally to the same shoreline or to new shores where they are cast on the driftlines. Many species of sedges such as *Carex arenaria*, *C. kobomugi*, *C. eriocephala* and grasses including *Ammophila breviligulata*, *A. arenaria*, *Leymus arenarius*, *L. mollis* and *Elymus farctus* may disperse as caryopses but their principle mode of short-distance dispersal is through water transport of rhizome fragments. Maun (1984) excavated plants of *Ammophila breviligulata* established along a 500 m long shoreline of Lake Huron and showed that the length of fragments varied from 10 to 80 cm and the number of nodes per fragment was 1 to 15, depending on the length of fragments. The most frequent length of fragment contained 2 to 5 nodes (Fig. 3.3). Most of the nodes on fragments, especially on longer fragments, bore roots but only a small proportion bore shoots. Establishment and survival of plants were more successful from rhizome fragments than seedlings (Fig. 3.4). For example, 79% of *Ammophila breviligulata* plants from rhizome fragments survived to the end of summer compared to only 4% from seedlings (Maun 1984). Harris and Davy (1986b) showed that shoots on multi-node fragments of *Elymus farctus* had greater chances of emergence than those on single node fragments. The vegetative regeneration of *Panicum racemosum* from rhizome fragments also showed similar results along a shoreline in southern Brazil (Cordazzo and Davy 1999). *Phragmitis australis*, a rapidly expanding alien species in North America, also invades shorelines by establishing from rhizome fragments.

An identical mode of vegetative regeneration is exhibited by dune species of tropical and subtropical shores. *Ipomoea pes-caprae*,

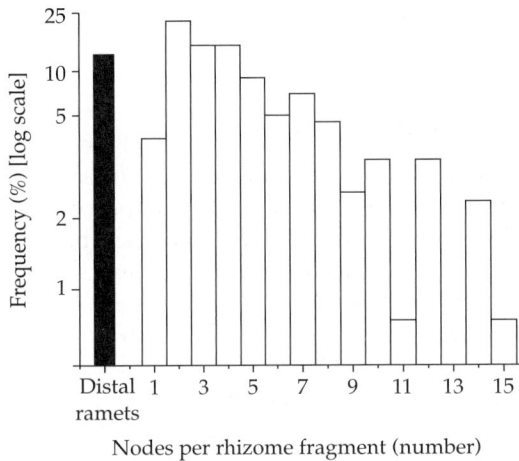

Figure 3.3 Frequency distribution of the number of nodes per rhizome fragment of newly established *Ammophila breviligulata* plants along part of the Lake Huron shoreline at the Pinery (after Maun 1984).

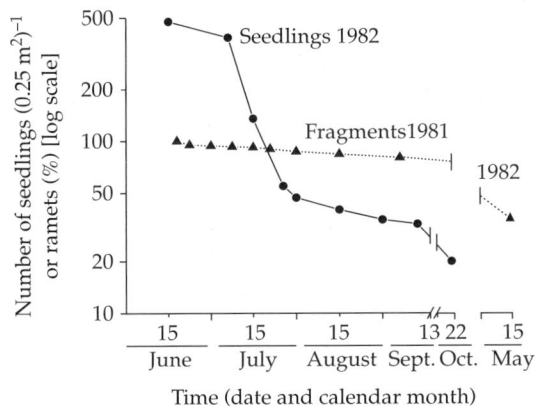

Figure 3.4 Survival of plants of *Ammophila breviligulata* derived from rhizome fragments (n = 100) and seedlings (n = 500) on the Lake Huron shoreline (after Maun 1984).

I. biloba and *Canavalia rosea* produce stolons that may be several metres in length. These are easily broken by wave action and transported to other shores where they establish new colonies. The creeping stems of *Calystegia soldanella*, a species of Antarctic origin (Ridley 1930), are much shorter than *I. pes-caprae*, yet

they can be broken by wave action and transported laterally. The fragments and branches of *Sesuvium portulacastrum* have a high survival rate during dispersal in seawater and readily establish when cast on shore (Ridley 1930). Plant fragments of *Honkenya peploides* and stem and root fragments of *Crambe maritima* (Scott and Randall 1976) can likewise be transported in water without loss of viability. Fruits, seeds and floating seedlings of mangroves, pioneers on the margins of estuaries and lagoons in tropical and subtropical regions, are well suited for long-distance dispersal and in viviparous species (*Rhizophora* spp.) viable seedlings remain buoyant in water for up to 50 days (Murray 1986b). The seedlings usually float horizontally until they get embedded in mud in shallow water. Detached branches of *Salix* and *Populus* are sometimes washed onshore in early spring and if buried by sand start to grow. Along the Indus River in Pakistan, mature plants of *Tamarix* may be carried away from river banks to other locations along the river and to the coast where they re-establish (Ridley 1930).

The colonization of these species along seashores by vegetative regeneration is advantageous for two reasons. First, the regeneration and establishment is higher from vegetative fragments than seeds. Secondly, because vegetative fragments have greater carbohydrate reserves, ramets produced from them grow and reproduce faster than seedlings. For example, *Crambe maritima* plants establishing from seeds may take 5–8 years to flower compared to only 1 year for plants establishing from fragments (Scott and Randall 1976). For both reasons, the major mode of artificial re-vegetation of denuded dunes and coastal strand is done through transportation of vegetative clumps, because they stabilize the habitat within a year (Jagschitz and Bell 1966).

A major disadvantage of vegetative regeneration is that the genetic make-up of the progeny is identical to that of parents. The outbreak of a disease or insect infestation can decimate a population within a short period of time. However, even though establishment from seeds is a stochastic event (Maun 1985; Lichter 1998), all strand species allocate some resources to sexual reproduction.

3.3 Dispersal in wind

Wind is perhaps the most efficient agent of dispersal along the coast and all species of the foredune complex use wind to disperse their disseminules. Since prevailing winds on coastal sand dunes are primarily from the direction of the sea, plants growing on the upper beach beyond the influence of waves or on the first dune ridge will disperse seeds towards the inland dunes. However, at certain times of the year the winds may be from the opposite direction so that seeds of some species do disperse towards the seashore. For example, along the Great Lakes, the establishment of *Populus* species on the mid-beach areas occurs from seeds that have dispersed from the first dune ridge (Poulson 1999). Several types of adaptations for wind dispersal, e.g. small size seeds, winged seeds, pappus around the seeds, catapult mechanisms and tumbling structures are exhibited by plants on the dunes, although some species appear to have none of these features. High wind velocity is also responsible for secondary dispersal of propagules.

3.3.1 Travel distances

The distance of disseminule travel depends on height of inflorescence above the sand surface, wind velocity and accessory structures that promote dispersal. Watkinson (1978) and Carey and Watkinson (1993) spray-painted mature infructescences of two annual dune grasses, *Vulpia fasciculata* and *V. ciliata*, and then measured the dispersal distances of single florets containing caryopses from the parents. They found that the mean dispersal distances of florets of *V. fasciculata* and *V. ciliata* were 7.0 ± 0.4 and 6.0 ± 0.2 cm, respectively. The direction of dispersal was non-random

and followed the prevailing winds. Physical disturbance caused by a person walking through the population increased the distance to 19.1 ± 1.5 cm. There was a significant positive correlation ($r = 0.62$, $P < 0.001$) between height of infructescences and dispersal distance (Fig. 3.5), but the greater the percentage cover of vegetation, the smaller was the distance travelled by the dispersal units.

In grasses, the disarticulation of the floret may occur either below the glumes or below lemma and palea and the dispersal unit is light enough to be blown by wind or washed by rain. Any bracts enclosing the caryopses can act as wings and aid in short-distance dispersal. Seeds of most species of the coastal foredunes, a large majority of the seeds, disperse close to the parents which is similar to those of other ecosystems. This applies to many foredune grass species: *Agropyron psammophilum, Ammophila breviligulata, Corynephorus canescens, Elymus canadensis, E. arenarius, E. mollis, Panicum virgatum* and *Uniola paniculata* fall into this category (Wagner 1964; Marshall 1967; Maun 1985). Similarly, in desert dunes generally the dispersal distances of grasses are short with most seeds falling close

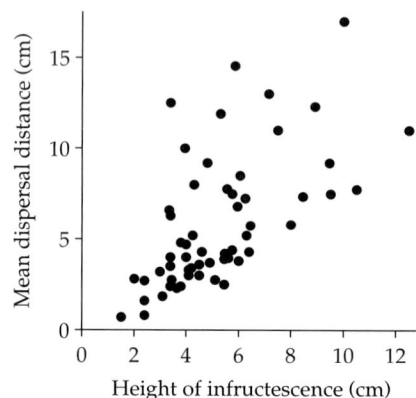

Figure 3.5 The relationship between mean dispersal distance of dispersal units and height of the infructescences of *Vulpia fasciculata* ($r = 0.62$, $P < 0.001$) on fixed sand dunes of Aberffraw and Newborough Warren in North Wales (after Watkinson 1978).

to parents (Kemp 1989), however, secondary movement by wind may be a major factor in dispersal to greater distances (Danin 1996).

Seed dispersal of *Lactuca virosa* and *Cynoglossum officinale* was studied on a sand dune system in the Netherlands (Boorman and Fuller 1984). Sets of compost-filled trays were placed to the north, south, east and west of plants at distances of 1, 2.5, 4.0, 5.5, 7.0, 8.5 and 10.0 m and seedlings were counted at regular intervals from September to May. Almost all seeds of *L. virosa* and *C. officinale* fell within a radius of 7.0 and 1.4 m, respectively. Seed release occurred during spells of windy dry weather and dispersal direction was strongly correlated with wind direction. In a coastal dune area near The Hague (the Netherlands), 50% of the seeds of *Cirsium vulgare* did not disperse farther than 1 m from the parent and 66% remained within 2 m while 11% dispersed outside the local population (Klinkhamer *et al.* 1988). Thus, in general, the dispersal distances follow a Kettle curve of dissemination of propagules from the parent (Fig. 3.6). The curve is explained by the equation as presented by Brewer (1994):

$$Y = ae^{-bx}$$

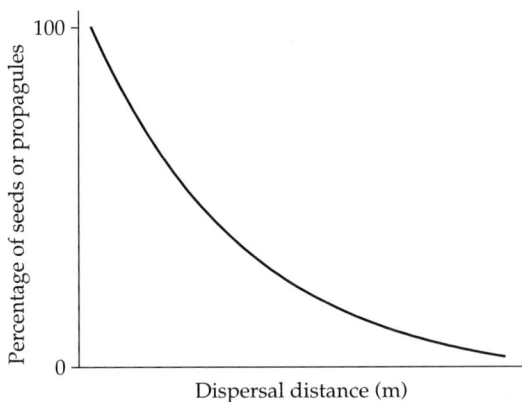

Figure 3.6 A Kettle curve showing common pattern of dispersal of seeds or other dispersing units from parent plants (after Brewer 1994).

where Y is number of dispersing units getting at least as far as distance x, a, the number of seeds at 0 distance, e, the base of natural log, b, the rate of decline of seeds with distance.

3.3.2 Role of micro-environmental factors

Dispersal distance of nutlets (seeds) of *Lithospermum caroliniense* from three mature plants growing just below the crest of the first dune ridge was studied on the Pinery sand dune system (Westelaken and Maun 1985b). At maturity in autumn the seeds of this species are dislodged from the inflorescence by wind. To determine the dispersal distance of seeds, three plants with mature seeds were spray-painted in July and dispersed seeds retrieved the next spring. Painting had the drawback of increasing the seed mass by about 1.7% but was within the 5% limit suggested as optimal in other studies (Watkinson 1978). For seed retrieval, an area of 16 m^2 (2 m on each side) around each plant was excavated to a depth of 5 cm. Sand was removed from $25 \times 50 \times 5$ cm sections, dried and sifted through a 1.5 mm nylon mesh. The location of each seed was plotted on a map and a diagram of general direction and distance of seed dispersal was drawn. More than 80–90% of the seeds were recorded within 1 m of the parent plants. However, a very small proportion had dispersed to 1.5–2.0 m. Seed density in the vicinity of adult plants ranged from 4.6 ± 1.4 to 13.7 ± 5.5 (SE) m^{-2}. Long-distance dispersal may also be facilitated in this species by adherence of seed-bearing calyxes to the fur of animals. The pattern of seed dispersal around the parent plants deviated significantly from a radially symmetrical distribution (Westelaken and Maun 1985b), suggesting that micro-environmental factors also strongly affected the direction of dispersal (Fig. 3.7). While wind velocity and direction were the principal factors, the direction and steepness of slope of the parental dune site and biotic factors such as the location of shrubs and trees also exerted a significant influence.

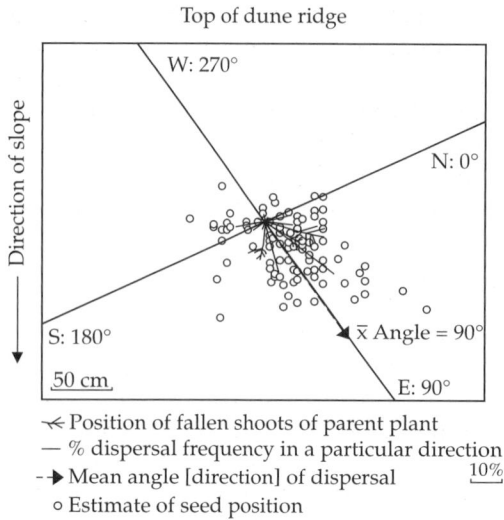

Figure 3.7 Pattern of seed dispersal, percentage frequencies of the circular distribution of seeds, mean angle of dispersal (with respect to N = 0°), position of fallen parent plant, and four compass points around the plant (after Westelaken and Maun 1985b).

3.3.3 Burial of whole inflorescences

High wind velocities sometimes bury whole plants or break off whole infructescences prior to seed dispersal of species such as *Carex arenaria* (Noble 1982) and *Corynephorus canescens* (Marshall 1967), *Ammophila breviligulata*, *Artemisia campestris* ssp. *caudata*, *Cakile edentula*, *Lithospermum caroliniense*, *Corispermum hyssopifolium*, *Elymus canadensis* and *Solidago sempervirens*. Seeds on these inflorescences usually germinate after burial and give rise to clumps of seedlings (Fig. 3.8).

3.3.4 Rolling and propelling over the sand surface (secondary dispersal)

Seeds and fruits of *Cakile*, caryopses of *Agropyron psammophilum*, *Ammophila breviligulata*, *Elymus canadensis*, *E. arenarius*, *E. mollis*, *Panicum virgatum*, *Lathyrus japonicus* and *Strophostyles helvola* do not possess any accessory structures for wind dispersal but strong winds can move them by rolling or pushing

Figure 3.8 Photographs of (a) seedlings of *Ammophila breviligulata* and (b) *Artemisia campestris* emerging from intact panicles buried in sand along the shoreline at the Pinery, Lake Huron (photographs by A Maun).

over the sand surface from the beach to other locations on the dune complex. Usually the transported distance is short unless the mature plants are situated on the windward sides of blowouts. Propagules may fall into depressions in the sand or, if they encounter an obstacle, accumulate on the leeward side. During transport, scarification of seeds may take place. A similar dispersal pattern was observed in the diaspores of *Stipagrostis scoparia* in the Negev desert (Danin 1996). Its diaspores are propelled over the sand surface by wind and spread over the leeward slopes where they are covered by sand.

3.3.5 Aerial transport by wind

Small disseminules such as viable spores of mycorrhizal fungi and small seeds of some plant species such as *Artemisia caudata* and the orchids may be lifted and carried by wind. The hairs surrounding the single seeds of *Asclepias viridiflora, Cirsium pitcheri, Salix* and *Populus* spp. help them to rise in height and drift in wind to considerable distances in open dunes. Some species of grasses such as *Andropogon scoparius, A. gerardii, Calamovilfa longifolia, Ischaemum anthephoroides, Phragmitis communis, Sorghastrum nutans* and *Stipa spartea* have hairs on glumes, calluses, spikelets, rachis or awns that may allow them to disperse farther in wind. Plumed seeds of *Solidago sempervirens* have been estimated to travel 2200 km to oceanic islands by wind (Ridley 1930). Seeds of *Pinus banksiana, P. longifolia, P. resinosa, P. strobus* and *P. sylvestris* that may grow close to the beach along certain coastal shorelines have wings that retard the speed of descent, thus allowing them to be carried considerable distances from parent trees.

3.3.6 Catapult mechanisms

Seeds of *Oenothera biennis, Strophostyles helvola, Ceanothus americanus* may be catapulted from capsules several metres from the infructescence at maturity when buffeted by wind. The terete capsule of *O. biennis* splits longitudinally from the top and the seeds are then released in both the windward and leeward direction from the parent when the tall raceme is bent by wind to the windward direction and then recovers towards the former position. In an experiment, the dispersal distance of seeds from each position within the pod of the legume *Strophostyles helvola* was measured in a lab (A. Maun unpublished data). Each seed within a pod was colour coded by injecting ink of a different colour with hypodermal syringes and pods were then placed singly in the centre of a large room and allowed to dehisce. The dispersal distance of each seed and its direction of movement on the compass were recorded. The pods dehisced along the sutures with each half corkscrewing outward in opposite directions. The mean dispersal dis-

tance of seed in positions 1, 2, 3, 4 and 5 from the proximal to the distal end was 33 ± 3 cm, 48 ± 10 cm, 84 ± 17 cm, 33 ± 25 cm and 163 ± 33 cm, respectively. The angle of dispersion from the pod for each seed was respectively, 321°, 307°, 230°, 33° and 34°. Thus, seeds at the distal end of the pod travelled further than others.

3.3.7 Tumbling

At maturity, the base of stems of tumbling species decays, breaks off or the inflorescence is torn off allowing the aerial portion of the plant to be rolled over the sand surface by strong winds. As these structures roll along, seeds are knocked loose and scattered. *Corispermum hyssopifolium, Cycloloma atriplicifolium* and *Salsola kali* are common plants of sea coasts and lacustrine shorelines of temperate regions that utilize this mode of dispersal. Plants of *Cakile* species, usually with lower fruits still attached, are sometimes uprooted by high waves and then tumbled inland by wind to a new location. Another example of tumble dissemination is provided by the dioecious grass species of *Spinifex* that are inhabitants of seashores of Australia, the Malay Peninsula, Ceylon (Sri Lanka), India, Siam (Thailand), Formosa (Taiwan) and China (Ridley 1930). Each spikelet of *S. squarrosus* bears a long radiating spike with long slender brackets ending in spines and the detached whole head and upper peduncle forms a sphere about 15 cm in diameter that tumbles swiftly along the sand in the wind, dropping seeds as it goes.

3.4 Dispersal by animals

Animals perform a useful function in the dispersal of many plants along sea coasts. Plants have various morphological features that aid transport by animals and produce edible parts that attract animals (Howe 1986). However, the relationship is of a diffuse nature because fruits of one plant species are often eaten by a number of different animals and/or bird species. Pair-wise co-evolution is unlikely due to variable spatial distribution, differential abundance of both plants and animals, seasonal production of fruits, and the times taken for evolutionary change by plants and animals (millions of years) do not coincide. Moreover, the animals adjust to whatever plants are available rather than the reverse (Howe 1986).

3.4.1 Disseminules ingested but not digested

A list of birds that consume fruits and seeds of coastal sand dune species is given in Table 3.3. Ridley (1930) may have assumed that fruit and seed consumption by birds is evidence of seed dispersal. This assumption may not always be true because some birds completely digest the seeds (Gillham 1970). Yet many birds are effective agents of seed dispersal of several plant species of foredunes, because seeds of many species are ingested but not digested and emerge unharmed and possibly even with higher germinability.

3.4.2 Disseminules cached but not recovered

Caching of seeds and fruits is an important activity of some animal and insect species for consumption later during inclement weather and in winter. For example, ants are frequently responsible for secondary dispersal of seeds of some species, such as *Panicum virgatum*, that are transported to their hills where some escape consumption and germinate under optimal conditions. Burrowing insects, namely white tiger beetles (*Cicindela lepida*) and digger wasps (*Microbembex monodonta*), can bury seeds as part of their digging activities.

Squirrels (*Sciurus carolinensis*) are important agents of dispersal of some foredune tree species. They collect acorns of *Quercus velutina*, *Q. rubra* and *Q. muehlenbergii* and then bury them in 3–5 cm of sand. Most of these fruits are removed later and eaten but some escape for various reasons and germinate the next spring. The establishment and further expansion of *Quercus velutina* in the sand plains of Connecticut was attributed only to the caching activity of squirrels (Olmsted 1937). The

Table 3.3 Name of bird species that may be responsible for the dispersal of fruits and seeds of plant species of coastal sand dunes (assembled from Ridley 1930; Guppy 1917; Van der Pijl 1982)

Blue jays *(Cyanocitta cristata)*. *Prunus virginiana, P. pumila, Vitis riparia, V. aestivalis.*

Snow buntings *(Plectrophenax nivalis)*. *Honkenya peploides.*

Bohemian waxwings *(Bombycilla garrula)*. Cedar waxwings *(B. cedrorum)*. *Arctostaphylos* spp., *Juniperus virginiana, J. communis, Vaccinium* spp.

Myrtle warbler *(Dendroica coronata)*. *Rhus radicans, Juniperus virginiana.*

Grouse family (Tetraonidae) Ring-necked pheasant *(Phasianus colchicus)*. *Carex arenaria* (Noble 1982). *Corema conradii, Empetrum nigrum, Arctostaphylos uva-ursi, Juniperus horizontalis, Symphoricarpos racemosus, Calluna vulgaris, Rosa blanda, Vaccinium* spp., *Shepherdia canadensis* and grasses.

Ravens *(Corvus corax)*, Starlings *(Sturnus vulgaris)*. *Symphoricarpos racemosus, Hippophae rhamnoides.*

Starlings *(Sturnus vulgaris)*. *Prunus virginiana, P. pumila, Rhus radicans, Rosa* spp.

Pine grosbeak *(Pinicola enucleator)*. *Vaccinium* spp., *Empetrum nigrum, Shepherdia canadensis, Carex* spp., *Gaylussacia baccata, Symphoricarpos racemosus.*

Pigeons *(Columba* spp.). *Quercus robur.*

Gulls *(Larus* spp.). *Empetrum nigrum* (Gillham 1970).

The robin *(Turdus migratorius)*. *Juniperus communis, J. virginiana* and *R. radicans.*

acorns in these caches numbered from 23 to 84 and their buried depth ranged from 5 to 10 cm. Excavation of many caches of acorns indicated that *Q. velutina* seedlings did not establish without being buried in sand and if the fruits were buried under >5 cm of sand, the seedlings were unable to emerge. Furthermore, caching of *Q. rubra* acorns by birds and rodents not only protected them from white-tailed deer but also increased their germination and seedling emergence (García *et al.* 2002). Chipmunks—*Eutamias quadrivittata* and *Tamias striatus*—also eat and cache seeds and fruits of several local species of other foredune and transition zone species such as *Quercus* spp., *Amelanchier laevis, Prunus pumila, P. virginiana, Juniperus communis, J. virginiana* and *Symphoricarpos racemosus.* The mouse, *Peromyscus maniculatus,* assists in the dispersal of *Cakile maritima* by carrying off the fruits to safer sites for consumption and then missing a very small proportion (only 0.002%) that germinate the next spring (Boyd 1991). Rodents, the major consumers of *Lithospermum caroliniense* nutlets, may occasionally aid in their dispersal by transporting seeds and not consuming them (Weller 1985). Seeds of *Carex arenaria* are dispersed by rabbits in Britain (Noble 1982).

3.4.3 Clinging fruits

The burrs of *Xanthium strumarium* bear hooked spines on their seed coat that cling to the fur of animals who transport them to other locations before detachment. *Cenchrus* species, occurring on seashores of islands in warm parts of the world, have hard acute spines that stick to the hair of animals (Ridley 1930). *Daucus maritima* has stiff spines that adhere to the feathers of sea birds.

3.4.4 Use as nesting material

Sea birds such as the booby (*Sula* spp.), use fragments of *Sesuvium portulacastrum* for their nests on shores where the plants take root and

start to grow (Ridley 1930). Much of the wide distribution of *S. portulacastrum* is a result of this mode of dispersal.

3.5 Human influence

Humans are the major transporters of species from one continent to another and many deliberate or accidental introductions of coastal sand dune species may be attributed to human actions. For example, *Ammophila arenaria* from Europe was deliberately planted along the sea coasts of Australia, New Zealand, South Africa, Argentina, the west coast of North America and the Middle East for stabilization of denuded sand dune habitats. The superior genetic, demographic, ecological, physiological and biotic traits of *A. arenaria* make it highly desirable for dune stabilization; however, the species has become invasive (an alien species that expands its range after its introduction and disturbs the ecological balance) and proliferated in the introduced habitats (Hertling 1997). For example, it has replaced native species along the Pacific coast of California and altered the topography of the foredunes and the dune complex (Barbour *et al.* 1976). There has also been a loss of species-rich indigenous communities consisting of *Leymus mollis–Abronia latifolia* and *Poa macrantha–Lathyrus littoralis* along the sea coast of Oregon (Wiedemann 1990). Similar impacts on the physiography of foredunes and community composition have been observed in Australia and New Zealand where *A. arenaria* has replaced the native communities dominated by *Spinifex hirsutus–Festuca littoralis* in Australia and *S. hirsutus–Desmoschoenus spiralis* in New Zealand (Kirkpatrick 1993; PN Johnson 1993). Invasiveness of this species has not yet been documented in South Africa, but caution is advised because of experiences in other regions (Hertling 1997).

Two naturalized species, *Cakile edentula* and *C. maritima,* were introduced along the Pacific coast of California. *Cakile edentula* var. *edentula* was introduced as a desirable beach plant from the Atlantic coast of North America around 1888. The species dispersed rapidly and within about 40 years occupied the sandy shorelines all along the Pacific coast (Barbour and Rodman 1970). Its congener, *C. maritima*, was introduced from Europe in 1935 for the same reasons and it also expanded very quickly and at present not only occupies all sandy shorelines of the Pacific coast, but is also displacing *C. edentula* at several coastal locations because of its higher rate of seed production and seedling establishment (Boyd and Barbour 1993). In sand dunes and beaches used for recreation, the seeds of *Carex arenaria* are dispersed by people visiting the beach (Noble 1982). The seeds of *Lithospermum caroliniense* are dispersed long distances by native Indians who used to extract a dye from its rootstock. Over the centuries, humans have also played a role in the dispersal of countless food plants. Along sea coasts, coconut (*Cocos nucifera*), is a highly desirable fruit for human consumption. Its original distribution in the tropics has been greatly expanded by artificial planting.

3.6 Summary

On coastal sand dune systems, water, wind, animals and gravity are the principal agents of dissemination. The first plants to evolve mechanisms of tissue aeration and buoyancy were hydrophytes, however, many plants living on foredunes by the sea are also well adapted to disperse in water. To use water as a medium for dispersal, a seed, fruit or a vegetative fragment must remain viable and resist decay caused by waterlogging, salt, micro-organisms and other hazards in transport. Plant species of coastal strands use several mechanisms for remaining buoyant in water, e.g. presence of air within the disseminule, buoyant tissue in seed coat, spongy albumin or mesophyll, hollow pericarp, lightweight cotyledons, corky covering on seeds or appendages and large surface to volume ratio. Some disseminules do not have any modifications but do dis-

perse on media such as pumice or driftwood. Some of the most successful species using seawater for dispersal belong to the genera *Cakile, Cocos, Ipomoea, Carex, Ammophila, Lathyrus* and *Honkenya*. Some, such as *Carex, Ammophila, Leymus, Elymus, Ipomoea* and *Canavalia* disperse both as seeds and vegetative fragments in seawater. Wind dispersal of coastal species is similar to that of land plants in other ecosystems. Disseminules are rolled and propelled along the sand surface, by suspension in air, borne aloft by hairs or wings or launched by catapult mechanisms and whole plants may be mobilized as tumbleweeds. Species with edible fruits are dispersed by birds that ingest but do not digest them, are cached but not recovered by rodents, transported to ant hills where they may escape consumption and travel on feathers or fur of animals.

Humans have also introduced coastal species to other seashores to stabilize damaged dune systems, but in some cases diseases and insects were also introduced and introduced species were so aggressive that they replaced indigenous taxa.

Arrival at a new site is of value to a species if it is able to successfully colonize which means that disseminules must not only germinate, but plants must establish and reproduce in the new location and build up a sustainable population. This is not always the case as many species may disperse and land in a new habitat but are unable to establish successfully.

The information presented here is based on existing data on dispersal along sea coasts at present, but this topic has not yet been thoroughly investigated and more effort is needed to examine the modes of dispersal of dune species. Another worthwhile objective would be to identify and determine the fates of propagules of plant species deposited on seashores around the world. We may then be able to better appreciate the most appropriate dispersal strategies and identify the factors limiting in establishment and survival.

CHAPTER 4

Seed banks

4.1 Introduction

The soil seed bank refers to a reservoir of viable seeds present on the soil surface or buried in the soil. It has the potential to augment or replace adult plants. Such reservoirs have regular inputs and outputs. Outputs are losses of seeds by germination, predation or other causes, while inputs include dispersal of fresh seeds from local sources and immigration from distant sources (Harper 1977). Since sand dunes are dynamic because of erosion, re-arrangement or burial by wind and wave action, efforts to find seed banks have largely been unsuccessful. Following dispersal, seeds accumulate in depressions, in the lee of plants, on sand surfaces, on the base of lee slopes and on the driftline. These seeds are often buried by varying amounts of sand. Buried seeds may subsequently be re-exposed or possibly lost over time. However, the existence of a seed bank can not be denied.

Plant species may maintain a transient or a persistent seed bank depending on the longevity of seeds. In species with transient seed banks, all seeds germinate or are lost to other agencies and none is carried over to more than one year. In contrast, in species with a persistent seed bank at least some seeds live for more than one year. The four types of seed banks described by Thompson and Grime (1979) provide useful categories for discussion of coastal seed bank dynamics of different species. Type I species possess a transient seed bank after the maturation and dispersal of their seeds in spring that remain in the seed bank during summer until they germinate in autumn. Type II species possess a transient seed bank during winter but all seeds germinate and colonize vegetation gaps in early spring. Seeds of both types are often but not always dormant and dormancy is usually broken by high temperatures in type I and low temperature in type II. Type III species are annual and perennial herbs in which a certain proportion of seeds enters the persistent seed bank each year, while the remainder germinate soon after dispersal, and Type IV species are annual and perennial herbs and shrubs in which most seeds enter the persistent seed bank and very few germinate after dispersal. Based on seed longevity a new system with three classes was proposed by Thompson et al. (1997) in which summer and winter dormant types I and II were classed as transient (<1 year), type II as short-term persistent (>1 to <5 years) and type IV as long-term persistent (>5 years). All three types of seed banks are found in coastal dune systems. An elegant synthesis of studies on seed banks, including seed ecology, ecological processes, factors affecting seed bank dynamics and evolutionary implications was presented by Leck et al. (1989). This chapter critically examines seed banks in foredunes and the dune complex and reviews seed bank dynamics in the disturbance-prone communities of coastal dunes.

4.2 Types of coastal soil seed banks

The presence of a transient or persistent seed bank is primarily dependent on the seed ecology of the species, its microhabitat on the sand dune, degree of disturbance, the length of time between disturbances, relative stability and stage during successional development of the dune system. Seed size is an important factor that is related to whether a species would possess a transient or a persistent seed bank. Usually species with a persistent seed bank have small seeds (Thompson and Grime 1979).

4.2.1 Transient seed bank

Many species of the foredunes possess a transient seed bank and can only survive in habitats with recurrent sand movement, high wind velocities and full sunlight. They are opportunistic, are dispersed by wind and water, may possess both vegetative and reproductive means of regeneration and require regular disturbance for survival. The energy required to develop adaptations for persistence of the seed bank will be too costly and will not increase their fitness. Based on observations over the last 30 years on the Pinery sand dunes, the persistence of fugitive annuals and monocarpic perennials on the strand is dependent on prolific and reliable seed production and effective dispersal. Seed production in these species has fluctuated over time but has never failed in these years. These plants are resilient and capable of high phenotypic plasticity in size which varies depending on the microhabitat of their occurrence on the shoreline. On beach sections used heavily for recreation by people, most of the seedlings are killed by trampling, but in early spring or next summer, the driftline is covered yet again with a large number of seedlings from seeds that have immigrated from nearby undisturbed beaches.

Planisek and Pippen (1984) examined sand samples (7.5 cm diameter × 11 cm deep) from the beach, foredunes, slack and dune forest in November 1981 along the eastern shore of Lake Michigan. The sifting of soil samples yielded no seeds from the beach and foredune which is rather surprising because *Cakile edentula* (a common plant on the beach) must have dispersed its fruits by November. The slack and forest contained 0.74 and 3.2 seeds per sample, respectively, but none of these seeds

was viable. It was concluded that coastal foredunes do not have a seed bank primarily because of substrate instability. Similarly, Looney and Gibson (1995) showed that in four habitats, the frequently disturbed beaches (extensive sand movement, salt spray), unvegetated sandy areas, early successional dredge spoils and the strand of a coastal barrier island of Florida, the seed banks of *Cakile constricta*, *Oenothera humifusa*, *Iva imbricata*, *Hydrocotyle bonariensis* and *Heterotheca subaxillaris* were transient and poorly developed. The dune annuals, *Cerastium atrovirens* (Mack 1976), *Vulpia fasciculata* and *V. ciliata* (Carey and Watkinson 1993) at Aberffraw, North Wales, also did not have any carry-over of seeds from one year to the next. This is rather surprising because annuals or biennials are more likely to have persistent seed banks in habitats with recurrent disturbance (Thompson *et al.* 1998).

Baptista and Shumway (1998) determined the seed bank composition of sand dunes with different disturbance histories at Cape Cod National Seashore and found that both the density of viable seeds in sand samples and species diversity were very low (Table 4.1). The seeds exhibited a clumped distribution, probably because of (i) accumulation of seeds in depressions following wind dispersal, and (ii) burial of complete infructescences of plants before the release of their seeds. A similar seed bank pattern was also observed in desert dunes by Kemp (1989), who concluded that the frequency of seed distribution was kurtotic, with a large number of seeds in some samples and a few or none in most and that the abundance of a species was not a good predictor of the abundance of its seeds in the seed bank.

Seed bank dynamics of *Cirsium pitcheri*, an endangered species in Canada and federally threatened species in the United States of America, was examined by Rowland and Maun (2001). They removed 700 cores of sand each 2.5–5.0 cm diameter × 35 cm deep located on transects perpendicular to the beach and extending to the leeward side of the second dune ridge. They then sifted the sand to

Table 4.1 Mean number of seedlings (±SD) emerging per core sample (n = 38) of 15 cm diameter and 10 cm deep from four sand dune sites within the boundaries of Cape Cod, Massachusetts, National Seashore (adapted from Baptista and Shumway 1998)

Name of species	Dune site			
	High Head	**Marconi Station**	**Coast Guard**	**Duck Harbour**
Ammophila breviligulata	0.03 ± 0.16	0.16 ± 0.82	0.16 ± 0.68	0
Artemisia caudata	0	0	0.03 ± 0.16	5.70 ± 19.4
Chenopodium rubrum	0	0.03 ± 0.16	0.05 ± 0.23	0.03 ± 0.16
Hudsonia spp.	0	0.05 ± 0.23	0	0
Solidago sempervirens	0	0.08 ± 0.27	0.03 ± 0.16	0.10 ± 0.50
Other species	0.03 ± 0.16	0.03 ± 0.16	0.05 ± 0.23	0.15 ± 0.43
Total	0.05 ± 0.23	0.34 ± 1.07	0.32 ± 0.84	5.97 ± 19.80

remove any seeds of *C. pitcheri*. In 1997 none of the samples contained seeds. The procedure was repeated during 1998 by increasing the diameter of the samples to one metre and locating the cores near existing populations. However, the number of samples was reduced to four. Only four seeds were recovered, one of which germinated, while the other three were dead. Thus, there were strong indications that the species only maintained a transient seed bank. Similarly, Rabinowitz and Rapp (1980) showed that *Andropogon scoparius*, a dominant grass of lee slopes of dune ridges, maintained a transient seed bank. Transient seed banks have also been reported in *Populus tremuloides*, *P. balsamifera* and *Salix* spp. that depend heavily on vegetative regeneration from root suckers to increase their population size. Bossuyt and Hermy (2004) showed that most of the early successional species of dune slacks did not form persistent seed banks and successional progression occurred primarily by dispersal of seeds.

In coastal sand dunes, persistence in soil is compromised in favour of other benefits such as survival from burial and seawater dispersal in the beach and strand habitat. A persistent seed bank may actually be a disadvantage in this habitat. Thompson *et al.* (1993) showed that size and shape of seeds determine the likeli-

hood of persistence in the soil. They examined 44 species of seeds and 53 species of fruits and concluded that seeds persisted longer than their fruits because they were more compact than fruits and that small seeds persisted longer than large seeds mainly because they were buried more easily. It appeared though that many other factors such as seed shape, germination physiology, defence against predators and pathogens, and burial depths modify the longevity of both large and small seeds in the seed bank. Nevertheless, careful studies are needed to examine the relationships of seed mass, seed shape and disturbance levels on the fate of seeds following dispersal of dune species of sea coasts using the radioactive tracer technique employed by Watkinson (1978).

4.2.2 Persistent seed bank

A persistent seed bank would include a body of reserve seed pool stored over more than one year. It would usually contain diverse genetic material from previous generations. Weller (1985, 1989) sowed nutlets of *Lithospermum caroliniense* in dunes of Lake Michigan and monitored seedling emergence. Quadrats with sand deposition ranging from 0–3 cm showed a seedling emergence of 55–73% in the first year, 1.6% in the second year and 0%

in subsequent years. In a similar experiment on Pinery dunes along Lake Huron, only 1 (due to a dry summer), 19 and 0% seedlings emerged in the first, second and third years, respectively (Westelaken and Maun 1985a). Excavation of seeds in the third summer showed that 13% of the seeds were still dormant and viable, 11% were non-viable, 15% had germinated and died, 21% were cracked but empty and the rest (21%) were missing. Both studies suggested that *L. caroliniense* maintained a short-term persistent seed bank in dune systems. Van Breemen (1984) showed that three species of monocarpic perennials—*Cynoglossum officinale, Echium vulgare, Anchusa arvensis*—growing in a sand dune system in the Netherlands maintained a small short-term persistent seed bank within the surface layer and partially in above-ground infructescences. Seeds of *Cynoglossum officinale* remained viable for at least 2 to 3 years (van Breemen and van Leeuwen 1983).

4.3 Soil seed bank dynamics

Seed banks are important in the dune complex because they provide insurance against the risks of frequent catastrophic disturbances caused by high wind velocities and violent wave action. The total amount of seed rain may range from very small to very large depending on the species, growing conditions during the growing season, soil fertility, and predation prior to dispersal (Chapter 3). However, the input of seeds to the seed bank may be very small because a large proportion (up to 99%) may be lost each year (Fig. 4.1) and losses may be substantial. For example, despite the large potential seed rain a very small fraction of seeds of three beach annuals, *Triplasis purpurea, Cenchrus tribuloides* and *Heterotheca subaxillaris*, was successful in seedling establishment, e.g. only 1% of the seed rain of *H. subaxillaris* was contributed to the transient seed bank (Cheplick 2005). Similarly, in a coastal sand dune, 98.4 and 75% of the seeds produced by *Lactuca virosa* and *Cynoglossum officinale* plants, respectively, failed to produce

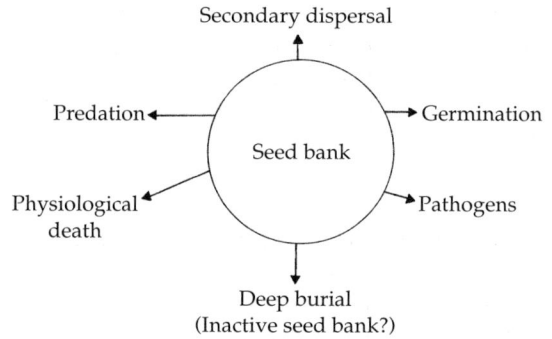

Figure 4.1 Seed bank dynamics: losses through germination, predation, pathogens, secondary dispersal, physiological death, and deep burial.

seedlings during the next growing season (Boorman and Fuller 1984). According to Kachi and Hirose (1985), 98% of the seeds produced by *Oenothera glazioviana* plants on the foredunes of Japan were lost within one year. The annual seed production of *Ptelea trifoliata* on Lake Michigan dunes was estimated at more than $330\,000$ seeds ha^{-1}, but only about 6% of these seeds successfully established seedlings (McLeod and Murphy 1977). A brief description of some of the main causes of seed loss is presented below.

All plants, including coastal sand dune species, experience a decline of a proportion of seeds by germination when temperature and moisture are optimal and by predation. Louda (1989) made several generalizations concerning the impact of predators on the evolution and persistence of seed banks in sand hills of Nebraska: (i) seed predation both in pre-and post-dispersal stages has a significant impact on input–output balance of seed banks, (ii) occurrence of a transient or persistent seed bank may depend on predator impact, (iii) predators are selective and cause differential losses among individuals, species and years, (iv) fugitive annual species depend heavily on large seeds for regeneration and persistence and predators prefer large seeds, and (v) environmental factors strongly influence the amount of seed loss to predators.

Seeds escape predation through dispersion and/or predator satiation. Few studies, however, are available on predation effects on the seed banks of strand and foredune species. Along the Pacific Coast, *Peromyscus maniculatus* consumes more than 99% of the fruits of *Cakile maritima* so that only a very small proportion are left to produce seedlings (Boyd 1991). Plant density and individual fitness of *Cirsium canescens*, a fugitive monocarpic perennial of Nebraska sand hills, are significantly reduced by predation (Louda 1989). In chaparral vegetation of California, the seed bank size of *Arctostaphylos*, a sand dune species, is very high but the rate of vertebrate predation of seeds in six months may be as high as 60%. However, 100% of the seeds of *Ceanothus crassifolius* are consumed by harvester ants (Parker and Kelly 1989). Ants remove a large number of seeds of *Panicum virgatum* to their colonies where many seeds are destroyed, but some do escape predation and some become part of the seed bank. In *Quercus velutina*, a major species of coastal sand dunes, large number of acorns are consumed by squirrels; however, because of predator satiation, especially in mast years, a small proportion of fruits do survive in acorn caches and contribute to the seed bank (Olmsted 1937). At the Pinery, sparrows, snow buntings, gold finches, warblers and starlings consume seeds of grasses and fruits of *Cornus stolonifera*, *Prunus pumila* and *P. virginiana* during their annual autumn migration, thus reducing the number of seeds that may have contributed to the seed bank.

On the foredunes a major seed loss occurs because of deep burial of seeds. The primary mode of burial is recurrent movement of sand by wind rather than cracks in soil, rain washes and earthworm activity as in some other ecosystems. Burial depths are highly variable depending on propagule size (large or small), dispersal (wind or water), microhabitat of the parent (beach or dune), wind velocity (high or low) and obstructions (vegetated, smooth or uneven with depressions) on the sand surface. Usually small seeds are more easily dispersed by wind and buried deeper than large seeds.

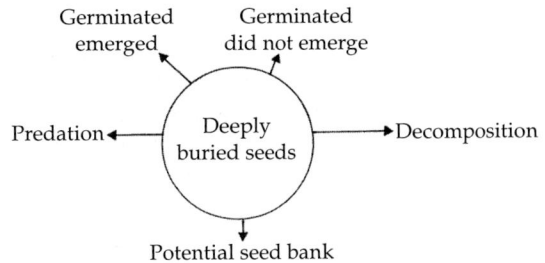

Figure 4.2 Fate of seeds buried deep in mobile foredunes along sea coasts.

Seeds of plants occupying the beach and foredune habitats are more prone to burial than those of less mobile dune ridges. Unlike other ecosystems, in sand dunes a large quantity of propagules may be present in deep locations.

What is the fate of such deeply buried seeds in sand dune systems? Maun (1985) marked ten microsites in which seeds of *Calamovilfa longifolia* had accumulated in autumn and measured the amount of natural sand accretion on these seeds during one winter. Seeds were buried to depths ranging from 1 to 29 cm during the winter. Excavated seeds in early spring were classified into four types: (i) those that germinated and emerged as seedlings in early spring, (ii) those that germinated but the seedlings were unable to emerge above the sand surface; such seedlings were decomposed by soil micro-organisms, (iii) those that did not germinate but were missing and probably consumed by seed predators or decomposed by fungi, and (iv) those that remained part of the seed bank especially at greater depths of burial (Fig. 4.2). Evidence for the first three types was shown by several other species, *Lithospermum caroliniense* (Weller 1985, 1989), *Cirsium vulgare* and *Cynoglossum officinale* (De Jong and Klinkhamer 1988) and six dune annuals (Pemadasa and Lovell 1975). Zhang and Maun (1990d, 1994) showed evidence that there was a potential for several other species to follow option (iv), however, the length of time the deeply buried seeds would survive in the seed bank is unknown. Olmsted (1937) excavated several caches of *Quercus velutina* acorns made by squirrels in a sand plain in Connecticut. In one cache, he discovered 27 acorns at a depth of about 10 cm. He found that 2 seedlings had emerged above-ground from this cache, 23 acorns had germinated but the seedlings had failed to emerge and two acorns had not germinated. In spite of remaining below-ground, all unemerged seedlings in the cache were still alive after 2.5 years.

In general, deeply buried or cached seeds do not contribute to an effective seed bank (Baker 1989). To be effective, at least a certain proportion of the seed population must be located at a depth that is appropriate for germination and seedling emergence. A similar conclusion was emphasized by Kemp (1989), who showed that effective seed banks of many dune annuals in North American deserts were formed by seeds located in the top 1 cm of soil and of shrubs approximately 4 cm. He suggested that seeds located deeper than 7 cm should not be considered an effective seed bank. However, in the Negev desert, Danin (1996) showed that the entire dune contained large quantities of seeds at all depths and that they were, in fact, part of an effective seed bank because constant erosion of sand decreased the depth at which buried seeds were located. When erosion removed a layer of sand, the seeds present in it were carried with sand to the leeward side of the dune and seeds located below the new sand surface were available for germination after a rainfall. Similar observations were made on sand dunes of the Pinery (Payne 1980). Thus buried seeds have the possibility of becoming an effective seed bank following erosion of sand. However, if these seeds were buried for a long time, they would probably deteriorate and be lost.

In desert dunes, there was high annual and seasonal variability in seed banks that depended on primary production and was related to variability in rainfall from year to year (Kemp 1989). Also, large transient seed banks were found in the vicinity of shrubs and trees during years of high rainfall, but their size decreased to very low levels in years of low rainfall. The seed banks of coastal and desert dunes are similar in that they are both highly heterogeneous spatially and the pattern of seed distribution is kurtotic. This patchiness in seed banks is related to patchiness in vegetation, seed dispersal close to the parent plants and accumulation of seeds in depressions in the sand surface.

4.4 Soil bud bank

All of the perennial species of foredunes reproduce both sexually and vegetatively through rhizomes, stolons or root suckers. For example, in the juvenile phase of growth, *Ammophila breviligulata* produces primary, secondary and tertiary rhizomes that bear one bud on each node and about 15 buds per metre of rhizome length (Krajnyk and Maun 1981a). On an intact rhizome, about 49% of the buds produce ramets, while the rest remain dormant. However, if a rhizome is fragmented into single node units, all buds are released from dormancy. Within an established population along the shoreline the number of buds on the landward end of the population was about 120 per 0.25 m^2 (Fig. 4.3) and decreased towards the lakeward end (Maun 1984). The

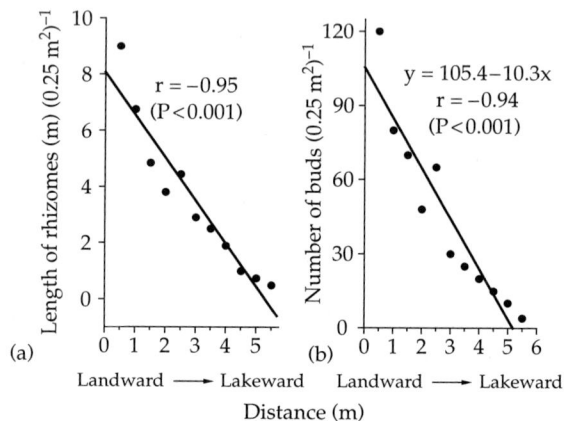

Figure 4.3 The number (b) of buds of *Ammophila breviligulata* per quadrat in contiguous quadrats (0.5 × 0.5 m) along a gradient from the landward end to the lakeward end of a beach along Lake Huron (adapted from Maun 1984).

large bud banks on rhizomes and stolons are advantageous to coastal dune species for recovery from herbivory or destruction from high wave action.

4.5 Soil seed banks of stabilized dunes

The transition of vegetation from graminoids in the open dunes to woody vegetation in the back dunes also changes the seed bank development. Looney and Gibson (1995) found well-developed seed banks with decreasing disturbance in dune woods, back slopes, marshes, wet swales and dry swales (low areas with freshwater lenses close to the sand surface) of a coastal barrier island off the north-west coast of Florida. Similarly, on stabilized Pinery sand dunes, species such as *Arctostaphylos uva-ursi*, *Viburnum* spp., *Rosa blanda*, *Shepherdia canadensis*, *Rhus aromatica*, *Ceanothus americanus* and *Symphoricarpos racemosus* maintained a persistent or short-term persistent seed bank.

Looney and Gibson (1995) also showed that species composition in the seed bank of coastal dunes clearly corresponded to that of the above-ground vegetation. Similar above- and below-ground correspondence was observed in tidal freshwater wetlands (Baldwin and DeRico 1999), although it may vary with the wetland and often with the site within a wetland (Leck 1989). However, in contrast, Thompson and Grime (1979) did not find any correspondence in ten ecologically contrasting habitats between species composition of above-ground vegetation and composition of seeds in the seed bank. Correspondence may therefore be habitat-dependent.

Examination of mature woodland communities in sand dune succession (Lichter 1998) reveals that many component species are represented in other coniferous and/or deciduous forest communities and their seed banks will probably have similar patterns (Pickett and McDonnell 1989). However, even though *Quercus velutina* and *Q. geminata* communities occur in well established habitats with no dis-

turbance, both species possess a transient seed bank (Olmsted 1937).

4.6 Seed bank potential

Zhang and Maun (1994) enclosed seeds of seven dune species of Lakes Erie and Huron in nylon mesh bags and buried them at 5, 10, 15, 25 and 50 cm depths in sand. Four bags of each species were retrieved after 167, 285, 372, 544 and 909 days and the seeds were categorized as germinated in the field + germinable in a greenhouse, ungerminated but viable, and dead. A significant proportion of seeds were lost from seed banks due to germination in all seven species, and interestingly this decreased with burial depth. The seed bank increased at greater depths of burial but the actual size was dependent on species, depth of burial, length of burial period and interactions between depth and burial length. The greatest statistical variation was accounted for by different species. After approximately 2.5 years of burial, seed loss during burial in most species was largely due to germination and in *A. breviligulata* and *C. longifolia* to both germination and death (Fig. 4.4). *Corispermum hyssopifolium* was unique in that both germination and mortality were very low. *Cirsium pitcheri* in contrast maintained a transient seed bank but when seeds were buried, 28–43% of the seeds remained ungerminated but viable after 13 months of burial at ≥15 cm depths (Rowland and Maun 2001), indicating the formation of a seed bank enforced by burial depth.

Pemadasa and Lovell (1975) showed that seeds of six dune annuals did not germinate as a result of burial and as the depth of burial increased, the number of ungerminated seeds increased. Similarly, seed germination of three foredune species, *Atriplex laciniata*, *Cakile maritima* and *Salsola kali*, decreased to <10% when buried at 16 cm depth (Lee and Ignaciuk 1985). Obviously deeply buried seeds would experience more moist soil, low oxygen, lower sand temperatures with fewer fluctuations, higher CO_2 levels, less light and greater accumulation

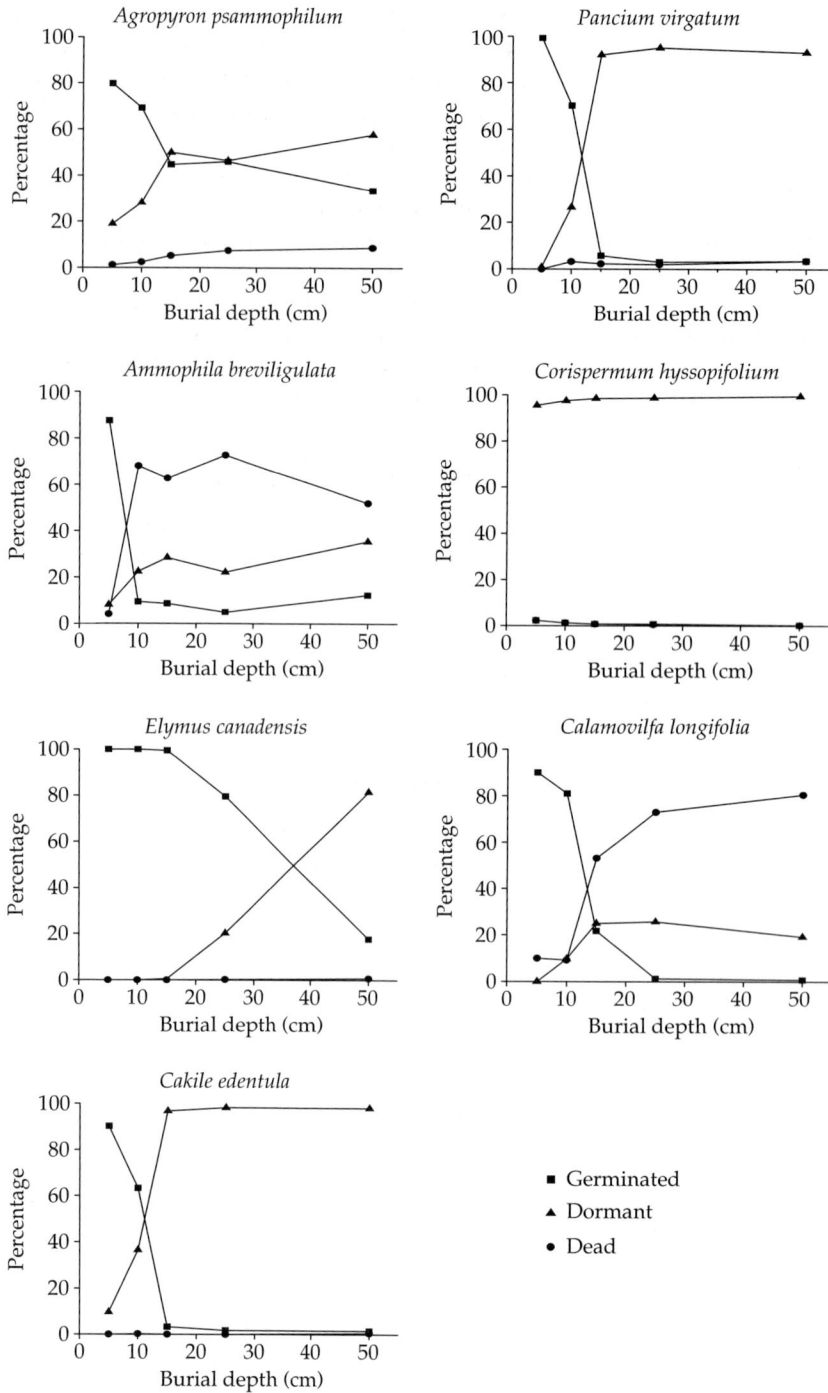

Figure 4.4 Percentage depletion (germinated, died, dormant) of seeds of seven dune species after 909 days of burial in sand at five different depths of 5, 10, 15, 25, 50 cm on a sand dune system along Lake Huron (adapted from Zhang and Maun 1994).

of toxins. Thus, there seems to be selection against premature germination of seeds at soil depths unsuitable for seedling emergence. This would be ecologically adaptive because an ungerminated seed lying deep in the soil would have the potential to produce a seedling at a later date provided erosion of sand from the surface reduced the depth of seed. One drawback of deep burial would be increased mortality of seeds with an increase in the duration of burial. *Calamovilfa longifolia* and *Ammophila breviligulata* seeds showed high mortality (up to 70%) due to decay, fungi, and bacteria. Further studies are needed to determine the dormancy mechanisms of deeply buried seeds.

Using regression equations for the relationship between time after burial of seeds and the percentage of seeds remaining in the seed bank (germinable + dormant/total buried seeds), the number of years required for depletion of all seeds were calculated (Table 4.2). Based on these calculations the seeds of *Agropyron psammophilum, Ammophila breviligulata, Calamovilfa longifolia, Elymus canadensis* and *Panicum virgatum* would be depleted completely in 3 to 5 years while the two annual species, *Cakile edentula* and *Corispermum hyssopifolium*, would be capable of remaining in the seed bank for a surprisingly long time: 10 and 20 years, respectively.

4.7 Above-ground seed banks

Several foredune species retain seeds in cones or inflorescences above-ground and release

Table 4.2 Regression equations showing the relationship between the mean percentage of potential seed bank after burial (germinable + dormant/total number of buried seeds) (Y) and the age of seeds (X) after burial for each species. The expected length of time (days after burial) by which all seeds (mean of the five burial depths) would have been depleted/died was calculated according to the fitted function for each species by setting Y = 0 (adapted from Zhang and Maun 1994)

Species (life form)	Function	Days (years) after burial when all seeds would have depleted/died
Agropyron psammophilum (P)	$Y = 88.90 - 2.43X^{1/2}$ ($r^2 = 0.87, P < 0.01$)	1338 (3.66)
Ammophila breviligulata (P)	$Y = 95.99 - 2.79X^{1/2}$ ($r^2 = 0.78, P < 0.02$)	1184 (3.24)
Cakile edentula (A)	$Y = 100.39 - 1.52X^{1/2}$ ($r^2 = 0.74, P < 0.03$)	4362 (11.95)
Calamovilfa longifolia (P)	$Y = 98.37 - 3.44X^{1/2}$ ($r^2 = 0.79, P < 0.02$)	818 (2.24)
Corispermum hyssopifolium (A)	$Y = 100.98 - 0.012X$ ($r^2 = 0.87, P < 0.05$)	8415 (23.05)
Elymus canadensis (B)	$Y = 96.05 - \ln X^{13.53}$ ($r^2 = 0.92, P < 0.01$)	1211 (3.32)
Panicum virgatum (P)	$Y = 94.33 - 0.053X$ ($r^2 = 0.80, P < 0.04$)	1780 (4.87)

P, perennial; B, biennial; A, annual.

Table 4.3 Number of seeds m^{-2} retained on inflorescences by ten plant species growing on foredunes of Lake Huron at Pinery Provincial Park and Lake Erie at Port Burwell Provincial Park (from Zhang and Maun 1994)

Name of species	Life form	Area m^2	Number of seeds $m^{-2} \pm$ SE
Agropyron psammophilum	P	5	23.25 ± 0.76
Andropogon scoparius	P	6	6.05 ± 0.62
Calamovilfa longifolia	P	900	0.05 ± 0.03
Elymus canadensis	M	300	0.32
Juniperus communis	P	900	14.55 ± 1.35
Juniperus virginiana	P	900	2.06 ± 1.13
Oenothera biennis	M	300	5.02
Rhus typhina	P	300	64.75
Rhus radicans	P	5	2164.80 ± 64.43
Xanthium strumarium	M	6	7.96 ± 0.18

M = Monocarpic perennial; P = Perennial.

them only in response to appropriate environmental cues. This strategy is adaptive because it augments the seed bank and spreads mortality risks of offspring over an extended time. For example, following a major destructive event, such as a coastal washover, an above-ground seed bank will be very useful in depositing seeds on newly exposed surfaces. Baker (1989), however, said that some of these seeds may fail to be an effective seed bank. Zhang and Maun (1994) collected panicles and reproductive parts from randomly located quadrats of various sizes (5 to 900 m^2) of ten species from foredunes of Lakes Huron and Erie in the spring of 1990 after the plants had over-wintered. The numbers of seeds retained by each species were highly variable (Table 4.3). All of the caryopses of the four grass species were contained on panicles of plants located on the leeward side of the dune ridges. The number of caryopses retained on the panicles ranged from only 0.05 m^{-2} for *Calamovilfa longifolia* to 23 m^{-2} for *Agropyron psammophilum*. More than 80% of the caryopses germinated under controlled conditions in a greenhouse while the rest were either dead or dormant.

Rhus radicans grew in clumps with local densities of 35 shoots m^{-2} and approximately 47% of its shoots bore seeds with retention of >2000 seeds m^{-2}. Although density of *Rhus typhina* was low it nevertheless retained about 65 seeds m^{-2} and all seeds of both *Rhus* species were dormant but viable. *Xanthium strumarium* retained about 8 seeds m^{-2}, half of which germinated while the rest were dormant but viable. Capsules on standing shoots of *Oenothera biennis* were retained over winter. Many of these capsules had dehisced and dispersed their seeds, but some of them had not opened and still contained up to 193 seeds per capsule. About 80% of these seeds germinated under controlled conditions but the rest were dead.

Juniperus communis occurred as multi-stemmed shrubs in some locations. The average number of seeds per female clump was 1455 ± 1351 or about 15 seeds m^{-2}. *Juniperus virginiana* trees retained on average about 514 ± 283 seeds per tree (n = 3). All seeds of both *Juniperus* species were dormant but viable. *Oryzopsis hymenoides*, a very common species of inland cold desert sand dunes, maintained a stable seed bank by retaining seeds on its old panicles and then releasing them gradually, thus continuously replenishing the seed bank (Kemp 1989). The annual species, *Corispermum hyssopifolium,* retained mature fruits on parent plants throughout the autumn and winter months along Lake Michigan and released them in spring (van Asdall and Olmsted 1963). The examples presented here are of temporary above-ground seed banks. In addition, long-term storage occurs in serotinous cones of species such as *Pinus banksiana* that occur on coastal sand dune systems along the Great Lakes and sea coasts (Poulson 1999).

4.8 Summary

Some studies on seed banks suggested that ever-shifting coastal dunes may not have seed banks. However, seed banks do exist and careful long-term studies are required to determine the mechanisms of seed dormancy, germination, and above all, the eventual destiny of seeds following their dispersal from parent plants. Based on the few studies, all three types, e.g. transient summer and winter dormant, short-term persistent and persistent seed banks are found in coastal dune systems. Several studies indicated that many of the annuals occupying the beach and dune habitats have transient (summer and/or winter dormant) seed banks, but in spite of the high seed output only a small fraction of seeds contributed to the transient seed bank. What happens to the rest of the seeds? Seed banks show considerable dynamism in dune systems because of germination, emigration, predation, physiological death and burial with one of the major physical parameters being burial in shifting dunes. Normally seeds buried deeply would

not form effective seed banks unless wind erosion brings them closer to the sand surface. This commonly occurs in mobile dunes, but deeply buried seeds may never be re-exposed by erosion and are lost. Emigration of seeds of beach species may occur by high waves, animal activity or bird dispersal and a major cause of loss for some species is predation by rodents, ants, birds and decomposition by pathogens. Under appropriate environmental cues a large proportion of seeds located at appropriate depths germinate and produce seedlings. The seed banks of woody plant communities of coastal dunes are similar to those of other coniferous and/or deciduous forest communities.

Controlled studies on plant species of Pinery dunes showed that a large proportion of artificially buried seeds remained ungerminated but viable, indicating that a seed bank was created by burial depth. There was also a substantial above-ground seed bank for a large number of dune species of the dune complex. Many foredune species also maintain a bud bank on rhizomes, stolons and root suckers that may be fragmented by high waves and produce new ramets.

More information on a number of topics on seed banks of coastal dunes will be helpful in understanding propagule banks of coastal dunes. First, a more careful study in specific dune habitats needs to be undertaken. Second, the fate of seeds from dispersal to their eventual loss needs to be monitored over time for each species. New methodology, perhaps similar to that of Watkinson (1978), more appropriate to a highly disturbed system with constant burial, erosion and sand movement, needs to be devised. Finally, the relationship between transient seed banks and recurrent disturbance must be resolved.

CHAPTER 5

Seed germination and seedling establishment

5.1 Introduction

For the transformation of a seed to a seedling complex physical and biochemical changes occur within a seed before germination can proceed. Germination is controlled by diverse seed dormancy mechanisms in plant species that delays germination until the conditions are most favourable for seed germination and seedling establishment (Thompson 1970). Baskin and Baskin (1998) identified four benefits for the evolution of seed dormancy in plants: (i) persistence in risky environments as seed banks, (ii) decreased intraspecific competition, (iii) improved chances of seedling establishment and (iv) increased fitness (seed production) of the individual and the species as a whole. They showed that seed dormancy may be caused by any one of physiological, morphological, physical, chemical and mechanical constraints or by a combination of more than one of these factors. For instance, seeds may possess an embryo with a physiological inhibiting mechanism, immature embryo, impermeable seed coat or may contain chemical inhibitors and hard woody fruit walls. In all of these cases seed dormancy is eventually broken by one or more of the following treatments: after ripening, heat treatment, cold temperature stratification, prolonged exposure to high temperatures, exposure to light, softening of seed coat by microbes or physical scarification, leaching of inhibiting chemicals, ageing of seeds and other subtle changes in the habitat.

In temperate North America with snow cover during winter months the seeds of a large majority of sand dune species—*Cakile edentula, Ammophila breviligulata, Calamovilfa longifolia, Iva imbricata, Croton punctatus, Uniola paniculata*—and others require cold stratification at <4°C for 4–6 weeks to break their dormancy requirements. Seeds of some species such as *A. breviligulata* and *U. paniculata* that require cold stratification at the northern end of their range lose this requirement in the south (Seneca 1972). At southern locations exposure to high temperatures may be required to fulfil the dormancy requirements. Winter annuals, *Vulpia ciliata, Cerastium atrovirens, Mibora minima* and *Saxifraga tridactylites*, that grow and mature their seeds in early summer on sand dunes at Aberffraw, North Wales, require exposure to high soil temperatures to overcome a state of dormancy in a certain proportion of seeds at the time of dispersal (Carey and Watkinson 1993; Pemadasa and Lovell 1975).

The second factor in seed germination is the environmental conditions to which a seed is exposed. Normally, for a non-dormant seed three basic requirements—availability of oxygen, proper temperature range and appropriate moisture conditions—must coincide for the onset of germination. Even though seeds of a species are not dormant or if the dormancy has been broken, they may be forced to remain dormant because of low or high temperatures and/or unfavourable moisture regimes. For instance, on Lake Huron sand dunes seeds of all species do not germinate after dispersal in autumn because of very low temperatures in autumn and winter that cause a thermal repression for 6 months, during which seeds undergo natural cold stratification and ready to germinate seeds accumulate. Seeds of some dune annuals require a prolonged period of continuous hydration before they will germinate (Pemadasa and Lovell 1975; Balestri and Cinelli 2004). Soil moisture is the key factor in the activation of enzymes, endosperm reserves and the onset of germination. Release from dormancy requires seeds of different species to undergo many different treatments (Zhang and Maun 1989a; Chen and Maun 1998). A student of seed dormancy mechanisms is referred to an elegant synthesis of studies on the ecology, biogeography, and evolution of seed dormancy by Baskin and Baskin (1998). In this chapter, a critical review of studies is presented on possible adaptations of seeds of dune plant species to germinate, emerge and then establish as seedlings in the harsh and unpredictable conditions of foredunes and the dune complex. Efforts are made to relate seedling death to possible causes of mortality even though it presents formidable problems.

5.2 Colonization

As shown in Chapter 4, many species are able to disperse successfully to a new site; however, dispersal will be adaptive only if the propagules are able to colonize the new site after arrival. For successful colonization in a new habitat, the propagules must be able to establish as seedlings or ramets. Even after landing safely on a distant shore, there is no guarantee of colonization and persistence in the new habitat. Hazards abound because of predation by crabs, rats, pigs, squirrels, insects (Ridley 1930), unsuitable climate, inappropriate substrate and unfavourable environmental conditions (Murray 1986a). For species such as walnuts (*Juglans nigra*) conditions may not be suitable for germination. For others, *Acer, Agropyron, Fraxinus, Pinus, Poa*, etc., germination conditions are favourable but seedlings are unable to survive because of burial by sand, sandblast and desiccation. Others grow to the

flowering stage but may be unable to set seeds because of self-incompatibility or absence of pollinators. For example, Baker (1955) showed that a single propagule of a self-compatible species would be sufficient to colonize and establish a sexually reproducing population. In contrast, at least two self-incompatible yet cross-compatible individuals will be required to establish a population.

In spite of all the hazards, there are many examples of successful colonization along coasts by species after long-distance dispersal in seawater (Ridley 1930; Murray 1986a). Outstanding examples are *Cakile* species that have evolved fleshy leaves and the corky fruit coat that adapted it to the coastal environment (Rodman 1974). Apparently, *Cakile maritima, C. edentula*, and *C. lanceolata* species of temperate regions evolved from the desert species, *Cakile arabica,* in the Mediterranean area along the shores of the ancient Tethys Sea. Guppy (1917) estimated that *Cakile* species have the potential to travel from 160 to 640 km on seawater. In fact, the first species to colonize Surtsey Island, formed 32 km off Iceland in 1963, was *Cakile edentula* (Thorarinsson 1966). Several other species such as *Crambe maritima, Ipomoea pes-caprae, Cocos nucifera* (Ridley 1930), some legumes (Murray 1986b) and mangrove species (Guppy 1917) have also expanded their distributional range by morphological adaptations for long-distance dispersal in the sea currents and then successful establishment. On newly formed islands, the majority of herbaceous plant species, and all tree and shrub species, arrived by sea and established there. Ernst (1908) estimated that within 23 years of eruption of Krakatoa Island 92 species of seed plants had established of which 30–72% had arrived by ocean currents, 10–19% by birds and 16–30% by wind. The relatively short period of time for re-vegetation could be attributed to its proximity to Java, Sumatra and other islands. Similarly, in Hawaiian and Galapagos islands a substantial proportion of species is believed to have arrived by flotation in seawater (Murray 1986b). Some species may, however, succeed only in specific habitats. For example, mangroves can only colonize mud flats while some littoral species such as *Ipomoea pes-caprae*, grasses, pioneer annuals and perennials will only grow on sandy or shingle beaches.

5.3 Seed mass and establishment

The frequency distributions of seed mass of some annuals, biennials and perennials of dune ridges are presented in Fig. 5.1. Some sand dune species have rather large seeds, others small.

Seeds of *Strophostyles helvola*, an indeterminate annual legume species that occurs on the mid-beach and driftline habitats of shorelines, have a rather large mass (Table 5.1). *Cirsium pitcheri*, a monocarpic perennial found on some sandy beaches and foredunes of the Great Lakes, also has large seeds as do other sea coast species of Europe and North America. Considerable overlap can be seen in the curves representing seed mass distribution of upper and lower fruit segments for all three annual species of *Cakile*. The mean values for upper fruits were significantly higher than that of the lower fruits.

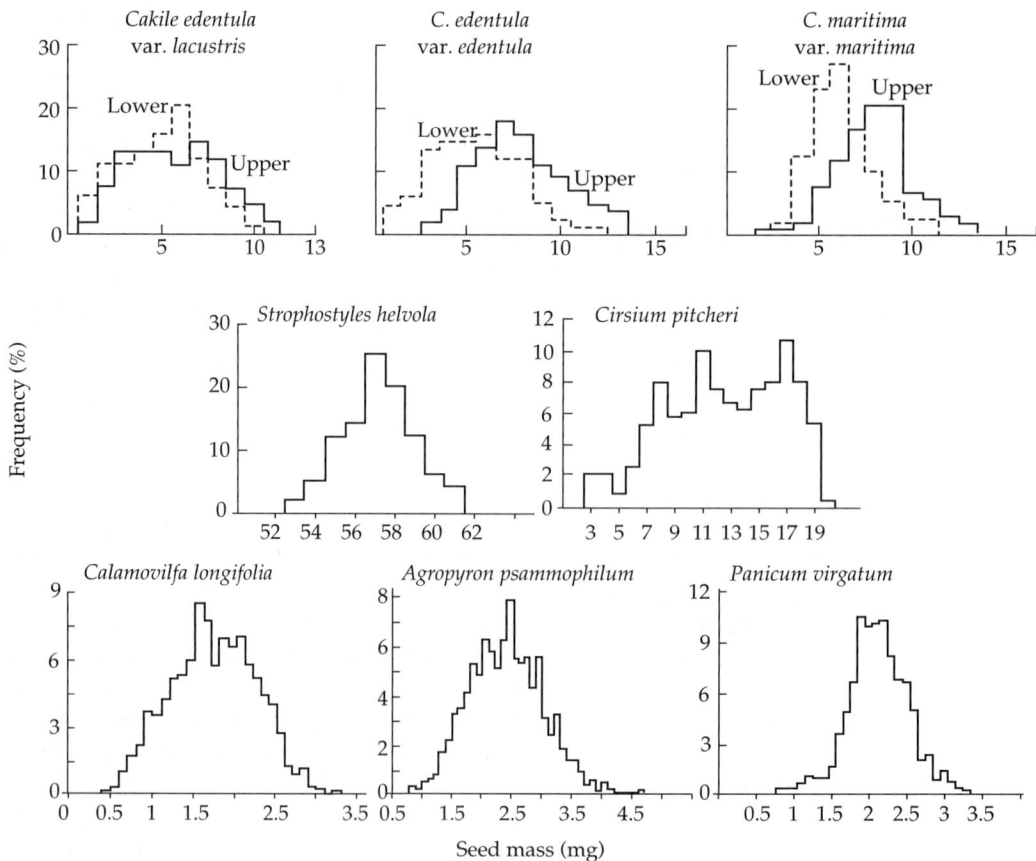

Figure 5.1 Bar diagrams of seed mass of upper and lower fruits of *Cakile edentula* var. *lacustris,* var. *edentula* and *C. maritima*. Also shown are seed mass distributions of annual species, *Strophostyles helvola*, the monocarpic perennial, *Cirsium pitcheri* and perennial grasses, *Calamovilfa longifolia, Agropyron psammophilum* and *Panicum virgatum* growing on the strand and dunes of Lakes Erie and Huron (after Maun and Payne 1989; Yanful and Maun 1996a; Maun *et al.* 1996; Zhang and Maun 1990b, 1991a, 1993).

Table 5.1 Mean seed mass of beach and dune species of some coastal dune systems

Species name	Morph	Seed mass (mg)	Authority
Beach and strand species			
Cakile maritima (A)	Upper	8.0 ± 2.0	Maun and Payne 1989
	Lower	6.2 ± 1.7	
C. edentula var. lacustris (A)	Upper	5.8 ± 2.5	Maun and Payne 1989
	Lower	5.0 ± 2.1	
C. edentula var. edentula (A)	Upper	7.9 ± 2.5	Maun and Payne 1989
	Lower	5.5 ± 2.2	
Strophostyles helvola (A)		57.20 ± 0.5	Yanful 1988
Cirsium pitcheri (M)		12.00 ± 4.3	Maun et al. 1996
Xanthium strumarium (B)			
Cornus stolonifera (P)		34.4	Mazer 1990
Calystegia soldanella		56.8	Salisbury 1942
Crambe maritima		20.0	Scott and Randall 1976
Mertensia maritima		1.67	Salisbury 1942
Honkenya peploides (on shingle)		14.4–17.4	Salisbury 1942
Ipomoea pes-caprae (P)			
Lathyrus maritimus (P)		0.023 g	Salisbury 1942
Rumex crispus var. trigranulatus		2.57	Cavers 1963
Dune species			
Agropyron psammophilum (P)		2.32 ± 0.02	Zhang and Maun 1990b
Calamovilfa longifolia (P)		1.68 ± 0.51	Zhang and Maun 1993
Carex eburnea (P)		0.354	Mazer 1990
Panicum virgatum (P)		2.10 ± 0.014	Zhang and Maun 1991a

P, perennial; B, biennial; A, annual; M, monocarpic perennial.

In contrast to these annuals and biennials, the perennial dune-forming grasses of dunes such as *Ammophila arenaria*, *A. breviligulata*, *Agropyron psammophilum*, *Calamovilfa longifolia* and *Panicum virgatum* have small seeds and the range of variability in seed mass within each species is quite high.

It is evident that seed sizes of plants in different dune habitats may vary by many orders of magnitude. Ecologically, large-seeded species tend to occupy closed low-light habitats while small-seeded species inhabit open high-light habitats of embryo dunes and open sand dune complexes (Mazer 1990). Similarly, Salisbury (1974) observed that pioneer species in early phases of dune succession have smaller seeds than the climax species. However, is there within-habitat segregation in seed mass in high-light habitats? Several studies show that the species of highly dynamic sandy and shingle beaches with persistent natural disturbance have larger seed mass than those of the dune complex (Table 5.1). Large seeds have an advantage over small seeds in this highly disturbed habitat for three possible reasons: (i) they emerge from greater depths of burial, (ii) they have better dispersal ability in water and (iii) they produce larger seedlings that establish more successfully. The only drawback may be greater predation because seed consumers may selectively choose large seeds (Louda 1989). Beach species also varied with respect to sand burial survival, extent of below-ground storage organs, canopy height >15 cm and capacity for seawater dispersal (García-Mora *et al.* 1999).

5.4 Seed germination in dunes

Sand dunes provide ideal conditions for germination provided ideal temperature and adequate moisture conditions are prevalent. Major problems faced by seeds are sand accretion and soil salinity, both of which have positive and negative effects on seed germination. Burial to an appropriate depth is beneficial for seed germination because it provides intimate soil contact, maintains high humidity around the seed, improves imbibition of moisture, protects it from surface predation, decreases evaporative seed surface and reduces chances of desiccation from heat and light. Surface-lying seeds of most foredune species, without a mechanism of burying themselves, are unable to germinate because of insufficient moisture and exposure to the dry environment. For example, acorns of *Quercus prinus* and *Q. rubra* failed to germinate or had lower germination in a sand dune if they were not buried (Olmsted 1937; García *et al.* 2002). Even under controlled greenhouse conditions with artificial watering, germination of surface-lying seeds of *Calamovilfa longifolia, Ammophila breviligulata, Elymus canadensis, Cakile edentula* and *Agropyron psammophilum* was significantly lower than seeds buried at shallow depths (Maun and Riach 1981; Maun and Lapierre 1986; Zhang and Maun 1990a, c). The only exceptions were small seeded species such as *Erigeron canadensis, Solidago sempervirens* (van der Valk 1974) and *Artemisia campestris* (Stairs 1986) that may be buried by extremely small amounts of sand.

Even if seeds germinate on the sand surface, their chances of seedling establishment are very low because their radicles are unable to affect speedy penetration of the sand surface and anchor the seedlings. To penetrate the soil the radicle must exert greater pressure than the resistance offered by the soil. Most of these seedlings die of desiccation and exposure to

high light intensity. However, because of chronic sand movement, the chances that a seed will remain unburied on the foredunes are very low indeed. The problem for seeds lying on the sand surface is only acute in stable transition zone dune habitats, usually behind the second or third dune ridges where the impact of wind and moving sand has been significantly reduced and the sand surface is covered by vegetation. Here, most species have evolved various mechanisms by which the seed may be able to bury itself by hygroscopic awns, orient itself in such a way that the embryo is in close contact with the soil, produce mucilage that increases soil contact or depend on disturbance of the substrate by rodents, burrowing insects, splashing of rain drops, hooves of animals and litter accumulation.

Burial also has negative consequences on seed germination. For species with a seed mass of <3 mg the highest seed germination was obtained at about 0.5–4 cm depths of burial. Up to a burial depth of about 10 cm seed germination of all test species except *Ammophila breviligulata* was not significantly reduced (Fig. 5.2a). However, seeds of some other species such as *Cirsium pitcheri* (Chen and Maun 1999) and *Leymus racemosus* (Huang *et al.* 2004) showed a significant ($P < 0.001$) decline in germination with an increase in burial depth. When the burial depth exceeded 15 cm seeds of almost all species started to exhibit dormancy.

In sand dunes, seed germination is strongly related to available moisture (García *et al.* 2002). Very small seeds (0.6 ± 0.01 mg/1000 seeds) of *Artemisia campestris* from the Pinery and *A. sphaerocephala, A. ordosica* species from Mu Us sandy land of China produced high germination at shallow depths of only 0.5 cm because they require small amounts of moisture for imbibition (Stairs 1986; Zheng *et al.* 2005). In contrast large seeds of *Strophostyles helvola* (65 ± 2.7 mg/seed) exhibited lowest germination at shallow depths of 2 cm because they required prolonged hydration

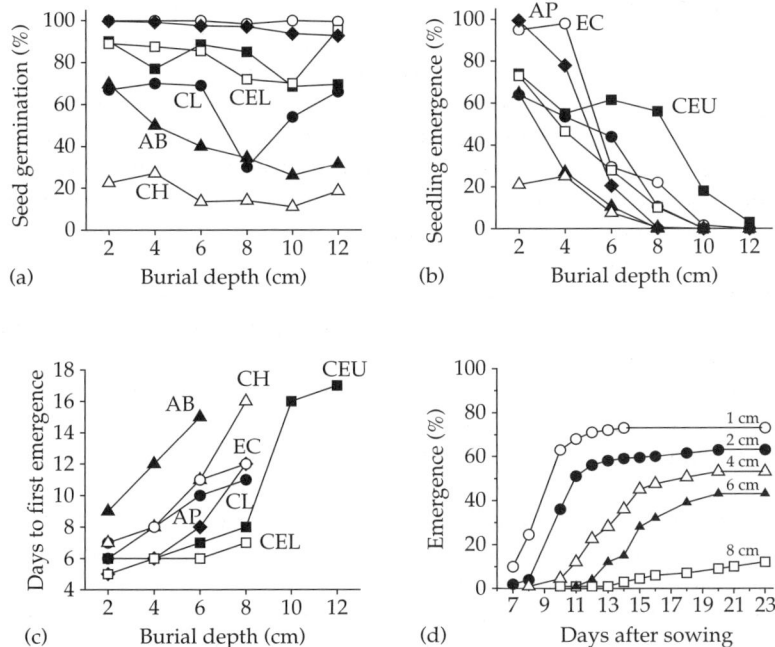

Figure 5.2 (a) Percentage of seed germination, (b) seedling emergence, (c) number of days to first emergence and (d) rate of emergence of *Calamovilfa longifolia* seedlings from various depths of burial in sand. Symbols: AB, *Ammophila breviligulata*; CH, *Corispermum hyssopifolium*; CL, *Calamovilfa longifolia*; CEL, *Cakile edentula* (lower fruits); EC, *Elymus canadensis*; CEU, *Cakile edentula* (upper fruits). (Maun and Riach 1981, Courtesy, *Oecologia*; Maun and Lapierre 1986, Courtesy, *American Journal of Botany*; Zhang and Maun 1990a, Courtesy, *Canadian Journal of Botany*).

and greater amounts of moisture for imbibition. As explained in Chapter 8, soil salinity above a certain level also has a strong negative effect on seed germination.

5.5 Seedling emergence

Emergence is strongly dependent on processes occurring after the germination of seeds because ideal conditions for germination are not necessarily ideal for emergence (García *et al.* 2002). Seedling recruitment may not occur because of desiccation of soil surface, high sand surface temperatures and the inability of the radicle to grow and establish contact with soil moisture. Zheng *et al.* (2005) tested the effects of three water supply regimes—5, 7.5, and 10 mm of water every third day—on six sand dune species and found that greater

water supply resulted in greater seedling emergence. As soon as burial exceeded a certain level specific for a species it became detrimental to the emergence of its seedlings. Studies on 15 dune species buried at different depths revealed that seedling emergence was negatively correlated with burial depth of a seed (Wagner 1964; Harty and McDonald 1972; Maun 1981; Maun and Riach 1981; Maun and Lapierre 1986; Weller 1989; Zheng *et al.* 2005). Based on similar results of many studies the following generalizations can be made.

1. Maximum emergence of all species occurred from shallow depths of about 0.5–4 cm (Fig. 5.2b). A significant decline in percentage emergence and rate of emergence of different species occurred as the depth of burial increased. For example, percentage emergence of *Panicum virgatum* showed an exponential response ($Y + e^{(4.785-0.046x)}$, $P < 0.01$, $r^2 = 0.423$) with little variation at shallow depths but a rapid decrease with burial depths exceeding 10 cm, while the rate of emergence showed a linear response with a constant decrease with increasing depths of burial (Zhang and Maun 1990c). Seedlings of most species failed to emerge from >8 cm depths. Studies on other species of sea coasts, e.g. *Spinifex hirsutus* in Australia (Harty and McDonald 1972; Maze and Whalley 1992b) and *Uniola paniculata* in North Carolina, USA (Wagner 1964) showed very similar results. Only in the case of *Cakile edentula* did 18% of the seedlings from the upper seeds emerge from 10 cm burial depth, mainly because of higher seed mass (Maun and Lapierre 1986). However, occasional seedlings of all species do emerge from greater than average depths. For example, even though the average emergence depth of *Spinifex hirsutus* seedlings on an Australian beach was 6.7 ± 2.4 cm some individuals emerged from 12.7 cm depths (Maze and Whalley 1992b). In fact the maximum depth of emergence (ED_{Max}) of occasional seedlings of a species was 30 to 100% higher than the depth from which 50% (ED_{50}) of the seedlings emerged (Fig. 5.3).

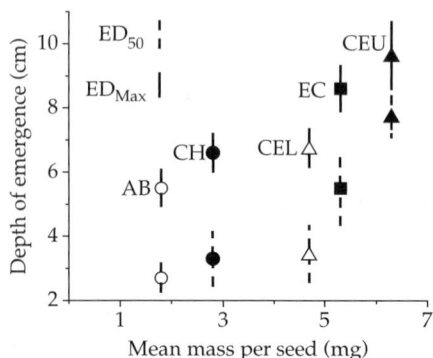

Figure 5.3 Maximum depth of emergence ($ED_{Max} \pm$ SD) and 50% depth of emergence ($ED_{50} \pm$ SD) of seedlings from seeds of different seed mass when they were buried in sand at different depths. For abbreviations of symbols see Fig 5.2. (Maun and Lapierre 1986, Courtesy, *American Journal of Botany*).

2. As expected there was a significant positive linear relationship between the depth of burial and the length of time required by the seedling to emerge above the sand surface (Fig. 5.2c). This effect was exhibited by all species even though the number of days required for emergence was different. Normally first seedlings emerged about 5–7 days after sowing from the lowest depth of burial, but it took 11–16 days to emerge from 10 cm depth. In grasses, the emergence occurred by the elongation of the first internode, but in dicots, the seedlings emerged by the elongation of the hypocotyl.

3. The cumulative rate of seedling emergence was the highest and fastest in all species at the lowest depth of burial and declined as the depth of burial increased. The emergence of all species reached a peak for each burial depth and then levelled off as shown for *Calamovilfa longifolia*, a common species of the Great Lakes dunes (Fig. 5.2d). All species followed the same pattern until no seedlings emerged beyond their limits of tolerance.

4. Seedling emergence was related to the energy contained in the endosperm or cotyledons of a seed rather than the embryo. Zhang and Maun (1989b, 1991b) found that removal of a portion of the endosperm of dune grasses, *Panicum virgatum* and *Agropyron psammophilum*, and seven other species increased percentage germination of seeds but the size of the seedling, rate of elongation of roots and shoots and seedling biomass were significantly reduced. Similar results were reported for *Leymus racemosus* seeds in semi-stabilized dunes in deserts of the Junggar Basin of Xinjiang, China (Huang *et al.* 2004). In general, seedlings of species with larger seed mass emerged from greater maximum depths of burial (Barbour *et al.* 1985) (Fig. 5.4a) and the depth from which 50% (ED_{50}) of seedlings emerged (Fig. 5.4b). This relationship was also true for the large and small seeds of a single species (Maun and Lapierre 1986; Weller 1989). Thus, seeds of larger size will have an advantage over small seeds under chronic sand burial conditions.

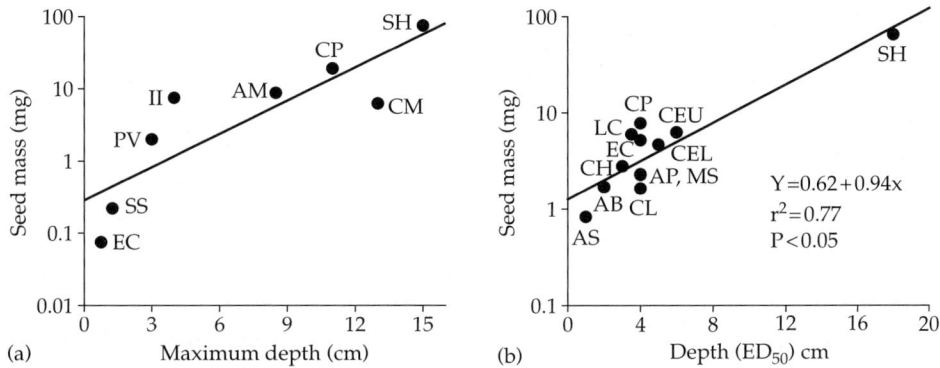

Figure 5.4 (a) The relationship between seed mass and maximum depth of emergence for successful establishment of seedlings of eight forbs (from Barbour *et al.* 1985) and (b) the relationship between seed mass and 50% depth of emergence for 12 species of grasses and forbs of sand dunes. Symbols: SH, *Strophostyles helvola*; CP, *Croton punctatus*; AM, *Abronia maritima*; II, *Iva imbricata*; CM, *Cakile maritima*; SS, *Solidago sempervirens*; EC, *Erigeron canadensis;* PV, *Panicum virgatum*; CL, *Calamovilfa longifolia*; AP, *Agropyron psammophilum*; AS, *Agriophyllum squarrosum*; MS, *Medicago sativa*; CEU, *Cakile edentula* (upper fruits); CEL, *Cakile edentula* (lower fruits); EC, *Elymus canadensis*; CH, *Corispermum hyssopifolium*; AB, *Ammophila breviligulata*; SH, *Strophostyles helvola*; LC, *Lithospermum caroliniense*. (AS, MS, from Zheng *et al.* 2005, Courtesy, *Annals of Botany*); (CL, from Maun and Riach 1981, Courtesy, *Oecologia*); (AP, from Zhang and Maun 1990a, Courtesy, *Canadian Journal of Botany*); (CEU, CEL, CH, AB, EC, from Maun and Lapierre 1986, Courtesy, *American Journal of Botany*); (SH, from Yanful and Maun 1996b, Courtesy, *Canadian Journal of Botany*); (LC, from Weller 1989, Courtesy, *Ecology*); (PV, from Zhang and Maun 1990c, Courtesy, *Holarctic Ecology*).

For example, the highest percentage emergence and rate of emergence were observed in large seeds of *Strophostyles helvola* (Yanful and Maun 1996b) and *Lithospermum caroliniense* (Weller 1989). The only exception was *Panicum virgatum*, a C$_4$ grass species with small seeds, in which some seedlings emerged from 16 cm depths (Zhang and Maun 1990c).

Since species with large seeds are more fit to emerge from greater depths of burial in sand there will be strong selective pressure for the evolution of species with larger seed mass in coastal foredunes. However, that is not the case. On the Pinery dunes, six species had seed mass of 0.0006–3.0 mg, two had >3.0–6.0 mg and one had >6.0 mg per seed. The most likely reason for the evolution of small seeds is the tremendous microhabitat variability in sand deposition. While some sites may be receiving large accumulations of sand a large majority has small amounts. Species producing a greater number of small seeds will be more fit to survive in microhabitats with low sand deposition than those with a small number of large seeds. Only in habitats with greater sand accretion will natural selection favour the evolution of large seed size. For instance, larger-seeded species such as *Cakile edentula, Strophostyles helvola* and *Lithospermum caroliniense* were prevalent in areas with larger chronic sand deposition on the strand and first dune ridge (Yanful and Maun 1996a; Weller 1989).

5.6 Patterns of seedling emergence

In coastal dune systems seedlings of most endemic species are abundant in open habitats on the beach, in blowouts and other bare areas between established plants in the foredunes, slack, dune ridges and stabilized areas behind dune ridges. A large proportion of seeds germinate during the first year after dispersal, with the possible exception of a few species that exhibit prolonged dormancy. The peak emergence is related to temperature in early spring. As soon as soil temperatures rise to the optimum for seed germination of a species, seedlings start to emerge. Since south slopes of a dune system warm up earlier in spring, the first flush of seedlings occurs over there. However, this response is short-lived because further rise in temperatures dries the top layers of sand with a corresponding decrease in further emergence. Emergence follows in other microhabitats and usually the period of seedling emergence in spring lasts about 4–6 weeks. Seedling flushes that occur later during summer in some species are related more to the frequency and amount of rainfall rather than temperature. Similarly, seedling emergence in autumn coincides with a decrease in temperatures and increase in rainfall. Three main patterns of seedling emergence were consistent from year to year despite variations in dates of snow melt and other weather conditions.

5.6.1 Entire emergence in spring

At the latitude of the Great Lakes, the emergence of most dune species occurred from late March to the end of May. It would occur earlier at more southern latitudes. Species that follow this strategy on the foredunes and dune complex are: *Cakile edentula, Strophostyles helvola, Elymus canadensis, Corispermum hyssopifolium, Salsola kali, Xanthium strumarium, Calamovilfa longifolia, Ammophila breviligulata, Lithospermum caroliniense, Cirsium pitcheri, Panicum virgatum, Quercus velutina, Q. rubra* and *Q. muehlenbergii.* All the herbaceous species reach maturity and disperse their seeds in summer and autumn months. At Holkam, North Norfolk and the Netherlands the bulk of the seeds of *Cynoglossum officinale* germinated very early in spring (Boorman and Fuller 1984). On the sandy coastal shores of north-west Europe, four strand annuals, *Atriplex glabriuscula, A. laciniata, Cakile maritima* and *Salsola kali,* that are restricted to a small belt on the driftline of high tides germinate in April–May which coincides with a decrease in salinity

by a factor of 10 and an increase in temperature from 10 to 20°C (Ignaciuk and Lee 1980). They do not germinate and emerge in autumn because of various bottlenecks such as wind or tide erosion of substrate, high salinity levels in autumn, winter and early spring and sand accretion.

5.6.2 Entire emergence in autumn

Species in this category germinate in late August and September when moisture is plentiful and the soil temperatures have decreased. They have endured high temperatures and drought conditions of June or July and their seeds did not germinate during the summer months, probably because of low moisture levels, high temperatures and dormancy mechanisms. Winter annuals, plants that germinate and establish in autumn and flower and mature seeds in spring, fall into this category. Pemadasa and Lovell (1975) followed the course of seedling emergence of artificially sown and naturally dispersed seeds of seven winter annuals in wet and dry slacks at Aberffraw, Anglesey, UK. These species dispersed their seeds and contributed to the seed bank from February to the end of July. They found that the emergence of seedlings of all species was very low until the end of September but a large increase occurred in October and November followed by a small increase in early December (Fig. 5.5). Weather fluctuations between years advanced or delayed peak emergence but the general pattern remained the same. Species differed slightly in the dates of peak emergence and cessation of emergence. They all started to emerge at the same time but *Aira caryophyllaea, Mibora minima,* and *Vulpia membranacea* were the first to reach maximum emergence (mid October), *Aira praecox* and *Cerastium atrovirens* were second (end of October) and *Erophila verna* and *Saxifraga tridactylites* were the last (mid November). In 1972, the sequences of emergence were still the same but the curves were shifted by about 3 weeks. *Arabis lyrata* followed a similar pattern along the Great Lakes;

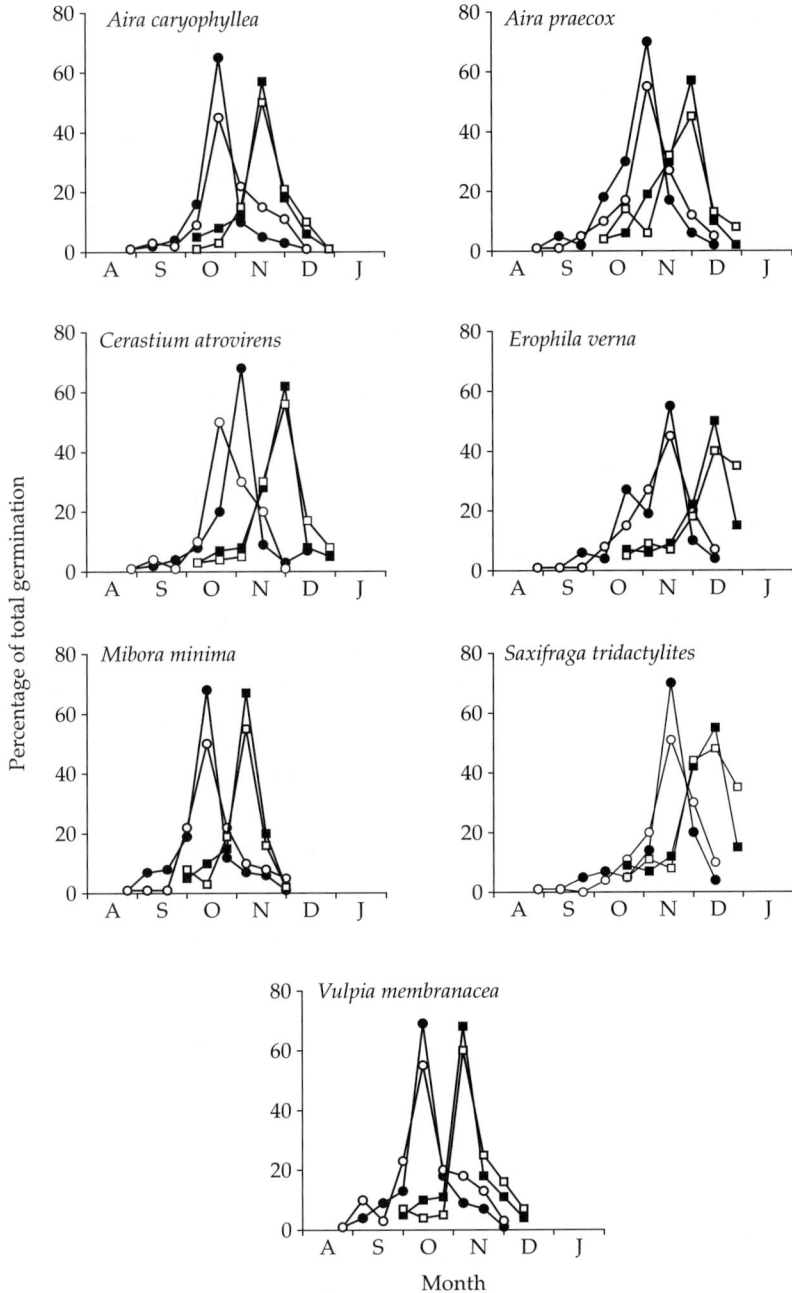

Figure 5.5 The percentage of total number of seedlings of seven winter annual species emerging from sand in dry slacks (open symbols) and wet slacks (filled symbols) from August to December at Aberffraw, UK during 1971 (circles) and 1972 (squares) (after Pemadasa and Lovell 1975, Courtesy, *Journal of Ecology*).

its seeds germinated and seedlings established before the onset of freezing temperatures and snow fall. The plants flowered in early spring and then dispersed their seeds from June to the end of July.

5.6.3 Species emerging in spring, summer and autumn

Artemisia campestris ssp. *caudata* is a facultative biennial whose seedlings emerged in spring, summer and autumn, but there were mainly two peaks of emergence; a large peak in spring and a small one in autumn (Stairs 1986). Some seedlings did emerge in midsummer after a warm soaking rain, however few if any of those seedlings survived. There was a tendency for seedlings to be aggregated both spatially and temporally. Variances in temporal distributions ranged from 0.3 to 5.75 m^{-2} in the dune ridge and 0.2 to 25.7 m^{-2} in the slack (Stairs 1986). There was a positive relationship between emergence of *A. campestris* seedlings and amount of rainfall in August–September. Kachi and Hirose (1990) reported a similar seedling emergence pattern in *Oenothera glazioviana* on a coastal sand dune system at Azigaura, Japan. Seedling emergence showed a large peak in autumn and a much smaller peak in spring that coincided with the yearly fluctuations in the amount of rainfall.

These differences in the periodicity of germination and emergence of different species in dune communities are evident but the relative merits of the different strategies are poorly understood. In general, the evolution of different strategies will be adaptive only if it will improve the long-term survival of the species. For example, if one cohort of seedlings fails others will allow the species to maintain its populations within the dune system. Do different cohorts of seedlings have different rates of survival? Stairs (1986) found no differences in survival rates of *A. campestris* seedlings emerging in spring and autumn. In contrast, rate of survival of *Oenothera glazioviana* seedling cohorts emerging in spring

(March–April) was 0.81–0.84, in summer (May to August) was 0.13–0.21 and in autumn (September–November) was 0.46–0.53 (Kachi and Hirose 1990). A demographic model predicted that seedlings emerging in September–November or in March–April would show greater fitness (higher number of seeds) per emerging seedling than those emerging in summer. According to Balestri and Cinelli (2004) *Pancratium maritimum*, a species common along the Mediterranean, distributes its germination and emergence to a period when moisture is plentiful and conditions are more favourable for seedling establishment and subsequent plant growth.

5.7 Survivorship and establishment

Once a seed overcomes the dormancy constraints, receives the appropriate germination cues and emerges above the sand surface as a seedling, it begins to cease its dependence on endosperm reserves and starts to receive energy from photosynthesis of young leaves. The establishment of the seedling is probably the most hazardous period in its life history. Is the period of maximum emergence related to maximum seedling establishment? According to Thompson (1970) the periodicity must be synchronized in the evolutionary process with the most favourable season when the chances of survival and reproduction of the seedling population are statistically the highest. Each microhabitat within the dune complex has its own unique limitations in the establishment of seedling populations.

5.7.1 Summer annuals

The annual species of coastal beaches possess a number of adaptive traits that enable them to withstand the hazards of the dune environment. The timing of germination is synchronized with the best period of emergence and growth. The fruits of *Cakile edentula* plants start to mature in mid-August and are dispersed in autumn. The winter and spring storms may move the buoyant fruits long distances and deposit them on the beach where they undergo natural stratification. The germination and emergence occurs very early in spring and by the middle of May, a large seedling population is visible on the driftline and mid-beach areas. However, the number of surviving plants shows a negative linear relationship with the passage of time. In a 1978 study by Payne and Maun (1984), none of the plants on the high beach survived before flowering but 5 and 14% of plants in the driftline and mid-beach flowered and produced fruits, respectively (Fig. 5.6a). The main reason for high mortality was a very dry summer with only 78, 21 and 66% of the normal precipitation during June, July and August, respectively. Survivorship may be drastically different between years. For example, during 1979 even the high-beach plants showed a survival rate of 0.10 but all the mid-beach plants died due to high levels of sand deposition by high waves (Payne and Maun 1984). The survivorship curves of seedlings of *Corispermum hyssopifolium*, another summer annual of Great Lakes foredunes, are quite different. The species possesses a C_4 photosynthetic pathway and is found in four major microhabitats of

foredunes, mid-beach, eroding surfaces of blowout, slacks and bare areas between plants on foredunes. The plants range in size from very small with single stems in blowouts to rather large branched individuals on the mid-beach. The large plants are usually located in areas with rich depositions of organic detritus by high waves. The seeds germinate in May of each summer as single seedlings or as clumps from seeds buried *in situ* on parent plants. Van Asdall and Olmsted (1963) established 10 plots of young seedlings at the end of May and followed their fate for 13 weeks. The survivorship curves (seedlings established and reproduced) in each of those plots is presented in Fig. 5.6b. In four of ten plots the survival was 100% and in three others it ranged from 83 to 91%. In plots 1 and 6, respectively, 65 and 31% of the seedlings established and reproduced. In plot 5 with 3907 clumped seedlings all of them died within the first seven weeks of their life. The probable reasons for mortality were competition and desiccation. The dead seedlings had very short root systems that were unable to reach moist layers of sand about 10 cm below the soil surface. Similarly, survival of *Strophostyles helvola* seedlings was very high (>95%) in some locations but it

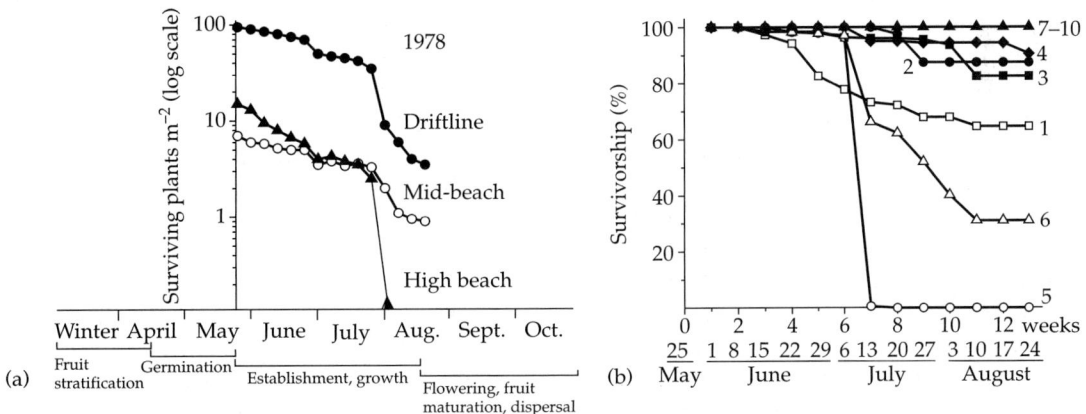

Figure 5.6 The survivorship of seedlings of (a) *Cakile edentula* on the beach of Pinery dunes during 1978 (after Payne and Maun 1984, Courtesy, *American Midland Naturalist*) and (b) *Corispermum hyssopifolium* in 10 plots located in different microhabitats on the foredunes and beaches of Lake Michigan after 13 weeks of growth (after van Asdall and Olmsted 1963, Courtesy, *Botanical Gazette*).

ranged between 67 and 79% in others (Yanful and Maun 1996b).

5.7.2 Winter annuals

Mack (1976) determined the survivorship of seedlings of *Cerastium atrovirens*, a winter annual in the sand dunes of Aberffraw, Anglesey, UK. He established seedling plots in September soon after their germination and then followed the fate of each seedling until their death or seed production. Few seedlings died in the early recruitment phase but a steady rate of mortality was observed from autumn to the end of winter. About 60% of the seedlings survived and reproduced successfully in spring. The major causes of mortality were desiccation, exposure of roots by erosion and excessive burial by sand. In another study on *Vulpia fasciculata*, a winter annual of sand dunes of Britain, Watkinson and Harper (1978) established twelve 40×40 cm plots on two sand dune systems and followed the fate of each seedling from August 1973 to July 1974. The total percentage mortality of seedlings ranged from 7 to 41%. In 9 out of 12 plots little or no mortality occurred during autumn and winter, however, a large number of plants died during the period of bolting in early summer. In the remaining three plots, the rate of mortality was higher in autumn and winter.

5.7.3 Biennials

Boorman and Fuller (1984) estimated that only a small fraction of the seed produced by *Lactuca virosa* and *Cynoglossum officinale* survived and produced seedlings during the next growing season. The subsequent survival of germinated seedlings to produce yearlings was estimated at 2.8% for *L. virosa* and 23% for *C. officinale*. A major cause of mortality was water deficit in the early summer months before the plants had developed a sufficiently deep and extensive root system. Kachi and Hirose (1985) determined survivorship and reproduction of *Oenothera glazioviana* in the coastal sand dune

system at Azigaura, Japan. Of the total seeds produced by a plant 98% were lost in the seed bank stage and only 2% produced seedlings. Half of the emerged seedlings died before establishment and the rest died after producing about 2000 seeds per plant. According to their calculations, the 10 000 established seedlings introduced initially, left behind 11 900 offspring. The overall survivorship of *O. glazioviana*, including the seed bank stage, approached a Deevey Type III curve.

Stairs (1986) randomly selected 20×50 cm plots in each of three sand dune habitats to determine seedling survival of *Artemisia campestris*. She mapped each seedling within each plot and monitored their survivorship at two-week intervals from spring 1979 to autumn 1981. The cumulative survivorship curves indicated that the mean survival time for seedlings in all habitats was less than a year; approximately nine months in the dune ridge and transition zone and half that in the slack. However, seedling survivorship was not normally distributed since most seedlings died soon after emergence: 50% of seedlings were dead after ten months in the dune ridge, eight months in the transition zone and after less than two months in the slack. The logarithmic plot indicated that the rate of mortality was greater in the slack than in the other habitats after the first month and also after the first year of life (Fig. 5.7). The overall survivorship was 0.49 in the dune ridge, 0.27 in the slack and 0.48 in the transition zone. The period immediately following emergence was the most vulnerable and all cohorts in the three dune habitats followed a Deevey Type III survivorship curve (Stairs 1986). The data clearly showed that differences between habitats were partly determined by the interaction between moisture availability and temperature and partly by between-habitat variability in physical factors. On the dune ridge there were fewer seedlings but their survival and relative growth were high primarily because of closer proximity to the Lake and nurse plant protection. In the slack sparse vegetation, coarse

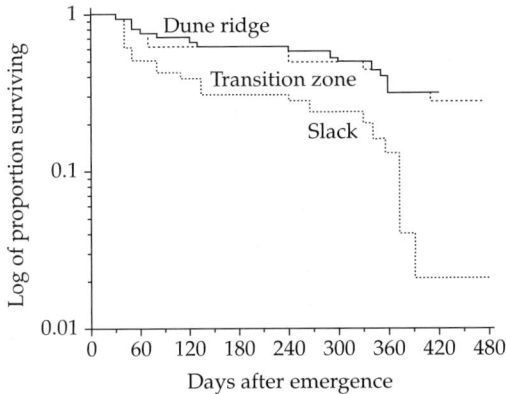

Figure 5.7 The log of cumulative proportion of seedlings of *Artemisia campestris* population surviving on the dune ridge, slack and transition zone for pooled data for 480 days in 1980 and 1981 (unpublished data from Stairs 1986 with her kind permission).

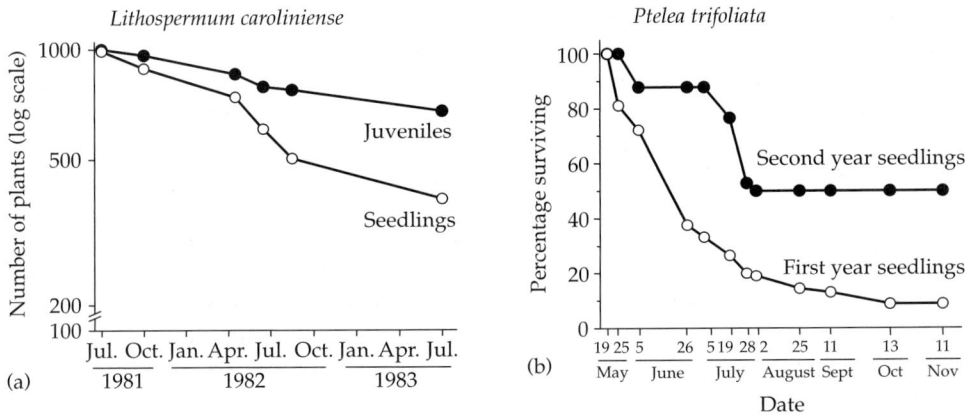

Figure 5.8 Comparison of the survivorship curves (a) of seedlings and juveniles (>1 year old) of *Lithospermum caroliniense* at the Pinery dunes (from Westelaken and Maun 1985a, *Oecologia*) and (b) first year and second year seedlings of *Ptelea trifoliata* on a sand dune along Lake Michigan (from McLeod and Murphy 1977, Courtesy, *American Midland Naturalist*).

texture of soil, severe drought caused by high soil and air temperatures and sand erosion led to high seedling mortality and rather poor seedling growth. The survivorship in the transition zone is higher, probably because of abundance of tall grass perennial vegetation acting as nurse plants, no sand movement, higher nutrient levels and greater soil moisture retention.

5.7.4 Perennials

Westelaken and Maun (1985a) subjectively selected forty plots (1 m × 1 m) and mapped the seedlings, juveniles and adult plants of *Lithospermum caroliniense* within each plot using a Cartesian coordinate system. At the end of the initial survey in 1980 the 40 plots contained 266 seedlings, 196 mixed age juveniles, and 43 mixed age adults. The survival of these plants was monitored for two years. The survivorship probability of juveniles was significantly higher (Chi-square = 36.62, P < 0.001) than seedlings 2 years after the initial survey (Fig. 5.8a). At the end of this period, 37% of the seedlings and 68% of the juveniles were still alive. According to the Mantel and Haenszel chi-square test, the two curves were significantly different. Similarly, along Lake Michigan, seedling survival was lower in the first year of life of seedlings than the second year and there were significant differences between locations: usually seedlings on leeward slopes located at greater distances from the Lake had lower rate of survival, probably because of diminished lake effect (Weller 1989).

The survivorship of *Calamovilfa longifolia* seedlings was 50% after the first 4 days of emergence in 1978 (Maun 1981). The high death rate continued throughout the summer and by 1 August only 2.5% of the seedlings were still surviving which decreased to 2, 0.9 and 0.5% by November 1978, May 1979 and November 1979, respectively. This high mortality was probably due to desiccation caused by a very dry summer in 1978. The survival rate of 1979 cohorts was much higher than those of 1978, probably because of greater and well distributed rainfall during the summer. The survival rate differed significantly among different microhabitats. For example, survivorship was the lowest on south slopes, intermediate within willow clumps and highest on north slopes. Survivorship also increased with age of the seedlings. McLeod and Murphy (1977) established two transects (1 × 20 m) to determine the growth and survivorship of newly emerged and 1-year-old seedlings of *Ptelea trifoliata* on Lake Michigan sand dunes. The fate of each seedling was followed after every three weeks from mid May to the middle of November 1973. The survival of first year seedlings was 10.5% by the end of the first growing season (Fig. 5.8b). The major causes of mortality were desiccation (63.2%), vehicular traffic (22.1%) and miscellaneous causes (14.7%). Mortality was rather high in June during a dry spell of three weeks when the seedlings were still young and had poorly developed root systems. In contrast, the survivorship of second year seedlings was 50% during their second growing season. The establishment of *Ammophila breviligulata* seedlings under natural

conditions is extremely rare except in localized microsites with abundant soil moisture such as moist dune slacks or driftlines (Maun and Baye 1989). Maun (1985) reported that all emerged seedlings of *A. breviligulata* died within about three weeks of their emergence and all new plants established from rhizome fragments. In contrast, on the Norfolk coast of England, caryopses and rhizome fragments of *Elymus farctus* were nearly of equal importance in the establishment of new clumps of tillers (Harris and Davy 1986b).

5.8 Limiting factors

In sand dune systems, the survivorship, establishment and growth of seedlings is influenced by a number of physical and biotic factors such as predation, disease, desiccation, competition, salt spray, nutrient deficiency, high soil surface temperatures and burial by sand (Laing 1958; Maun 1994; Payne and Maun 1984; Watkinson and Harper 1978). Payne and Maun (1984) followed the fate of 691 *Cakile edentula* seedlings from emergence until maturity by mapping them and examining them every 3 to 6 days. They showed that 31, 9 and 5% of plants were killed by desiccation, insect damage by a cutworm, and burial and erosion, respectively (Fig. 5.9). A large proportion (48%) of seedlings disappeared between readings and the reasons for their mortality could not be ascertained. Only 7% of the plants reproduced successfully. Generally, the mortality of seedlings is very high but it varies between life forms, physiological pathway (C$_3$ or C$_4$), emergence date, years and microhabitat. Since there are also large yearly fluctuations in seedling establishment depending on life form, genetic constitution of seedlings, harshness of microenvironment, weather conditions and biotic interactions, comparisons of specific values for different species must be made with caution. A detailed account of some of the major causes of mortality is presented below.

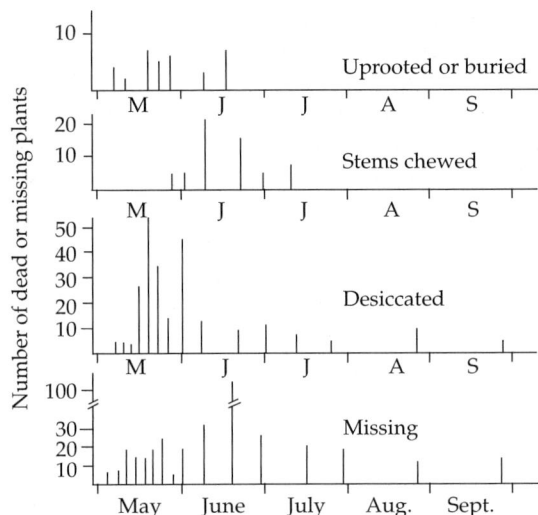

Figure 5.9 Number of dead and missing plants of *Cakile edentula* from May to September 1978 caused by various factors on the Pinery beach along Lake Huron (after Payne and Maun 1984, Courtesy, *American Midland Naturalist*).

5.8.1 Nutrients

The primary cause of low seedling establishment, high phenotypic plasticity, and low productivity may be deficiency of nutrients, and large within- and between-habitat variability (Holton 1980; Maun 1985; Hawke and Maun 1988; Houle 1997). Kachi and Hirose (1983) compared the relative importance of the three macronutrients for the growth of *Oenothera erythrosepala* and found that under field conditions plant growth was controlled more by nitrogen rather than phosphorous and potassium. Artificial fertilization significantly increased seedling establishment of *Ammophila arenaria* (van der Putten 1990). However, the addition of fertilizer to artificially planted grass populations did not improve seedling survival but significantly increased the growth of survivors (Maun and Krajnyk 1989), especially in early spring when plentiful moisture was available. Similarly, McLeod and Murphy (1983) showed that biomass was significantly increased by the

application of complete fertilizers and the greatest reduction in biomass occurred when plants were deprived of nitrogen. Seedlings of annual species growing on the driftline and mid-beach areas such as *Salsola kali*, *Cakile* species, *Corispermum hyssopifolium*, *Atriplex glabriuscula*, *A. laciniata* and others are not colonized by arbuscular mycorrhizal (AM) fungi and depend heavily on washed-up detritus consisting of macroalgae, *Fucus vesiculosus*, *Ascophyllum nodosum*, *Alaria* and *Laminaria* spp., fish kills, insects, rainfall input, aerosol spray and other debris (Ignaciuk and Lee 1980; Payne and Maun 1984). Since this material is randomly distributed, it creates large nutrient variability in microsites. Seedlings growing in the vicinity of rich microsites have high seedling survival and produce high seed output compared to plants growing in nutrient-poor microsites, thus creating very high plasticity in these plant populations (van Asdall and Olmsted 1963; Hawke and Maun 1989). Leguminous species growing on the strand such as *Strophostyles helvola* and *Lathyrus japonicus* have an advantage over other species because they can fix their own nitrogen.

As one moves from the beach to inland dunes perennial species become abundant. Their seedlings depend primarily on soil organic matter, colonization by mycorrhizal fungi, aerosol spray and input from rainfall. Perhaps the most important source of nutrient acquisition, especially phosphorous, by these plants in coastal dunes is their association with AM fungi in the root zone (see Chapter 9 for more details). Another factor in foredunes is the rather low quantities of essential nutrients required by dune species that may be satisfied by the amounts deposited from salt spray. For example, Hawke and Maun (1988) showed that *Oenothera biennis* was less exacting in its requirements of N, P and K than *Corispermum hyssopifolium* and *Cakile edentula* and all three species not only required rather low concentrations of macronutrients but also exhibited flexibility in allocation of resources and mobility of nutrients within the plant.

5.8.2 Soil temperature

As reported earlier (Chapter 2), sand surface temperatures can rise to high levels during summer, similar to those of the desert environment, because sand is a poor conductor of heat. Warming up of surface layers has both useful and deleterious effects. In early spring, the sand surface warms up faster thus allowing a rise in temperature to the optimum for seed germination of plant species. Seedlings and tillers of many species emerge from the soil very early in spring when soil moisture is plentiful. However, what started out as a benefit becomes a hazard as the temperatures of the sand surface rise to levels above 50°C and soil surface is desiccated. For example, surface temperatures on bright sunny days in June and July were 55–60°C in sand plains of Connecticut (Olmsted 1937), 48–53°C at the Pinery along Lake Huron (Baldwin and Maun 1983), above 45°C for 4 hrs day^{-1} on Lake Michigan sand dunes (McLeod and Murphy 1977), 42°C at Blakeney Point (Salisbury 1934), and fluctuated during a single day between 10–35°C at the northern site (38°20' N lat.) and 10–50°C at the southern site (34°02' N lat.) along the Pacific coast of California (De Jong 1979). High temperatures of this magnitude kill young seedlings of plant species by girdling the epicotyl or hypocotyl and destroying the phloem and xylem of seedlings. For instance, seedlings of C. longifolia exhibited very high mortality from 29 May to 2 June 1978, which coincided with a heatwave that raised sand surface temperatures to about 45°C and scorched the seedlings at the point of contact with sand.

5.8.3 Desiccation

The potential of the environment to induce water loss from the plant is called drought. Plants on sandy coasts frequently experience drought stress and in order to survive and reproduce under such conditions the plant must develop adaptations (heritable changes in function) to overcome low moisture conditions. If a plant is able to survive, grow and reproduce under water-deficit stress conditions, it is called a xerophyte. Drought resistance in plants may develop in two ways: (i) by utilizing adaptations that exclude moisture stress from its tissues (avoidance) or (ii) by developing adaptations to withstand the penetration of water stress (tolerance). A majority of higher plants avoid drought stress by maintaining high vapour pressure in their tissues by conserving water through morphological adaptations such as increase in water absorption, pubescence on leaves, thick cuticle, fleshy leaves or by lowering their metabolic rate to extremely low levels. In contrast, lower plants such as lichens, mosses and seeds tolerate drought by desiccation and decreasing their vapour pressure to very low values. However, they recover normal metabolism as soon as they rehydrate after a rainfall.

There is evidence that a major hazard for seedling survival in coastal sand dunes is desiccation caused by low soil moisture and high temperatures. For instance: (i) watering at weekly intervals of artificially planted seedlings of A. breviligulata on the first dune ridge resulted in 92% survival (Maun and Krajnyk 1989), (ii) seedling establishment of A. breviligulata was higher in soils with finer sand texture owing to the positive relationship between grain diameter and soil moisture-retention capacity (Jagschitz and Bell 1966) and (iii) safe sites for seedling establishment of Ammophila arenaria and A. breviligulata were found only in moist locations, especially in moist slacks (Huiskes 1977; Maun and Baye 1989). Sandy substrates are very well drained with low moisture-holding capacity. Even though annual precipitation in temperate latitudes and longitudes is about 80–90 cm per year, dry spells of weather are very common. A large number of seedlings are killed by dehydration during such spells, especially in their very young age. For example, 36% of 4-week-old seedlings of *P. trifoliata* died of dehydration during a dry spell of 9 days on Lake Michigan dunes (McLeod and Murphy 1977). Several studies clearly show that 20–80% of the seedlings may succumb to desiccation in different microhabitats of foredunes (Laing 1958; Maze and Whalley 1992b; Payne and Maun 1984; van Asdall and Olmsted 1963; Weller 1989). Seedlings establishing successfully possess several adaptive traits that allow them to survive in this habitat.

1. *Optimal season of emergence.* Seedling emergence period coincides with the highest available soil moisture levels. In general along all coastal locations the season of germination, emergence and establishment are synchronized with the highest rate of survival to maturity (van Asdall and Olmsted 1963; Wagner 1964; Weller 1989). The timing of germination provides the maximum chances of survival.

2. *Fast growth of root systems.* Seedlings of species that can extend their root system to relatively moist layers in the soil within a short period of time have better chances of survival. Bright sunshine dries the top layer of sand to a depth of about 5–10 cm (Laing 1958) which acts as a protective layer over the sand surface, thus conserving moisture in the deeper layers of sand. Ranwell (1960) showed that within the first four days of germination the roots of *Euphorbia paralias* had extended to a depth of 5–6 cm and after 14 days had grown into the zone of permanently moist sand at a depth of 10–15 cm. Westelaken and Maun (1985a) excavated surviving seedlings of *Lithospermum caroliniense* on different dates during summer and showed that they allocated a larger proportion of their assimilates to the growth and development of their root system than leaves and stems. Following their emergence in early May the seedlings had extended the tips of their roots to a depth of about 12 cm (zone of relatively greater moisture content) within 30 days (11 June), 14 cm by 1 July and

19 cm by 8 September. Similarly, the roots of *Ptelea trifoliata* seedlings completed 67% of their total growth within the first month of their life on Lake Michigan sand dunes when the rains were more frequent and soil moisture levels were relatively high (McLeod and Murphy 1977). Usually June and July are the warmest months and soil water potentials can go down to levels below the survival limits of seedlings. De Jong (1979) found that along the Pacific coast of California water potentials can decline to −30 to −85 bars at 10 cm depth in southern locations and −30 bars in the northern locations, however, the potential was much higher (−15 bars) at 30 to 100 cm depths of sand. Thus, in this habitat the roots of seedlings will have to grow to a depth of 30 cm to reach the relatively higher moisture levels.

3. *Nurse plants.* Seedlings growing in the vicinity of shrubs may benefit from the significant modifying effect of shade cast by them on the sand surface. This positive influence of other plants (that are potential competitors) is caused by decreased insolation in the shaded habitat, lower evaporation, less substrate disturbance, lower surface temperatures and probably higher soil nutrient availability (Blake 1935; De Jong and Klinkhamer 1988). For instance, McLeod and Murphy (1977) showed that under the crown of *P. trifoliata* the solar intensity was reduced by 64 to 70% and sand temperatures were <30°C. In sheltered habitats the survival of *Ptelea trifoliata* seedlings was 24.5% compared to 0% in open habitats (Fig. 5.10). Similarly, rank correlation analysis revealed that the establishment of *Cirsium vulgare* and *Cynoglossum officinale* seedlings was significantly higher in sheltered compared to open plots (De Jong and Klinkhamer 1988). On a subarctic coastal dune on the east coast of Hudson Bay (55°17′ N, 77°45′ W), *Honkenya peploides* facilitated the recruitment of *Leymus mollis* seedlings by trapping and retaining seeds and acting as a nurse plant that improved seed germination and seedling emergence (Gagné and

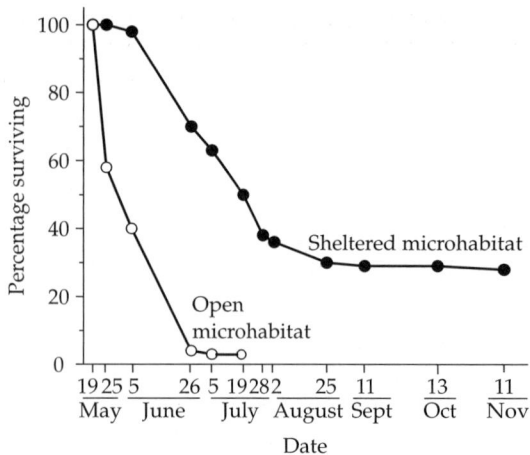

Figure 5.10 Survivorship curves of *Ptelea trifoliata* seedlings during the 1973 growing season in open and sheltered microhabitats on the foredunes of Lake Michigan (after McLeod and Murphy 1977).

Houle 2001). In the foredunes of the Pinery along Lake Huron a large number of mature seedlings can be found in close proximity to trees and shrubs and on north slopes. Similarly, seedlings of *Uniola paniculata* (Wagner 1964) and *Ammophila breviligulata* (Laing 1958) growing under shade showed higher survival than those growing in the open. However, what started out as a benefit in seedling establishment may not last long because of below-ground competition with roots of nurse plants.

4. *Avoidance.* Seedlings of *Lithospermum caroliniense* avoid desiccation owing to dense hair on the leaf surface. Rolling of leaves by *Ammophila arenaria* reduces moisture loss by transpiration and thick cotyledonary leaves of *Cakile edentula* retain moisture for a longer period of time. All these adaptations improve seedling survival.

5.8.4 Salt spray and soil salinity

Salt spray and soil salinity may, under certain environmental conditions, exert an influence on seed germination, seedling emergence and

their establishment. For the effect of salt as a factor on sea coasts refer to Chapter 8.

5.8.5 Density-dependent processes

Some studies have shown that competition was not a major factor in the survival of seedlings (Maze and Whalley 1992b), mainly because they were growing at low density in bare areas with little or no intra- or interspecific competition with neighbouring plants. It was rather difficult to show evidence of density dependence because of the confounding influences of physical and biotic factors such as desiccation, burial by sand, erosion of substrate and predation. As shown earlier, seedling survival is strongly depended on stochastic events. However, seedlings growing in clumps, usually in sites of parents of the previous year or in association with neighbours of other species, may experience both inter- or intraspecific competition. Payne and Maun (1984) compared survival of *Cakile edentula* seedlings growing in 19 clumps with those growing singly. Only the seedlings in large clumps (30.5 cm diameter) had a significantly higher mean mortality rate than the adjacent controls with low density. The survival in smaller and medium-sized clumps did not differ significantly. Keddy (1981) showed that seedlings of *Cakile edentula* clearly indicated a significant ($P < 0.001$) effect of intraspecific competition on the gradient from the seaward to the landward end of the beach. On the seaward end and middle of the beach, density dependence did not affect survival, primarily because of low seedling density, but on the landward end of the beach as the seedling density increased, the proportion of surviving plants significantly decreased. However, the density-independent mortality stayed constant in all three habitats on the beach.

Zhang and Maun (1991a) sowed seeds of *Panicum virgatum* at different densities in 100 cm^2 plots in an artificially created sandy

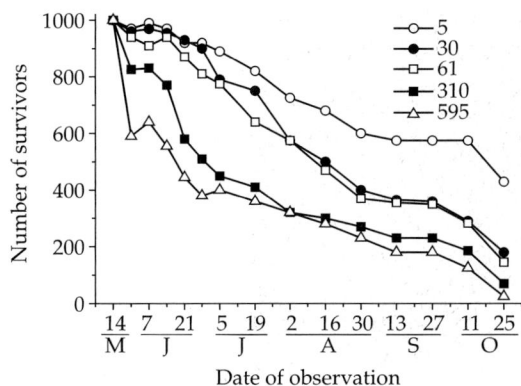

Figure 5.11 Survivorship curves of *Panicum virgatum* seedlings at different densities in an outdoor garden. The seedling survival curves were standardized to start at 1000 for comparison of all densities (after Zhang and Maun 1991a).

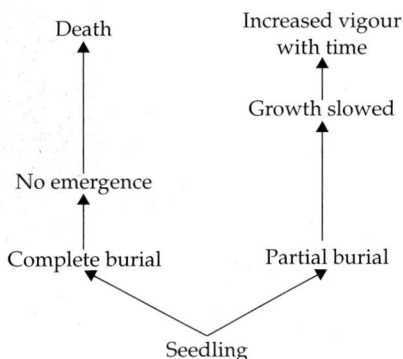

Figure 5.12 Possible fates of seedlings of sand dune species following partial or complete burial in sand (after Maun 1998).

substrate in an experimental field garden. Upon germination, the survival of seedlings growing at five different densities, 5, 30, 61, 310, 595 per plot, was monitored for five months. At low seedling densities of 61 or less the mortality rate was constant but at high densities greater seedling mortality occurred at early stages of growth (Fig. 5.11), suggesting a significant density-dependent effect. Regression analysis revealed a significant correlation ($Y = -0.59 + 1.09x$, $r^2 = 0.999$, $P < 0.001$) between seedling density and mortality. The slope of the line did not differ from 1, suggesting that mortality was proportional to seedling density. In the early stages of growth the seedlings were uniform in size but within about four weeks some seedlings in each plot started to grow faster than others and a size hierarchy became evident. The number of large seedlings was significantly higher in high-density plots. According to De Jong and Klinkhamer (1988), clustering of >15 seedlings dm^{-2} decreased survival and seedling mass of *Cynoglossum officinale*. On the coastal dunes at Aberffraw, Wales, the growth of several winter annuals was suppressed by *Festuca rubra* (Pemadasa and Lovell 1974).

Cheplick (2005) showed that *A. breviligulata* exerted a strong negative influence on the distribution, density and seed production of *Cenchrus tribuloides* and *Heterotheca subaxillaris*. The evidence could be seen by the greater abundance of *Triplasis purpurea* and *C. tribuloides* in bare areas not under the influence of *A. breviligulata*.

5.8.6 Sand erosion or accretion

The major physical forces on the strand were wave action and wind velocity that undercut, overwashed, eroded, or deposited large quantities of sand, gravel and detritus. Storms of high intensity may sometimes completely bury the flora and create a primary bare area. Similarly, high wind velocity may erode sand from the bases of seedlings, which always leads to toppling of seedlings and death because of desiccation of their root system (Wagner 1964; Maun 1981). A model of the possible fate of partially or completely buried seedlings is presented in Fig. 5.12. The survival depends on its size or age, the relative depth of burial, and the energy contained in its leaves, roots and stems. In general, the greater the depth of burial the lower was the survival of seedlings. With few exceptions complete burial of seedlings was almost always fatal (Zhang and Maun 1991a, 1992;

Maun *et al.* 1996) unless they were re-exposed within a week (Harris and Davy 1987). Sykes and Wilson (1990b) suggested that mortality was caused by the inability of seedlings to tolerate darkness and/or to grow up through the sand. In their study 29 New Zealand sand dune species exposed to darkness survived for periods ranging from 19 to 140 days. They concluded that with the exception of some species darkness was an indicator of survival under sand burial conditions (r = 0.607, P < 0.001). The suggestion is probably only partially correct: in spite of arguments in favour of some similarities between darkness and burial direct comparisons between the two are unwarranted because the two environments are not analogous (Harris and Davy 1988).

Zhang and Maun (1991a) tested the effect of post-emergence burial on the survivability of 1-week-old *Panicum virgatum* seedlings emerging from seeds sown at different depths of burial on a Lake Erie strand. In spring when seedlings were one week old they were divided into four groups on the basis of their depth of emergence and then artificially re-buried to depths ranging from <1 to 6 cm. As the depth of emergence increased from 1 to 11 cm, the survivorship of seedlings to withstand post-emergence burial decreased significantly (Table 5.2). For example, seedlings emerging from 3–4 cm depths in spring could withstand up to 6 cm of burial while those emerging from 10–11 cm survived only 1–2 cm of burial. The main reason for poor performance of seedlings emerging from greater depths was their vigour relative to seedlings emerging from shallow depths. They had used up all their resources for emergence from a greater depth of burial and were unable to recuperate within one week of their life. For example, Yanful and Maun (1996b) clearly showed that the emergence of *Strophostyles helvola* seedlings and their survival following burial were dependent

Table 5.2 Number of surviving seedlings of *Panicum virgatum* that emerged in spring from seeds buried at different depths and were subsequently buried to various depths by natural sand accretion (Zhang and Maun 1991a)

Emergence Depth (cm)	Post-emergence burial depth (cm)					
	<1	1–2	2–3	3–4	4–5	5–6
Group I						
1	1	1	0	0	0	0
Group II						
2	9	25	10	2	0	0
3	17	28	10	2	3	1
4	24	34	6	3	3	2
5	11	23	10	0	1	0
Group III						
6	13	34	7	0	0	0
7	15	32	12	0	0	0
8	9	24	1	0	0	0
9	3	3	1	0	0	0
Group IV						
10	2	11	0	0	0	0
11	0	1	0	0	0	0

on the seed mass from which a seedling had originated. Larger seeds produced larger seedlings and more robust seedlings and emerged from greater depths of burial than those from smaller seeds.

5.8.7 Biotic agents

Many insect species, fungi, birds and other animals consume parts of seedlings and contribute to their demise (see Chapter 10 for more details).

5.8.8 Human disturbance

Coastal strands provide ideal locations for human recreation. Most of the sandy beaches are overcrowded during summer. Construction of tourist resorts, cottages and roads by the sea coasts has virtually destroyed many foredunes. Other areas, for example sea side parks and conservation areas, are also so heavily used that seedlings of annuals growing on the beach are killed by trampling, vehicular traffic, enhanced erosion and burial. On the foredunes of Lake Michigan, 22 to 37% of the seedlings of *Ptelea trifoliata* were lost due to vehicular traffic (McLeod and Murphy 1977). Trampling by people kills seedlings and adult plants, thus creating bare sand surface on paths that become susceptible to wind erosion and significant reduction in cover of vegetation.

5.8.9 Unknown causes

Fenner (1987) pointed out that large seedling mortality occurs in arable habitats for which the cause of death cannot be assigned. Sand dune systems are no exception.

5.9 Summary

Species of coastal dune complex produce large numbers of viable seeds that are dispersed to various microhabitats. Usually seeds accumulate in depressions where they are buried to various depths by sand. Burial to an appropriate depth is useful for the germination of a seed and the emergence of seedlings. However, as soon as burial exceeds a certain threshold level, the seedlings are unable to emerge from the soil. The plant species employ different patterns of seedling emergence with entire emergence in spring or autumn, a major flush in spring and a small one in autumn or emergence throughout the growing season. In all cases it seems to be synchronized with optimal temperatures, high amounts of soil moisture and/or high rainfall. The relative merits of each strategy are poorly understood but all studies agree that each strategy is adaptive and maximizes the chances of overall establishment of seedlings, reproductive maturity and long-term survival of the species. However, seedling establishment is highly variable from year to year depending on the weather conditions and other limiting factors during the growing season. There is strong evidence that even under the most unfavourable conditions in some years a certain proportion of seedlings of almost all species are able to establish in safe sites. The level of success is variable among summer annuals, winter annuals, biennials and perennials depending on the weather and microhabitat conditions.

Seedlings must cope with many limiting factors: low nutrient availability, sand accretion, erosion, sandblasting, salt spray, high soil temperatures, desiccation and harmful biotic agents. After its emergence a seedling requires proper nutrient levels for its establishment. Since dune systems lack essential macronutrients, detritus washed up on the driftline, atmospheric inputs, nutrients in salt spray and association with arbuscular mycorrhizal fungi provide the much-needed required elements. High soil temperatures and desiccation take their toll on a certain proportion of seedlings but avoidance mechanisms such as optimal season of emergence, fast growth of the root system, survival in the shade of nurse plants and seedling morphology improve their survival. The two most predictable environmental stresses on sea coasts are salt spray and burial in sand. Seedlings exhibited a high degree of tolerance to both factors by employing both tolerance and avoidance strategies. Although complete burial was almost always fatal, the seedlings not only tolerated partial burial but also showed marked stimulation in overall growth. The main mechanism for withstanding salt spray was by an increase in leaf thickness, decrease in ion uptake, sequestering salt in vacuoles or excreting salt through salt glands. However, certain limiting factors such as insect and disease damage, inter- and intraspecific competition, herbivory and trampling by humans can not be avoided. Nevertheless even a very small proportion of seedling establishment is sufficient for the overall survival of the species.

CHAPTER 6

Burial by sand

6.1 Introduction

In coastal dune systems, plant communities are fundamentally the product of interaction between disturbance of the substrate, impact of high wind velocities, salt spray episodes, sand accretion levels and other factors of the environmental complex. Burial by sand is probably the most important physical stress that alters species diversity by eliminating disturbance-prone species (Maun 1998). There is a close correlation between sand movement and species composition, coverage and density (Moreno-Casasola 1986; Perumal 1994; Martínez *et al.* 2001). Sand accretion kills intolerant species, reduces the relative abundance of less tolerant species and increases the abundance of tolerant species. It filters out species as the level of burial starts to exceed their levels of tolerance. For example, lichens and mosses are the first to be eliminated, then the annuals and biennials and finally the herbaceous and woody perennials. Again within each life form and genus there are significant differences in survivability. Burial imposes a strong stress on production by altering normal growth conditions and exposing plants to extreme physiological limits of tolerance.

Do plant communities occurring in different locations within a dune system correspond to the amount of sand deposition? Several studies (Birse *et al.* 1957; Moreno-Casasola 1986; Perumal 1994) show that the species composition and their distribution are strongly related to the long-term average sand deposition. The evolution of a plant community in coastal foredunes requires frequent

and persistent predictable burial events specific to a particular coast. In a large majority of sea coasts burial occurrences are of relatively low magnitude and species occupying the coasts are well adapted to withstand the stress imposed by burial. This recurring event within the generation times of plant species allows them to acquire genes of resistance over time and evolution of adaptations to live in this habitat. A prerequisite to survive in this habitat happens to be the ability to withstand partial inundation by sand. To survive the dynamic substrate movement a plant species must be a perennial, be able to withstand burial, endure xerophytic environment, spread radially and vertically, and adapt to exposure on deflation and coverage on burial (Cowles 1899). Nevertheless, in some very high-energy coasts few plants are able to establish and survive because wind velocities are too strong and substrate movement too frequent.

In this chapter a critical review is presented of the ecological impact of burial by sand as a stress on the physiological ecology, growth, maximum limits of tolerance and adaptive mechanisms of seedlings and adult plants exposed to regular sand accretion in dune systems.

6.2 Impact of burial in other ecosystems

Burial is not unique to sand dune systems because it occurs in several other ecosystems as well. It invariably has a strong impact on community composition, density and distribution.

6.2.1 Wetlands

In wetlands under normal sedimentation over thousands and tens of thousands of years, plants are well adapted to the prevailing sediment inputs (Walker 1970). However, a change in sedimentation rate imposes stresses on the system and alters species composition through natural selection. An increase in burial by floating plant debris or wrack brought in by a high tide has altered the composition of the plant communities in New England salt marshes (Brewer *et al.* 1998). Burial caused significant mortality of *Juncus gerardi* and *Spartina patens* and increased bare areas which allowed establishment of *Salicornia europaea* and *Aster tenuifolius* and increased expansion of *Distichlis spicata*. Clear cutting of forests for timber and intensification of agriculture also increase the soil and silt levels in streams and rivers that empty into lakes and oceanic deltas. In some Oxbow lakes in Central Alberta, sediment loads in streams have increased by at least four to six times (van der Valk and Bliss 1971). As a result, the normal ecological balance of wetland communities has been significantly altered.

6.2.2 Coral reefs

The flow of sand, silt and clay sediments into the river deltas of sea coasts in tropical and subtropical regions destroy corals by blocking sunlight and smothering polyps. Dodge *et al.* (1974) found an inverse relationship between rate of resuspension of sediments and average growth rate of coral in Discovery Bay, Jamaica. Similarly, extensive coral mortality occurred in reefs of Castle Harbour, Bermuda because of burial and increased turbidity caused by dredging from 1914 to 1943 (Dodge and Vaisnys 1977). Major losses occurred to corals older than 20 years while younger ones survived. The primary reasons for mortality and reduction of coral growth were impairment of light intensity and quality, reduced photosynthetic activity of algae symbionts and drastic reductions in calcification rate of corals. Experimental suspension of peat particles over reef-building coral, *Montastrea annularis*, decreased production, reduced chlorophyll content by 22% and increased the rate of respiration (Dallmeyer *et al.* 1982). Similarly on rocky seashores, community structure showed a strong relationship with sand inundation at San Nicolas Island, California (Chapman 1992). Long-lived mussels dominated rock pinnacles not subject to sand inundation while opportunistic species of green algae, perennial sea grasses and large sea anemones dominated on sand scoured surfaces.

6.2.3 Unpredictable disturbances

Burial also occurs occasionally by infrequent disturbances, disasters or catastrophic events such as long drought periods, volcanic tephra deposits and lava flows, hurricanes, tsunamis, mudslides and collapse of talus slopes that dump huge loads of sediments on plant communities. Such events are unpredictable and devastate ecosystems whose recovery to former levels or even new levels may take many years. For instance, the seven-year drought from 1933 to 1940 in the prairies of Midwest Canada and the United States caused major alterations in species composition (Weaver 1968). Three factors, severe drought, high winds, erosion and burial, combined to reduce coverage of less resistant species. Severe drought curtailed growth, reproduction and lateral spread of plant species. High-velocity winds created dust storms that covered vegetation with thin blankets of silt and buried plants in drifts of loose earth. Vast areas of vegetation were killed by layers of dust of 1–4 cm in depth. Farm soils ploughed up for agriculture were blown away as dust because there was no vegetation to stabilize the surface. Even natural prairie that had not been ploughed was weakened by differential mortality of susceptible species and subsequent development of bare areas. There was a major change in the composition of vegetation from a true prairie to a mixed prairie; the dominant grass of true prairie, *Andropogon scoparius* (C_4), was significantly reduced because of relatively short roots and replaced by *Agropyron smithii* (C_3) mainly because of its sod-forming capability, production of vertical rhizomes, deeper root system, ability to initiate growth very early in spring when moisture was plentiful and continued growth rate in autumn. At the end of the long drought period in 1941, the recovery of vegetation started but it took 20 years for the re-establishment of the true prairie to its former levels (Weaver 1968).

6.2.4 Volcanic eruptions

Catastrophic volcanic eruptions occur infrequently and release large amounts of chemically active and impervious tephra into the atmosphere which settles and buries plant communities. Harper (1977) suggested that since catastrophes are rare events they are of little selective consequence to fitness in succeeding generations. Antos and Zobel (1984, 1985) examined the effects of tephra released by the eruption of Mount St Helens, Washington, on the recovery of vegetation. All plants showed no effects by 2.3 cm of burial but bryophytes were killed at 4.5 cm, herbaceous plant species richness was reduced by 32% at 7.5 cm and all herbaceous plants were killed at 15 cm depth of tephra. Woody shrubs and trees had greater survival than herbaceous plants. Erosion of ash at certain locations by rainfall events significantly improved survival.

6.3 Alteration of soil micro-environment

Sand deposits alter soil micro-environment (Fig. 6.1) and both beneficial and harmful changes are experienced by completely or partially buried plants. Suddenly, the aboveground environment of the plant has been transformed into a dark suffocating milieu. Since temperature decreases with soil depth, the roots are experiencing much lower soil

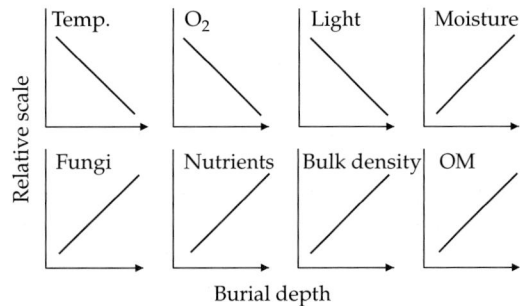

Figure 6.1 Changes in temperature, moisture, bulk density and other environmental factors in the soil environment of a plant buried in sand (assembled from Baldwin and Maun 1983).

temperatures. Under arctic and subarctic conditions, burial by sand may also create snow lenses beneath the surface that linger into summer (Marin and Filion 1992; Filion and Marin 1988) and significantly delay emergence and growth of buried plants in spring. The roots are still anchored in the old soil with higher organic matter content compared to low organic matter in the new sand, probably with a slightly different pH. An overburden of sand increases soil compaction and bulk density of the soil in the root zone. The amount of moisture in the root zone also increases from about 2% at the surface to about 7% at 100 cm depth (Fuller 1912; Oosting and Billings 1942; van der Valk 1977; Baldwin and Maun 1983).

Another major change is in the concentration of oxygen in the root zone because burial by sand impairs the normal diffusion rate of oxygen to the roots, hence lowering the oxygen content (Bergman 1920; Kurz 1939; Bertrand and Kohnke 1957). Since oxygen diffuses into the soil from the surface, the level decreases with increasing depth of soil (Armstrong 1979).

6.4 Burial and plant growth
6.4.1 Seedlings

Burial in sand is a major factor affecting the growth of seedlings in sand dunes. The initial

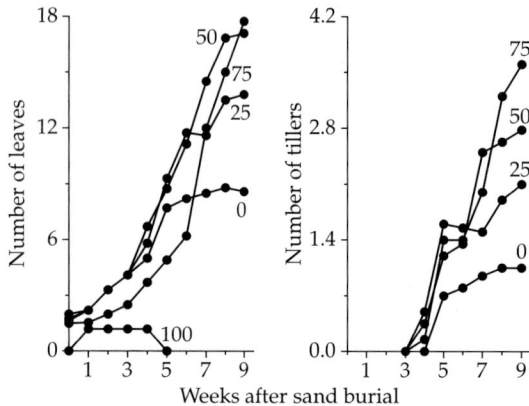

Figure 6.2 Number of leaves and tillers of *Agropyron psammophilum* seedlings buried at 0, 25, 50, 75 and 100% of their height when they were one week old (after Zhang and Maun 1990a).

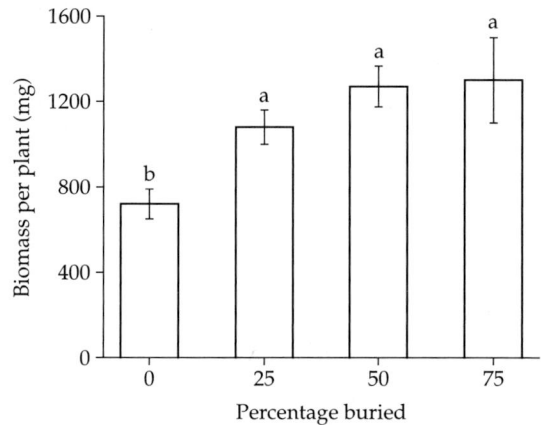

Figure 6.3 Mean biomass ($\pm$SE) per plant (mg) of *Agropyron psammophilum* buried at 0, 25, 50, 75 and 100% of their height when they were one-week-old seedlings (after Zhang and Maun 1990a).

response of seedlings to partial or complete burial in sand is reduction in growth and vigour (Maun *et al.* 1996). However, this negative response is short-lived. The plant adjusts to the new micro-environment, starts to recover and may eventually show growth stimulation. One-week-old seedlings of *Agropyron psammophilum* buried at 25, 50, and 75% of their height in sand took 6 to 8 weeks to recover but then started to show luxuriant growth and surpassed control plants in the number of leaves, tillers (Fig. 6.2) and total biomass per plant (Fig. 6.3) (Zhang and Maun 1990a). The plants buried to a greater percentage of their height took longer to recover and surpass the control. Similar increase in the numbers of leaves and tillers was observed in seedlings of *Uniola paniculata* (Wagner 1964). In another experiment, Zhang and Maun (1992) partially buried 4-week-old seedlings of *Cakile edentula* to 7 cm depths (63–93% of plant height) in sand and then removed the sand after 1, 2, 3, 4 and 5 weeks. The plants responded positively by a significant increase in relative growth rate (RGR) even 1 week after burial. There was a significant linear relationship ($r^2 = 0.64$, $P < 0.01$) between RGR and percentage burial. Burial also showed a significant increase

in the number of flowers, fruit biomass and total biomass but the mean seed mass was not affected (Table 6.1). Similarly, six tropical sand dune species, *Chamaecrista chamaecristoides*, *Palafoxia lindenii*, *Trachypogon gouinii*, *Canavalia rosea*, *Ipomoea pes-caprae* and *Schizachyrium scoparium*, exhibited an increase in leaf area and biomass following burial in sand (Martínez and Moreno-Casasola 1996). In general except for *T. gouinii* buried plants allocated more biomass into leaves and above-ground stems. Enhancement in biomass occurs primarily due to increased root mass (Cheplick and Grandstaff 1997; Perumal and Maun 2006).

Another important consideration is the rate at which plants are being buried by sand. For example, how does gradual burial in sand compare with sudden deposit of the same magnitude? Seedlings of *Cirsium pitcheri* were exposed to one time burial of 4 and 8 cm and 4 repeated burials of 1 or 2 cm every 8 days (Maun *et al.* 1996). Gradual burial permitted faster recovery and exhibited enhanced growth earlier than one time burial. It also showed greater stimulatory effects than single burial treatments. This may not be surprising because repeated burial was more akin to natural burial events and hence the plants

Table 6.1 Reproductive output and biomass per plant of 20-day-old *Cakile edentula* plants buried at 0, 3, 6 cm depths in washed sand (adapted from Zhang and Maun 1992)

Burial depth (cm)	Number of flowers	Fruit biomass (mg)	Total biomass (mg)	Seed mass (mg)	
				Upper	Lower
0	10.6a	136.9a	315.8a	5.53a	4.55a
3	15.8b	188.6b	396.8b	5.46a	4.44a
6	16.7b	211.9b	447.7b	5.51a	4.53a

Means in each column followed by the same letter are not significantly different at $P < 0.05$ according to Tukey's test.

were better adapted to it. A study by Lee and Ignaciuk (1985) also showed that *Cakile maritima* and *Atriplex glabriuscula* exhibited the most enhanced growth with gradual weekly sand deposition of 4–8 cm, whereas *Atriplex laciniata* and *Salsola kali* produced maximum dry matter at 8–12 mm of burial per week.

6.4.2 Adults: herbaceous species

The response of adult plants to burial is similar to that of seedlings, but because mature plants have greater energy reserves they can withstand greater depths of overburden. A large majority of foredune perennial grasses are so well adapted to this habitat that they actually require burial to maintain high vigour. For example, vigorous stands of *Ammophila breviligulata* (Eldred and Maun 1982), *Calamovilfa longifolia* (Maun 1996), *Agropyron psammophilum*, *Panicum virgatum* (Perumal 1994), *Ammophila arenaria* (van der Putten *et al.* 1988), *Spinifex sericeus* (Maze and Whalley 1992a), *Uniola paniculata* (Wagner 1964), *Festuca rubra* var. *arenaria* (Anderson and Taylor 1979) and other foredune species growing along sea coasts, are prevalent in foredunes receiving sand accretion rates of about 1–25 cm per year (Maun and Baye 1989). Enhanced vigour in these plants is manifested as an increase in biomass per shoot (Eldred and Maun 1982; Disraeli 1984). Similar increases in below- and above-ground biomass and percentage cover were observed

in several dune grasses and forbs under various artificial sand accretion rates of 1–15 cm (Perumal 1994). Nevertheless, responses to burial are rate dependent and above a certain critical level plants are negatively affected. Burial depths of 5–20 cm generally maintain or enhance shoot density of dune grasses while deeper deposits precipitate a temporary setback. The greater the depth of burial the greater was the reduction in reproduction of *Ammophila breviligulata* plants (Seliskar 1994) and shoot density of *A. breviligulata* and *Calamovilfa longifolia* (Maun 1998).

Sykes and Wilson (1990a) exposed 2-month-old plants of 30 New Zealand sand dune species to four burial treatments according to the proportion (0, 66, 100 and 133%) of plant height in a greenhouse and recorded their biomass after 15 weeks of recovery. At 66% burial, 6 of 30 species showed a significant decrease in biomass, 23 were unaffected and 1 showed a significant increase. At 100% burial, 20 of 30 species and at 133%, 28 of 30 species recorded a decrease in biomass and the remaining species were not affected. Effects are thus dependent on extent of burial and the plant species.

6.4.3 Adults: woody species

Burial may stimulate the growth of woody plants as well. Shrubs of a tropical coastal foredune were more tolerant of burial than herbaceous, erect and creeping species (Martínez and Moreno-Casasola 1996). Burial by sand increased the size of growth rings of *Betula papyrifera* and *B. glandulifera* (Hermesh 1972) and *Thuja occidentalis* on Crystal Lake dunes in Michigan (Wolfe 1932). However, in arctic and subarctic locations there was actually a decrease in width of growth rings of trees following burial, primarily because the long-lasting snow lenses beneath sand delayed and curtailed growth (Marin and Filion 1992). Shi *et al.* (2004) experimentally buried 1-year-old seedlings of *Ulmus pumila*, an endemic species of the Hunshandak Sandland of China, and later measured net photosynthesis, growth and biomass allocation. The seedlings were buried at 33, 67, 100 and 133% of their height. All seedlings survived partial burial of 33 and 67%, while 70% of seedlings buried to 100% survived after 6 weeks and only 20% survived the highest treatment. The surviving seedlings exhibited an increase in net photosynthetic rate, transpiration, water use efficiency and in total biomass. Although the total leaf area was reduced by burial, the increase in net photosynthetic rate of remaining leaves compensated for the lost tissue, thus balancing the carbon and resource requirements of plants. McLeod and Murphy (1983) found no significant differences in growth between partially buried and undercut (sand removed) seedlings of *Ptelea trifoliata* except that the buried seedlings had 3.6 leaves compared to 2.6 in sand removal treatments. The partially buried seedlings were also slower to wilt when water was withheld and 28.5% produced adventitious roots on the cotyledonary nodes. *Pinus sylvestris*, however, instead of producing new roots on the stem elongates its roots upwards into the burial deposit against the force of gravity (Gaël' 1975).

6.5 Physiological ecology

Complete burial of plants drastically reduces net photosynthesis of plants. Harris and Davy (1988) showed that the net CO_2 uptake in the second leaf of *Elymus farctus* plants was reduced to 30% of the control after 2 days of complete

burial and essentially terminated after 5 days. The dark respiration rate was also reduced due to curtailment of light, probably an adaptation for energy conservation under stressful conditions. Seven days after emergence from burial the relative growth rate rose to the same level as controls and full photosynthetic competence was established within 24 hrs. Burial caused a reversal of normal source–sink relationship whereby water-soluble carbohydrates, N, P and K were translocated from roots and stems of buried plants to the leaves. Stored energy reserves were thus shifted to the unburied portions of the plant. Maun *et al.* (1996) also showed that *Cirsium pitcheri* seedlings emerged from burial by re-allocating energy from their roots to aboveground parts. Similar results were obtained by Brown (1997) who showed that three species, *Sarcobatus vermiculatus*, *Chrysothamnus nauseosus* and *Distichlis spicata*, of Mono Lake desert in California shifted their biomass from below-ground to above-ground parts of plants. However, Sykes and Wilson (1990a) did not record a shift in the root to shoot ratio in 23 out of 29 New Zealand species. Brown (1997) noted, however, that the root to shoot ratio in 21 species had indeed decreased, but the differences were not significant. Another reason may be that Sykes and Wilson (1990a) allowed their plants to recover and grow for 15 weeks following burial during which the energy resources used to recover from the episode by 23 species of plants may have been replenished and the normal source to sink relationship re-established.

Thus, there is strong evidence for a shift of resources from storage tissue and buried parts of plants to the meristematic tissues. When a plant is completely buried under sand and photosynthetic function ceases, its metabolism shifts from an energy-producing to an energy-consuming state and it must mobilize stored energy reserves. The major source is energy in its storage tissues: roots, stems, corms, rhizomes or stolons and carbohydrates in the buried green leaves. Other sources may be root connections, in some cases via AM fungi, with other plants in the community or clonal integration.

Yuan *et al.* (1993) partially buried 8-week-old plants (15–20 cm tall) of *Ammophila breviligulata* and *Calamovilfa longifolia* to 0, 3, 6, 9 and 11 cm depths in sand in a greenhouse. Measurements of net CO_2 uptake of fully developed third leaf from the top were commenced 10 days after burial and continued at weekly intervals for 11 weeks. The mean rate of CO_2 uptake increased with sand burial in both grasses (Fig. 6.4) and the total biomass, by the end of the experiment (75 days after burial), was also enhanced (Fig. 6.5). Burial increased leaf thickness of *A. breviligulata* slightly (not significant) but a significant increase was observed in *C. longifolia*. In another experiment (Perumal and Maun 2006) plants of six species, *Strophostyles helvola, Cirsium pitcheri, Elymus canadensis, Oenothera biennis, Agropyron psammophilum* and *Panicum virgatum*, were buried at 33 and 66% of their height in sand and grown in a growth chamber at light intensities of 500, 1000, 1500 and 2000 μmol m^{-2} s^{-1} at a temperature of 25°C. There was a significant increase in CO_2 exchange rate and leaf

thickness of buried plants at all light intensities over control in all six species. When buried plants were exposed to a constant light intensity of 1500 μmol m^{-2} s^{-1} and various temperatures of 20, 25, 30, and 35°C there was an increase in CO_2 exchange rate in C_3 species up to a temperature of 30°C; however, it continued to increase even at 35°C in the C_4 species, *P. virgatum*.

In a field experiment, natural populations of two grass species, *Ammophila breviligulata* and *Calamovilfa longifolia*, were buried to 0, 20 and 40 cm depths (Yuan *et al.* 1993). Upon emergence, photosynthetic rate and water potential were determined every 3–7 days until the middle of August. The net CO_2 uptake was significantly higher than controls on 12 out of 20 dates of measurement in *A. breviligulata* and 6 out of 15 dates in *C. longifolia* (Fig. 6.6). The overall mean CO_2 uptake of *A. breviligulata* for the whole summer was significantly higher in both burial treatments than control; however, in *C. longifolia* even though uptake values were higher they were not significantly different. The leaf water potential of buried plants of both species was not significantly different from controls. A problem with measurements

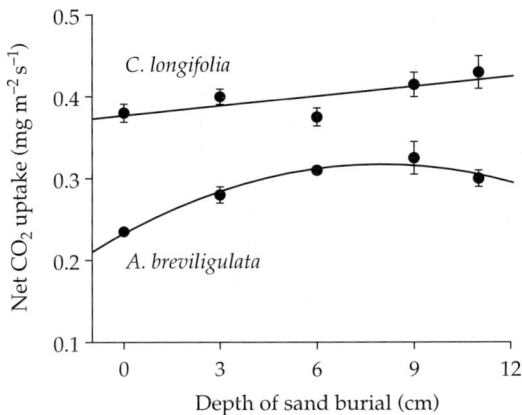

Figure 6.4 Mean ± SE net CO_2 uptake (mg^{-2} s^{-1}) by leaves of *Ammophila breviligulata* and *Calamovilfa longifolia* over the whole experiment as affected by burial of seedlings to various depths in sand in a greenhouse (after Yuan *et al.* 1993).

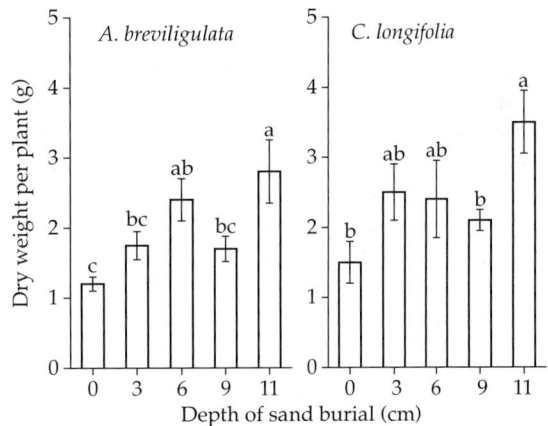

Figure 6.5 Mean ± SE dry weight of seedlings of *Ammophila breviligulata* and *Calamovilfa longifolia* buried to different depths in a greenhouse. Bars within each species with a different letter are significantly different (P < 0.05) according to Tukey's test (after Yuan *et al.* 1993).

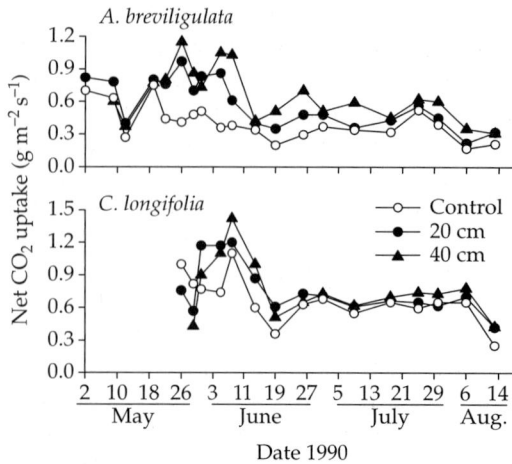

Figure 6.6 Mean net CO_2 uptake (mg^{-2} s^{-1}) by leaves of *Ammophila breviligulata* and *Calamovilfa longifolia* plants after emergence from burial depths of 0 (○), 20 (●) and 40 cm (▲) in sand under field conditions during summer 1990. For visual clarity standard error bars are not presented (after Yuan *et al.* 1993).

of CO_2 exchange rate (CER) in the field was the high variability caused by changes in micro-environmental conditions of temperature, light, wind velocity and other environmental factors. To overcome the problem, Perumal and Maun (2006) measured the effects of partial burial on the CER of six common sand dune species in a growth chamber as well as in the field and greenhouse. All test species, *Agropyron psammophilum, Oenothera biennis, Strophostyles helvola, Cirsium pitcheri, Elymus canadensis* and *Panicum virgatum*, showed an increase in CER under controlled conditions in the growth chamber, thus confirming that growth was actually stimulated by partial burial. There was also a predictable increase in biomass, leaf area and leaf thickness that enhanced the total area of mesophyll thus increasing the capacity for more CO_2 absorption and photosynthesis. Growth stimulation lasted for 2–3 weeks in the field, 3 weeks in the greenhouse and 4 weeks in the growth chamber.

Kent *et al.* (2005) determined the physiological response of a community of plants to artificial burial. They removed turf from four Machair subcommunities: (i) foredune grassland (containing 16 species), (ii) slack (24 species), (iii) 3-year fallow grassland (20 species) and (iv) unploughed grassland (21 species). Five species were common to all four subcommunities, seven to three subcommunities and 12 to two subcommunities. Sixteen species were each present in only one of the four subcommunities. Turves were grown under controlled conditions for 10 weeks before burial at 2 cm depth with sterilized sand and after 2 or 6 weeks the burial deposit was removed. Following exposure, the plants were allowed to adjust to incident light for 16–20 hours or 40–44 hours before net photosynthetic measurements were recorded. The net photosynthetic rate of buried turf was significantly lower than control and the rates of CO_2 exchange were not significantly different between the four subcommunities. Plant recovery of treated turf was also very slow following the burial episode. The lowest rate was recorded in slack turf probably because several of the component species were aquatic and more adapted to waterlogging than to burial. In this community there was no evidence of stimulation of net photosynthesis in partially buried plants.

In a similar experiment Owen *et al.* (2004) subjected turf from the same four subcommunities to two burial treatments, one time burial of 5 cm or five burial events of 1 cm depth each and then followed the survival, growth and percentage frequencies of species after the burial treatments. The burial events did not alter species richness but reduced the relative abundance of species. The overall frequency of plant species in turves was drastically decreased by a single application of 5 cm; however, steady recovery continued subsequently. In the intermittent burial treatment even though the recovery response was similar initially, later instalments of sand produced erratic responses until a sharp decline occurred following the last application. Some component species showed no effect, others

Table 6.2 Mean density (± SE) of grass shoots per 0.5 × 0.5 m before (September 1989) and after (September 1990) the sand burial treatments in *Ammophila breviligulata* and *Calamovilfa longifolia* plants under field conditions (adapted from Yuan *et al.* 1993)

Depth (cm)	A. breviligulata		C. longifolia	
	Before	After	Before	After
0	41 ± 4	42 ± 3	24 ± 1	21 ± 3
20	43 ± 2	24 ± 2	32 ± 5	19 ± 3
40	34 ± 4	5 ± 2	23 ± 5	5 ± 1

declined, recovered or died and some new species were recruited from the seed bank. Some species, under natural field conditions, would probably have dispersed into the community as well.

Under natural conditions, some plant species gain benefits from the burial episode at the expense of others because both intra- and interspecific competition decrease with an increase in the depth of burial (Moreno-Casasola 1986; Perumal 1994). For example, the mean shoot density of both *Ammophila breviligulata* and *Calamovilfa longifolia* decreased significantly following burial treatments (Yuan *et al.* 1993) (Table 6.2). Susceptible species are eliminated and that creates more space for the tolerant species. The released space may also allow new species to disperse into the habitat and gain a foothold. The decomposing remains of susceptible species also provide a source of nutrients for the tolerant species. The plant may also get a brief respite from some injurious soil organisms while it moves its roots into the new soil and new shoots into the aerial environment. In a way it gets a new start in life.

6.6 Tolerance limits

Sand dune plants have widely different tolerance limits depending on the life form, developmental stage, rate and season of burial, the amount of stored energy reserves and evolutionary history. Poikilohydric organisms such

as fungi, diatoms, green algae and cyanobacteria secrete mucilaginous polysaccharides and create microbial crusts on the sand surface that can tolerate only small amounts of sand burial (Danin 1996). The tolerance levels of lichens and bryophytes, for example, ranged from about 1–5 cm of sand deposition per year (Birse *et al.* 1957) and species of mobile dunes are generally more tolerant of burial than those of more stable inland habitats. *Ceratodon purpureus* and *Ditrichum flexicaule*, mosses of open dynamic foredunes, emerged from greater depths (7 cm) compared to other inland species (1–3 cm) (Martínez and Maun 1999). The same is true for herbaceous perennials. Stored energy dictates the tolerance limits of a species. Thus seeds are less tolerant of burial than seedlings, annuals are less so than biennials, while biennials are in turn less tolerant than perennials. Dominant grasses on heavily accreting beach and foredune habitats and a few trees and shrubs are able to survive the highest amounts of sand accretion per year with maximum values ranging from 40–100 cm of burial (Mueller 1941; Ranwell 1958; Rihan 1986; Wagner 1964; Maun and Lapierre 1984; Maun 1996; Houle 1997). Some woody species of shrubs and trees, e.g. *Prunus pumila*, *Rosa rugosa*, *Cornus stolonifera*, and *Hypericum kalmianum*, are also found on actively accreting dune sites despite 15–45 cm of sand deposition annually (Cowles 1899; Belcher 1977; Maun 1993). Similarly, two species, *Sarcobatus vermiculatus* and *Distichlis spicata*, occupying areas of high sand accretion at Mono Lake, California, tolerated greater sand deposition than *Chrysothamnus nauseosus*, a species of microhabitats with low sand accretion (Brown 1997). Nevertheless, the optimum levels for best growth, reproduction and replenishment of storage tissues are approximately half the maximum values.

6.6.1 Natural levels of burial

Many scientists have measured yearly rates of sand deposition on sea coasts (Maun

Table 6.3 The amount of net change (accretion or erosion) per year at different locations along coastal shorelines

Location	Plant species	Sand accretion per year (cm)	Authority
North America			
Pacific coast	*Ammophila arenaria*	57	Barbour *et al.* 1985
	Abronia umbellata	44	Barbour *et al.* 1985
	Elymus mollis	32	Barbour *et al.* 1985
Atlantic coast	Dune community	30	van der Valk 1974
	Ammophila breviligulata	−78 to 102	Zaremba and Leatherman 1984
	Sand fence	Variable	Zaremba and Leatherman 1984
	Panicum amarum	5–120	Woodhouse 1982
Gulf of Mexico	*Calamovilfa longifolia*	80–120	Dahl *et al.* 1975
Lake Huron	*Ammophila breviligulata*	8	Maun 1985
Lake Michigan		30	Olson 1958a
Europe			
Atlantic coast, England (Newborough, Warren)	*Ammophila arenaria*	30–90	Ranwell 1958
Iceland South coast	*Leymus arenarius*	19–34	Greipsson and Davy 1996

1998) and the mean values range from 8–120 cm (Table 6.3). The extent of dune accretion depends on the amount of sand cast onshore by waves each year, the frequency and velocity of winds above 4 m s^{-1} and the general climate of the region. Another factor of major consequence to a plant is the rate at which it is being overwhelmed by sand. The rate depends on many environmental factors such as wind velocity, duration of high winds, sand texture, relative humidity, soil moisture, intensity and seasonality of storms, aspect and location on a sand dune (e.g. crest, windward or lee slope) and the resistance offered by plants. Sand movement on coasts is primarily related to the frequency and duration of extreme winds rather than to the average wind velocity (Bagnold 1960). In tropical regions, because of high frequency of rainfall, sand becomes wet and more coherent and does not drift as easily as dry sand. As a consequence dune systems are lower in stature than those along sea coasts in warmer more dry regions with long drought periods as in Australia, Asia and Africa. Burial events in different world regions

may be regular, seasonal or sudden and unpredictable. In temperate latitudes most of the sand accretion occurs in late autumn to winter and early spring. Along the Norfolk coast the movement was greatest from February to May (Harris and Davy 1986a). At most other latitudes the burial or erosion are also regular annual events. The rest of the year, the winds above the threshold for movement are relatively minor to be of significance. When burial and erosion occur regularly, vegetation is well adapted, but in some parts of the world these events are unpredictable. Violent wave storms and hurricanes penetrate inland as overwash and dump large loads of suspended sediments on top of plants on the summit and lee of dune ridges. Recovery of plants from such heavy loads may take a long time. One 'north-easter' along the North Carolina coast in 1978 penetrated 250 m inland and eroded −0.78 m from the seaward end and deposited +1.02 m on the leeward side of the dune field (Zaremba and Leatherman 1984). The dune system was shifted landward and vegetation was either eroded or covered by large amounts of sand.

Major storms also erode material from one area and deposit it on another and adjustment and recovery from such storms is time-consuming and unpredictable.

6.6.2 Length of time for emergence

The length of time required by a plant to survive a burial episode and emerge above the surface is species-specific and depends on developmental stage, burial depth and weather conditions during the period of recovery. The more quickly a plant emerges, the greater are its chances to re-establish at the new sand surface. Generally the length of burial period critical for survival is limited by the energy reserves available to maintain buried photosynthetic tissue. Newly emerging seedlings of *Elymus farctus* survived complete burial for only one week (Harris and Davy 1987), while those buried for two weeks succumbed. However, some species have more energy reserves than others. For example, *Rumex alpinus* emerged from a burial depth of 20 cm after four months mainly because of large carbohydrate reserves (Klimeš *et al.* 1993). In temperate latitudes, plants undergo winter dormancy and emerge from burial in spring by utilizing stored energy reserves in their underground organs (Maun 1984; Nobuhara 1973; Nobel *et al.* 1979), but if more burial occurs soon after emergence, the chances of survival are limited because their energy reserves have already been utilized (Maun and Lapierre 1984; Maun 1998). Normally, however, sand storms are rare during the summer growing season (Davidson-Arnott and Law 1990) and if the burial is shallow the plant has no difficulty emerging. However, if burial during winter is deep the plants may either fail to emerge or it may take a long time for them to emerge in summer and the remaining length of the growing season may not be long enough for the plant to develop photosynthetic tissue, establish new roots, buds and rhizomes close to the new sand surface and store enough energy reserves to survive the next winter (Maun 1996). These plants will be vulnerable to winter kill. Weather conditions during the summer also affect survival, with mild rainy weather being more conducive to emergence and survival.

6.7 A general model of sand accretion and plant response

We have seen that an advancing dune kills vegetation in its path. However, death rate depends on the rate of advance, dune height and above all the plant species (Cowles 1899). Maun (1998) showed that plants exhibited three types of responses to burial (Fig. 6.7):

1. *Negative inhibitory response.* Any plant growing in the path of moving sand accumulates sediment on its leeward side. Plants not adapted to live under these conditions are soon killed. For example, seeds of many genera, namely *Quercus, Juglans, Acer, Fraxinus, Thuja* and numerous other herbaceous plants and woody shrubs are washed up to the driftline in early spring. They may germinate but their seedlings are very short-lived because of burial, sandblast and other unfavourable environmental constraints.

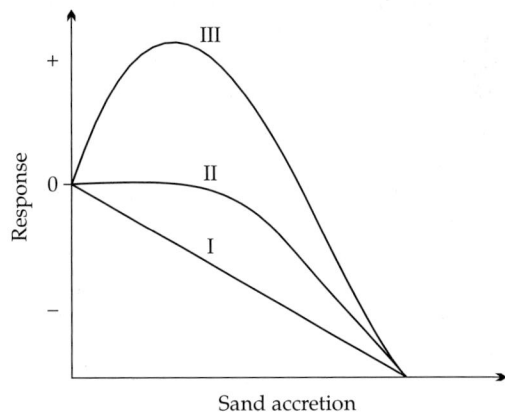

Figure 6.7 A model showing three possible response curves of plant species to burial in sand dunes (adapted from Maun 1998).

Figure 6.8 South slope of a blowout at the Pinery, showing a tall tree of *Pinus resinosa* that has died from burial in sand (photograph by A Maun).

2. *Neutral and then negative response.* Adult plants show little or no visible response initially but as the level of burial increases they start to show symptoms of injury and are eventually killed. This response is particularly evident in tree species such as *Quercus velutina*, *Pinus resinosa* and *P. strobus* which are not adapted to habitats with sand deposition. They may survive small amounts of burial for a few years but die long before they are completely buried (Fig. 6.8). However, there is wide variation, even within a species, in the length of time taken by individual trees to succumb.
3. *Positive stimulatory response.* As shown earlier, all species of the dune complex exhibit stimulation in growth especially following partial burial, although the relative amounts of burial required for growth stimulation of different species may vary by many orders of magnitude. Nevertheless, their response curves are very similar. However, every species has a maximum tolerance limit beyond which it starts to show a negative response and is eventually killed.

6.8 Modes of emergence from burial

Seedlings, herbaceous and woody species employ one or more of the following means to emerge from the burial deposits.

6.8.1 Elongation of first internode, coleoptile or hypocotyl

Following germination, the seedlings of grasses may emerge by elongation of the first internode (Maun 1981; Maun and Riach 1981) or growth of the coleoptile. Dicot species of foredunes, *Cakile edentula*, *Lithospermum caroliniense*, *Cirsium pitcheri*, *Salsola kali*, *Corispermum hyssopifolium* and *Xanthium strumarium* emerge from the soil by the elongation of hypocotyl.

6.8.2 Vertical growth of apical meristem

A large number of herbaceous annuals, biennials and perennials commonly grow vertically from shoot apical meristems (Fig. 6.9). For example, buried rosettes of *Artemisia campestris*, *A. borealis*, *Oenothera biennis* and *Cirsium pitcheri* produce new growth from the apical meristem, emerge from the soil and produce a rosette (Hermesh 1972; Rowland and Maun 2001) as well as new tillers (Maun *et al.* 1996) on the new soil surface. Other species, e.g. *Lotus corniculatus*, *Galium verum*, *Rumex acetosella* and *Thymus serpyllum* produce a number of etiolated lateral shoots from the crown that emerge from the sand surface and finally generate new leaves (Farrow 1919). In the *Poaceae*, the apical and axillary meristems of the clump of shoots likewise elongate, grow through the overburden and produce new sets of shoots. All perennial species (Table 6.4) use this strategy; however, some are more adept at growing vertically than others. Of the two *Ammophila* species, *A. arenaria* forms tussocks with many ramets which grow vertically, while *A. breviligulata* responds by growing from vertical shoots as well as buds on horizontal rhizomes. Apparently, the vertical growth response is mediated by an increase in ethylene levels and darkness (Voesenek *et al.* 1998).

6.8.3 Vertical growth of rhizomes

Many species of plants growing in the actively depositing sites on sand dunes expand into

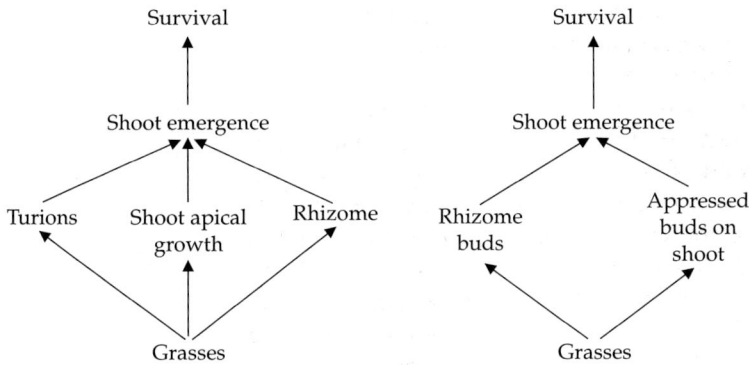

Figure 6.9 The mode of emergence of grass species from burial episodes in foredunes by vertical growth of rhizomes and apical meristems (modified from Maun 1998).

Table 6.4 Dominant perennial plant species in sand-accreting habitats along coastal, lacustrine and desert sand dune systems around the world

Location	Plant species and authority
Japan	*Ischaemum anthephoroides, Carex kobomugi* (Nobuhara 1973)
Australia	*Spinifex hirsutus, Spinifex sericeus, Ammophila arenaria* (introduced) (Hesp 1991); *Spinifex sericeus* (Maze and Whalley 1992a)
Europe	*Ammophila arenaria* (Ranwell 1958); *Calamovilfa baltica* (Rihan 1986); *Elymus farctus* (Harris and Davy 1986a); *Agrostis vulgaris, Festuca ovina, Festuca rubra, Ammophila arenaria, Galium verum, Rumex acetosella, Thymus serpyllum* and *Lotus corniculatus* (Farrow 1919)
Gulf coast, California	*Elymus (Leymus) mollis, Abronia umbellata, Ammophila arenaria* (introduced), *Ambrosia chamissonis* (Barbour *et al.* 1985)
Mono Lake Desert, California	*Chrysothamnus nauseosus, Distichlis spicata, Sarcobatus vermiculatus* (Brown 1997)
Gulf of Mexico	*Ipomoea pes-caprae, Ipomoea stolonifera, Croton punctatus, Palafoxia lindenii, Chamaecrista chamaecristoides* (Moreno-Casasola 1986)
Great Lakes, North America	*Ammophila breviligulata, Calamovilfa longifolia* (Olson 1958a; Maun 1985, 1996); *Agropyron psammophilum, Panicum virgatum* (Zhang and Maun 1990a, 1990c)
Lake Athabasca, Canada	*Elymus mollis, Deschampsia mackenzieana, Festuca rubra, Calamagrostis neglecta, Bromus pumpellianus, Juncus baltica* (Hermesh 1972)
North Atlantic, North America	*Ammophila breviligulata* (Woodhouse 1982)
South Atlantic, North America	*Uniola paniculata* (Wagner 1964)
Iceland south coast	*Leymus arenarius* (Greipsson and Davy 1996)
Hudson Bay, Canada	*Elymus (Leymus) mollis* (Houle 1997)

the new sand surface by producing rhizomes that grow horizontally, but as soon as they are buried the terminal apices of rhizomes change their orientation from horizontal to the vertical and grow upwards until they emerge above the sand surface (Fig. 6.9). Many of the grass species listed in Table 6.4 produce such rhizomes.

6.8.4 Growth of shoots on stolons

When buried by sand, the dormant buds on stolons of species such as *Spinifex sericeus*, the robust strand species of Australia, New Caledonia and New Zealand, and *Ipomoea pes-caprae*, the cosmopolitan stoloniferous plant of tropical and subtropical coastlines, grow upwards until they emerge from the sand surface.

6.8.5 Activation of dormant buds

The rhizomes, shoots and stolons of dune species on the strand support a large reservoir of dormant buds that are activated upon burial and develop vertical shoots that emerge from the sand surface.

6.8.6 Elongation of leaf petioles

Leaf petioles of some plant species, *Cirsium pitcheri* (Maun *et al.* 1996) and *Hydrocotyle novae-zelandiae* (Sykes and Wilson 1990b) elongate until they emerge from burial. A similar response is also seen in *Rumex* species following waterlogging in the flood plain of rivers (Voesenek and Blom 1989).

6.8.7 Clonal integration

Of all the aspects of species resistance to environmental stresses on the strand, clonal integration probably has the greatest relevance to the successful colonization of plant species on the beach strand. It is useful to buried units of an integrated clone because the resources such as water, nutrients, oxygen, hormones and photosynthates may be shared by translocation from unburied parts, thereby increasing the chances for survival. An integrated clone is better able to withstand environmental stresses imposed by pathogens, herbivory, salinity (Baye 1990), competition (Alpert 1999), nutrient limitations (Dong 1999), low moisture (Dong and Alaten 1999) and other stresses (D'Hertefeldt and Falkengren-Grerup 2002; D'Hertefeldt and van der Putten 1998; Amsberry *et al.* 2000). Amsberry *et al.* (2000) showed that the expansion of *Phragmitis australis* in New England marshes is aided by clonal integration because severing of interconnected rhizomes drastically reduced the growth, survivorship, photosynthetic rate and colonization potential of the species. Dong and Alaten (1999) showed that in semi-arid dunes of Inner Mongolia clonal integration of *P. villosa* improved plant survival when the soil was infertile, dry and disturbed. The severing of rhizomes of *P. villosa* reduced growth of rhizome segments compared to controls (Dong and Alaten 1999). Ramets, receiving less water, benefited by connection with well-watered parent ramets that provided strong support to the newly formed ramets of *P. villosa* in this mobile dune system.

Plants with an ability to produce shoots on their roots such as *Populus balsamifera* or with a rhizomatous growth habit, e.g. *Carex arenaria* (Noble *et al.* 1979), *C. kobomugi* (Nobuhara 1973), *Ammophila breviligulata*, *Leymus mollis* and many others are better adapted to live in dynamic sandy substrates because the response is not only that of an individual, but of a reservoir of dormant buds, ramets and shoots linked below the sand surface. Even if burial kills part of population the survivors re-occupy the newly created bare areas by expanding plagiotropic rhizomes. Integrated apices may obtain energy from leaves, stems, roots and rhizomes of buried parts, as well as from unburied ramets. Yu *et al.* (2004) tested this hypothesis using *Psammochloa villosa* in a sand dune in Inner Mongolia and showed that clonal integration promoted emergence from burial deposits through carbon transfer from connected ramets and rhizomes that were not buried. This allowed the buried part of the clone more time to endure the darkness and altered environment and more energy to emerge from the overburden. However, a price is paid by the donor because it shares resources and sacrifices its own performance. Nevertheless, clonal integration serves as a powerful tool of natural selection that provides a net benefit to the growth of the whole clone.

6.8.8 Increase in the number of nodes and length of internodes

The emergence of grasses and other herbaceous species following burial occurs by an increase in the number of nodes and/or increase in the length of internodes. For example, the shoots of *Calamovilfa longifolia*, *Ammophila breviligulata*,

Agropyron psammophilum and *Panicum virgatum* emerged from sand by an increase in the number of nodes and the length of internodes (Maun 1996; Maun and Lapierre 1984; Perumal 1994). Similarly, *Spinifex sericeus* grew through the accumulated sand by producing long internodes (Maze and Whalley 1992a). In the Negev desert *Cyperus macrorrhizus* produced 10–30 mm long internodes under burial conditions compared to only 2–3 mm in stable sand (Danin 1996). Many dicot species of coastal and inland dunes such as *Hudsonia tomentosa*, *Stellaria arenicola*, *Empetrum nigrum* and *Cakile edentula* also form long internodes while growing upwards through the burial deposits (Hermesh 1972; Zhang and Maun 1992). Following burial a 30-fold increase in internode length of *Rumex alpinus* was recorded by Klimeš *et al.* (1993).

6.8.9 Response of woody species

Upon burial trees and shrubs may exhibit three types of responses to survive the burial episode: (i) development of roots and shoots, (ii) development of roots only and (iii) no root or shoot development (Fig. 6.10).

1. *Development of roots and shoots.* Some woody species respond by producing new buds on the buried trunk that develop into roots and shoots. The new roots anchor the tree and exploit resources of the burial deposit while shoots rise above the soil surface close to the parent trunk. With continued sand accretion more roots and shoots are produced and a small grove may develop from a single tree (Hermesh 1972). Many species of *Salix, Betula, Populus* and others (listed in Table 6.5) are able to withstand rather large depositions of sand regularly and maintain their trunks and secondary shoots above the sand surface (Fig. 6.11a). Most of these species establish in moist areas in slacks.

2. *Development of roots only.* Secondary rooting occurs in several tree species, *Larix laricina, Picea glauca, P. mariana* (Hermesh 1972), *Juniperus communis, J. virginiana* and perhaps many others (Fig. 6.11b). Their survival is dependent upon the rate of burial. Gradual burial allows trees time to develop new adventitious roots and survive the burial episode but a faster rate of burial usually results in mortality of trees or shrubs in this category.

3. *Development of neither roots nor shoots.* Many tree species—*Pinus banksiana, P. resinosa, Quercus velutina*—produce neither roots nor shoots when buried yet survive for considerable lengths of time using the original root system. *Pinus banksiana* persists much longer

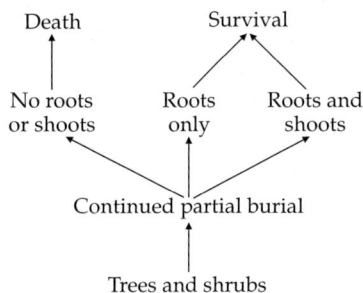

Figure 6.10 Three types of responses shown by woody species to burial episodes: development of roots and shoots, development of roots only, and no root or shoot development (modified from Maun 1998).

Table 6.5 Types of responses shown by tree species to burial by sand in sand dune systems

1. Species that produce both roots and shoots
Salix adenophylla, S. brachycarpa, S. glaucophylla, S. pyrifolia, S. salicicola, S. turnorii, S. tyrrellii, Betula papyrifera, B. resinifera, B. ×sargentii (Hermesh 1972; Rowe and Abouguendia 1982; Karpan and Karpan 1998), *Tilia americana, Ulmus americana, S. adenophylloides* (Cowles 1899), *Magnolia grandiflora* (Kurz 1939), *Salix exigua, S. interior, Cornus stolonifera, Prunus pumila* (Maun 1998), *Populus balsamifera, S. cordata* (Dech and Maun 2006).

2. Species that produce roots only under gradual burial conditions
Larix laricina, Picea glauca, Picea mariana (Hermesh 1972), *Juniperus virginiana, J. communis* (Dech and Maun 2006).

3. Species that produce neither roots nor shoots
Quercus velutina, Q. rubra, Pinus banksiana, P. resinosa, P. strobus (Dech and Maun 2006).

(a)   Roots and shoots   (b)   Roots only

Figure 6.11 (a) Exposed roots and shoots of a birch tree along the banks of William River in the Athabasca sand dunes and (b) exposed roots of white spruce in Thomson Bay on the south shore of Lake Athabasca (from Karpan and Karpan 1998 with permission).

than other species and produces a large crop of cones before dying (Hermesh 1972). There is a wide range of variation of individual trees because of genotype variability. Some individual trees die soon after being covered by a small amount of sand while others in a population can withstand large amounts for a longer period (Cowles 1899), but complete burial by an advancing dune kills the whole forest in its path (Fig. 6.12a). Continued movement of the sand field eventually exposes ghost forests on the windward side in which all trees have been killed by burial (Fig. 6.12b). Under desert conditions *Retama raetam*, a very common shrub of the Negev desert, does not produce roots on the stem when covered by sand but it withstands cycles of burial and deflation up to

a depth of 1–2 m without producing any roots in the sand deposit (Danin 1996). However, complete burial is fatal.

All these conclusions are based on field observations of trees exposed to burial events in coastal and desert dunes. Dech and Maun (2006) tested these hypotheses in an experimental field study in which saplings of seven tree species of coastal sand dunes of Canada were artificially planted on a gentle windward slope of the second dune ridge and artificially buried in sand to depths of 25, 50 and 75% of tree height. The responses of these trees to burial were monitored for two years after which the saplings were excavated and assessed for morphological and growth responses. They classified the responses into three categories:

Figure 6.12 (a) An easterly migrating dune encroaching on a forest at Athabasca sand dunes and (b) previously buried by sand these uncovered trees, once part of an ancient forest, overlook Thomson Bay (from Karpan and Karpan 1998 with permission).

negative response, neutral and then negative and positive response. *Pinus strobus* and *Picea glauca* showed a pronounced decline in growth. Both species were intolerant of burial and did not form adventitious roots in the burial deposit. They relied on their original root system to overcome the stress. The neutral and negative response was observed in *Thuja occidentalis, Juniperus virginiana* and *Picea mariana* where the species maintained a steady state up to a burial depth of 25% of the tree height, after which decline in biomass was evident. All these species produced adventitious roots but were unable to alter the source–sink relationship and overcome the stress. Two species, *Salix cordata* and *Populus balsamifera*, showed a positive response to burial by an increase

in total biomass, plant height and extensive production of adventitious roots. There was an increase in allocation of resources to the growth of new shoots and adventitious roots and a decrease to the original root system, thus altering the source–sink relationship to the benefit of the trees.

6.8.10 Possible role of adventitious roots and role of oxygen in survival

Why does a tree die when all of its photosynthetic area is still intact? Many authors have hypothesized that the deposition of sand on top of a plant smothers or chokes roots because of a lack of oxygen in the root zone (Bergman 1920; Kurz 1939; Daubenmire 1974). Although no experimental evidence is available to substantiate it, this hypothesis is plausible. Indeed roots covered by sand must experience a drastic reduction in oxygen content because of an increase in moisture level and decreased diffusion from the surface. Survival depends on the ability to cope with low oxygen levels. Both avoidance and tolerance strategies are utilized. Tolerance may be achieved by functioning under reduced oxygen levels temporarily until a new set of adventitious roots is produced on the buried stem close to the sand surface. All woody (Dech and Maun 2005) and herbaceous plants adapted to live on the beach and open dunes possess this capability.

The primary function of adventitious roots is to acquire O_2 for the proper functioning of the root system and secondarily to anchor the plant at the new sand surface. Over time as the old roots become non-functional the adventitious roots assume the function of moisture and nutrient absorption as well. Another possibility is that the original roots may continue to function, perhaps less efficiently, under anaerobic conditions. According to Bergman (1920) plant roots start to show visible symptoms of injury when oxygen levels drop to 5–10%. In dune systems because of good drainage conditions and relatively high diffusion capacity the oxygen levels may remain within the natural range for considerable periods of time. Another factor affecting oxygen tension in soil is rainfall and its frequency. If rain occurs regularly, the oxygen dissolved in it replenishes some of the oxygen and helps to meet the requirements of plant roots. Thus even though plant species may persist for long periods they are unable to develop adventitious roots.

The ability to grow under low oxygen levels also depends on a number of other important morphological and physiological adaptations (Kurz 1939; Kozlowksi 1984). In some species of *Betula, Prunus* and *Cornus* the number of lenticels may increase. Some species such as *Salix* and *Betula* may develop aerenchyma tissue in response to low oxygen levels following burial. Finally, as mentioned earlier clonal integration may also play a role in the transfer of oxygen from the unburied portions of the rhizomes or roots of clonal plants to the buried parts deprived of oxygen. This happens in some species of trees, e.g. *Populus balsamifera*, and many species of grasses, e.g. *Elymus farctus* (Harris and Davy 1986b), *Ammophila breviligulata* (Maun 1984) and *Calamophila baltica* (Rihan 1986).

6.9 Summary

Along sea coasts and shorelines of lakes strong wave action deposits sand on the beach that is later blown inland by wind velocities above a certain threshold value depending on sand texture. Dunes are formed by the marked surface roughness created by plant species and consequent sand deposition around the plant. Such burial events alter the physical microenvironment of vegetation and in order to survive a plant must make significant physiological adjustments to overcome the episode. Depending on the wind velocity, duration and amount of available sand plants may be partially or completely buried. There is ample evidence that gradual and partial burial of plants shows stimulation in net CO_2 uptake, relative growth rate and total biomass of plants. However, as the level of burial increases, the

plant requires an initial period of recovery during which it must mobilize its resources from storage organs to adjust to the environmental stress. Thus it takes time for the plant to start showing an enhancement of vigour following a major burial event. The length of time required by a plant to recover depends on depth of burial, stage of plant growth and species-specific response. Plants with slow rates of recovery are less fit to occupy rapidly accreting habitats.

Different life forms have widely different limits of tolerance. For example, poikilohydric organisms that create microbial crusts over sand surfaces can tolerate less burial than lichens and bryophytes which are less tolerant than angiosperms and gymnosperms. Seeds are less tolerant than seedlings, annuals less so than biennials and biennials less than perennials. Grasses and some woody species of trees that occupy the beach and foredunes are able to survive the highest amounts of yearly sand accretion. Usually, the rates of sand deposition along sea coasts range from 8 to 120 cm. In a large majority of cases burial occurs regularly only in certain parts of the year. The rest of the year sand accretion is relatively small. However, violent storms and hurricanes may suddenly and unpredictably dump large amounts of sand on plant communities and cause significant devastation, recovery from which may take many years.

Plants exhibit three types of responses to burial: a negative, a neutral and then negative or a positive stimulatory response depending on the plant species. Species adapted to live in accreting habitats emerge from burial by using different techniques. Germinating seedlings may elongate their first internodes, increase coleoptile length or hypocotyl. In rhizomatous grasses and stoloniferous species, vertical growth of the apical meristem of shoots, rhizomes and stolons, activation of dormant buds on roots, stolons and rhizomes, elongation of petioles and clonal integration, make it possible for them to emerge above the sand surface. In stoloniferous species such as *Ipomoea pes-caprae* and rhizomatous grass *Ammophila breviligulata* large colonies of shoots are integrated by stolon or rhizome connections that allow the whole population to survive burial episodes. The unburied part of the clone compensates for the negative impact on another part of the clone.

Tree species that occupy the foredunes may survive burial episodes by producing new roots and shoots, roots only with no shoots or shoots on buried stems. Most successful are species that produce both roots and shoots in the burial deposit. Those that produce only roots require gradual rather than sudden large burial events. The third category of trees on coastal dunes cannot withstand burial and primarily occupy dunes at a later stage in succession. The major impact of burial of trees is the suffocation of their root system. The lack of oxygen to the root system kills roots and the tree eventually dies.

CHAPTER 7

The *Ammophila* problem

7.1 Introduction

Even a cursory look on foredune plant communities shows vigorous dense stands of dune species in areas with moderate recurrent sand accretion levels specific for each plant species (Disraeli 1984; Maun and Baye 1989; Maun 1998). The phenomenon has been well documented in species of *Ammophila arenaria* (Carey and Oliver 1918; Tansley 1953), *Corynephorus canescens* (Marshall 1965), *A. breviligulata* (Eldred and Maun 1982) and *Calamovilfa longifolia* (Maun 1985). Burial has a positive influence on growth and flowering of plants and debilitated populations of foredune plant species can be rejuvenated by sand deposition (Maun 1998).

Clear evidence of this phenomenon was presented by Maze and Whalley (1992a), who examined population dynamics of *Spinifex sericeus* in five zones receiving different amounts of sand deposition on a coastal dune system of Australia: the sea side of the first dune ridge, crest of first dune ridge, swale, *Acacia* thickets and stable hind dunes. In the very dynamic area on the sea side or toe of the first dune ridge (high beach) with regular burial or erosion of up to 1 m or more the plants produced very vigorous stolons with long internodes. On the crest of the dune ridge with sand deposition of about 17.5 cm per year even though plants had fewer stolons, they responded to burial by growing upwards with long internodes. In *Acacia* thickets in spite of very little sand deposition, plants were vigorous with little or no dead material, produced stolons and grew upwards with some long and some short internodes, probably because of greater nitrogen content in the soil.

However, in the swale (slack) with little or no sand deposition, plants showed strong clumping tendency with very short internodes, a large amount of dead material on the surface and very low vigour. Unburied nodes usually died. Similarly, in the stable sand dunes with little or no sand deposition debilitated low-vigour clumps with very few stolons were abundant. Another example of this decline was presented by Martin (1959) on a shoreline along the Atlantic coast of North Carolina. He measured deposition and deflation of sand on two transects and showed that as one moved inland from the shoreline the total deposition of sand decreased (Fig. 7.1). The dune species, *Ammophila breviligulata* and *Carex kobomugi*, were very vigorous in areas with average sand deposition of about 17 cm (transect 1) or 28 cm (transect 2) per year in the first 40 m from the nip—the beginning of the primary foredune. As the depth of burial declined from 41 to 60 m the density of shoots decreased. Farther inland (61–110 m), deflation exceeded sand deposition and both species degenerated and exhibited a significant decline in vigour. Several authors show that the decline is manifested by a decrease in density, net CO_2 uptake, photosynthetic efficiency, tillering, rhizome production, flowering and biomass per unit area (Eldred and Maun 1982; Maun and Baye 1989; Yuan *et al.* 1993; Hope-Simpson and Jefferies 1966; Perumal and Maun 2006).

Marshall (1965) called this the '*Ammophila* problem'. As shown above, this debilitation of populations is not confined to *Ammophila* alone because it occurs in populations of all perennial plant species adapted to live in the sand-accreting habitats of foredunes (Maun

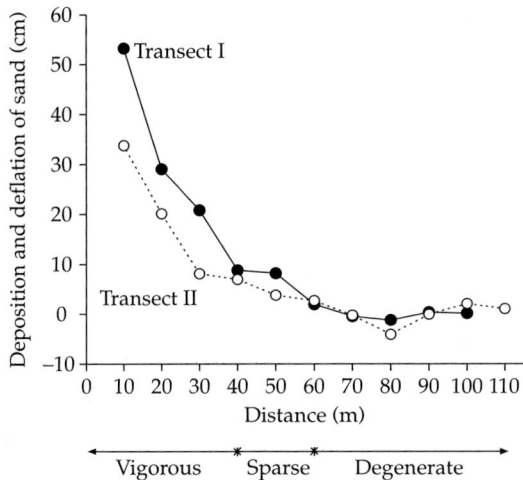

Figure 7.1 Annual deposition and deflation of sand (cm) on stakes installed at 10 m intervals on two transects from the nip (the beginning of primary foredune) to 110 m inland. The relative vigour of *Ammophila breviligulata* at different distances is indicated below the figure (from Martin 1959).

1998). Dune ecologists have pondered over the possible causes of this decline for over a century and many possible explanations have been proposed to explain this phenomenon (Table 7.1). Most of them are based on subjective visual field observations or deductions and have not been experimentally tested. Only recently experimental studies on the population dynamics and demography of a few dune species have yielded tests of some of the hypotheses. Some of the major explanations are elaborated in the following pages.

7.2 Degeneration response

7.2.1 Nutrient deficiency

Willis (1965) suggested that the major cause of decline of *Ammophila arenaria* was deficiency of three macronutrients—nitrogen, phosphorous and potash—that are lacking in coastal sand dunes. He tested the hypothesis

Table 7.1 Explanations for the differences in vigour of *Ammophila breviligulata* and *A. arenaria* plants on sites with sand accretion and no sand accretion (an extension of Table 1 in Eldred and Maun 1982)

Explanation	Authority
On sites with no sand accretion:	
(a) Dead and decaying organic matter accumulates	Waterman 1919; Farrow 1919; Wallén 1980
(b) Interspecific competition increases	Tansley 1953; Salisbury 1952; Willis *et al.* 1959a, b; Marshall 1965; Watkinson *et al.* 1979; Huiskes and Harper 1979
(c) Plants are older in age	Westgate 1904; Wallén 1980
(d) Soil acidity increases	Salisbury 1952
(e) Toxic substances accumulate	Tansley 1953
(f) Site of adventitious roots decreases	Halwagy 1963; Marshall 1965; Willis 1965
(g) Reduced number of adventitious roots	Willis 1965; Marshall 1965
(h) Roots lose cortex	Marshall 1965
(i) Harmful soil organisms	Van der Putten *et al.* 1988; De Rooij-van der Goes *et al.* 1995b
(j) Aeration of roots and rhizomes decreases	Tansley 1953; Salisbury 1952; Szafer 1966
On sites with sand accretion:	
(a) Promotion of adventitious root growth	Willis 1965; Marshall 1965
(b) Increase in soil moisture and decrease in temp.	Olson 1958a; Willis 1965; Marshall 1965
(c) Addition of nutrients by accreting sand	Willis 1965; Houle 1997
(d) More nitrogen fixing bacteria	Wahab 1975; Ralph 1978; Pugh 1979
(e) Mycorrhizal fungi exploit soil resources	Little and Maun 1996
(f) Multifactor hypothesis	Maun 1998
(g) Genotypic differentiation	Laing 1967; Gray 1985

in a field and a glasshouse experiment where he incorporated a mixture of major nutrients (N, P, K, Mg) in soil of a debilitated stand of *A. arenaria* at Braunton Burrows on six occasions over a period of two years. In the glasshouse he planted young seedlings and gave similar fertilizer treatments. The amount of supplied nutrients was much higher than could possibly be contributed by sand accretion. In both experiments there was a marked improvement in plant density, shoot height, vigour, tillering and total biomass, however, the plants did not flower. Based on these results, Willis (1965) hypothesized that burial by sand acted as a natural fertilizer by increasing soil volume and contributing small amounts of nutrients. Similarly, Boudreau and Houle (2001) reported that fertilization of debilitated plants of *Ammophila breviligulata* increased shoot density, shoot and root mass but did not increase rhizome mass. However, roots were less efficient in utilizing soil nutrients than naturally occurring vigorous stands. Pemadasa and Lovell (1974) also reported a significant increase in number of tillers, leaf area, number of leaves and biomass per plant of dune annuals sprayed with complete nutrient mixture and grown under both glasshouse and field conditions at Aberffraw dune system. Several recent studies have confirmed that indeed there was a significant increase in plant growth, productivity and community structure in response to the addition of nutrients in low latitude (Olff *et al.* 1993; Seliskar 1994) and high latitude subarctic sand dune systems (Houle 1998).

Hope-Simpson and Jefferies (1966) questioned this hypothesis and argued that the presence of nutrients in sand could not be a factor in debilitation of stands because even burial of moribund *A. arenaria* by leached sand showed a stimulatory response. Similarly, burial of plants with fresh sand (with no nutrients) from the lake enhanced plant growth (Perumal 1994). Baye (1990) gave further evidence that the stimulation response shown by fertilization was much different than the

one produced by burial in sand. Both burial and nutrient addition increased growth, however the effects were not identical because burial promoted flowering, nutrient addition did not. Soil analysis by Boudreau and Houle (2001) indicated that in debilitated stands root mass density was higher and nitrogen content was three times (0.0066%) that of vigorous stands in the foredunes (0.0022%), and concluded that the most logical explanation was efficiency of root absorption rather than nutrient deficiency.

7.2.2 Competition increases

Several studies on succession in coastal sand dunes argued that vigorous growth of grasses in accreting habitats of foredunes was caused by the effective elimination of competition (Salisbury 1952; Huiskes and Harper 1979). Conversely, Marshall (1965) reported that as sand deposition declines and the habitat stabilizes other plant species invade and *A. arenaria* populations start to decline because of interspecific competition. According to Halwagy (1963), in stable areas on sand banks and islands in the Nile river near Khartoum, Sudan, decline in growth of *A. arenaria* and other species such as *Tamarix nilotica* was effectively caused by increased competition following the establishment of other plant species. It is assumed that other plant species must lower the growth and vigour of *A. arenaria*. However, this assumption may not be true for two reasons. First, in many areas the debilitated populations of *A. arenaria* and *A. breviligulata* in North America did not contain any other species (Hope-Simpson and Jefferies 1966; Baye 1990; Poulson 1999). According to Poulson (1999) such debilitated monospecific stands of *A. breviligulata* had not changed in more than 10 years along Lake Michigan. He argued that the species actually retarded natural sand dune succession by not allowing other species to gain a foothold. Second, experimental studies by Houle (1998) clearly showed that heterospecific removal of either *Elymus mollis* or *Lathyrus japonicus,* in a subarctic plant community, failed to show any significant positive or negative effects on either of the two test species.

7.2.3 Decline in root production and decrease in rooting depth

Olson (1958c) suggested that the inherent tendency of *Ammophila breviligulata* to elongate its internodes is not overcome completely even when there is no sand deposition. This elongation brings the growing point and adventitious roots into the dry surface sand layers ranging from 5–10 cm, which is unfavourable for the production of new nodes. Similar conclusions were also drawn by Willis (1965) for *A. arenaria* in which old roots were not replaced by new roots in stable areas with no sand accretion. Thus, the potential sites on nodes for the formation of adventitious roots decreases. Marshall (1965) also reported that the site of adventitious root production of *A. arenaria* below the sand surface decreased from 24 cm in embryo dunes to 12, 5 and 2 cm on first, second and third dune ridges, respectively. Similarly, in a foredune along Lake Huron the average depth from which roots of *A. breviligulata* originated was 12.9 ± 4.7 cm with a range of 5.8–23.7 cm in different areas of the population depending on the rates of sand accretion (Eldred and Maun 1982). The roots often grew upwards towards the sand surface from rhizome nodes. According to Marshall (1965) continued development of new roots is required to support the growth of increasing number of tillers in *Corynephorus canescens*. He assessed the efficiency of old (nine months in age) and young (new) root systems by using a split root technique and showed that new roots were more efficient in acquiring moisture and nutrient resources than old roots. In the absence of sand deposition the uptake function of old roots declined.

However, Hope-Simpson and Jefferies (1966) showed that in *A. arenaria*, even though old roots lost cortex with age, they were fully functional in the transport of water from the soil. The wiry decorticated roots frequently terminated in a fleshy fully decorticated distal end. Similar observations were made by Weaver (1968) in other dune grasses of the tall grass prairie, especially *Calamovilfa longifolia*. In stable habitats, *A. breviligulata* plants continue to produce rhizomes even in the absence of sand deposition and may thus effectively lower the site of adventitious root production (Maun 1984). Hope-Simpson and Jefferies (1966) argued further that the *Ammophila* problem occurs only in sand dunes. They planted small tufts of *A. arenaria* in a calcareous garden soil and found that plants continued to show vigorous growth, tillering and flowering for two years without showing any signs of deterioration. How many roots are produced on each node? There is some confusion on this question. Purer (1942) recorded 5–6 roots per node on planted *A. arenaria* while Waterman (1919) found only two roots per node in *A. breviligulata*. Hope-Simpson and Jefferies (1966) showed that *A. arenaria* initially produced one or two roots which later increased to a maximum of four.

7.2.4 Accumulation of organic matter

Waterman (1919) observed that roots of plants of *Ammophila breviligulata* exhibited little response to dead and decaying organic matter and did not survive in sand possessing appreciable quantities of humus. Similarly, the extension of roots of seedlings was inhibited by the presence of decaying plant parts. This observation was experimentally tested by Zaremba and Leatherman (1984) who excavated sand to a depth of 20 cm, horizontally placed genetically identical *A. breviligulata* fragments on the sand surface and then covered them with 15 cm of three types of organic drift (*Ascophyllum* algae, bay drift, *A. breviligulata* debris) and beach sand. All four treatments were then covered with 5 cm of sand. The litter of any type did not influence the number of leaves and tillers. However, algal litter produced

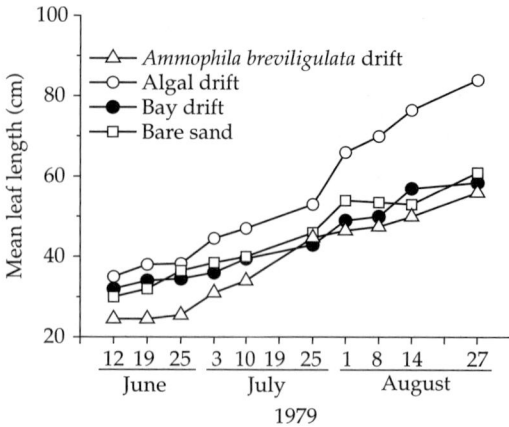

Figure 7.2 Comparison of mean leaf length per tiller of *Ammophila breviligulata* grown in four habitats containing algal litter, *A. breviligulata* litter, Bay drift litter (varied flora) and bare sand (adapted from Zaremba and Leatherman 1984).

significantly longer leaf length compared to bay drift, bare sand and *A. breviligulata* debris which did not differ from each other (Fig. 7.2). The most probable reason for highest leaf lengths in algal litter was the higher concentration of nitrogen compared to other types of litter. Thus, the addition of organic matter to sandy soil did not inhibit the growth of plants. On the contrary it stimulated growth and relative growth was dependent on the C:N ratio which ranged from 15:1 in algae to 75:1 in *A. breviligulata* litter.

7.2.5 Harmful soil organisms

Van der Putten *et al.* (1988) argued against Marshall's (1965) hypothesis of reduced function of old roots by physical ageing and suggested an alternate explanation: that the malfunctioning of roots in the absence of sand deposition is caused by a build up of populations of harmful micro-organisms. They conducted two greenhouse experiments to test this hypothesis. In the first experiment, they grew seedlings of *A. arenaria* in sand collected from the sea floor and from the rhizosphere of a debilitated stand of *A. arenaria*. They found

that the growth in the rhizosphere sand was significantly lower than sea sand and showed that stimulation in vigour occurred because fresh sea sand did not contain harmful soil pathogens. In the second experiment, they sterilized both types of sand to kill insects and micro-organisms and then compared the growth of *A. arenaria* seedlings in sterilized and unsterilized sand. The seedlings growing in sterilized rhizosphere sand and sea sand produced significantly higher biomass than unsterilized rhizosphere sand. There were no significant differences in biomass between sea sand, sterilized sea sand and sterilized rhizosphere sand. Based on the results of these two experiments they hypothesized that under natural conditions burial by fresh sand allowed plants to escape harmful soil micro-organisms (the escape hypothesis) and exhibit high vigour.

This was later confirmed in a field experiment by De Rooij-van der Goes and colleagues (1995b) who buried plants in a debilitated population of *A. arenaria* with 20 cm of root zone sand, sterilized root zone sand and beach sand and compared it to non-buried control. Plants buried with beach sand produced significantly greater above-ground biomass than control and root zone sand, however there were no significant differences between beach sand and sterilized root zone sand (Table 7.2), thus confirming the results presented by Van der Putten *et al.* (1988). However, there were no significant differences in the dry weight of roots produced in the deposited layer (Fig. 7.3). She (De Rooij-van der Goes *et al.* 1995c) then analysed the soil and found 47 species of fungi and 10 genera of plant parasitic nematodes in the rhizosphere of vigorous and debilitated stands of *A. arenaria* in Dutch foredunes. According to the Canonical Correspondence Analysis, 10-year-old stands of *A. arenaria* contained significantly different communities of soil organisms than 3-year-old stands. In another study De Rooij-van der Goes (1995a) showed that artificial inoculation of seedlings by single species of fungi did not precipitate a

Table 7.2 The above-ground living biomass (g m^{-2}) of a degenerating *A. arenaria* population buried with unsterilized root zone sand, sterilized root zone sand and beach sand, compared with control (modified from De Rooij-van der Goes *et al.* 1995b)

Sand source for burial	Deposited layer (g)	Above-ground (g)	Total (g)
Control	—	150 ± 28.6b	150c
Unsterilized root zone sand	120 ± 21.4b	150 ± 14.3b	270b
Sterilized root zone sand	135 ± 35.7ab	185 ± 57.1ab	320ab
Beach sand	180 ± 32.1a	250 ± 42.9a	430a

Means in each column followed by a different letter are significantly different at P < 0.05 according to Tukey's test.

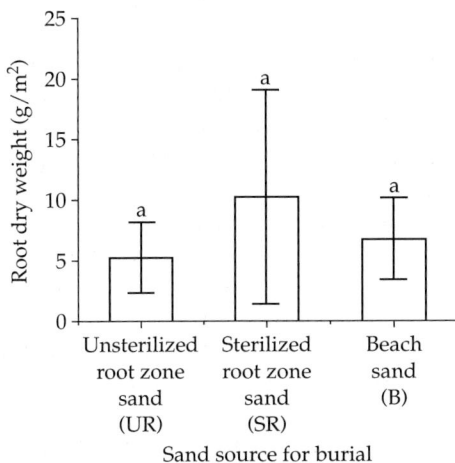

Figure 7.3 The dry weight of roots in the deposited layer of a degenerating *A. arenaria* population buried with beach sand, root zone sand and sterilized root zone sand. There were no significant differences between treatments at P > 0.05 (redrawn from De Rooij-van der Goes *et al.* 1995b).

decline but combining all fungi led to a reduction of 80% in the growth of *A. arenaria* seedlings. Similarly large numbers of nematodes (80 times normal) were required before a significant reduction in growth became evident. In conclusion, the decline in *A. arenaria* populations was not caused by a single well-defined patho-system but rather by a combination of many different soil organisms (De Rooij-van der Goes 1996).

There are several drawbacks in the proposed escape hypothesis. First, nematodes can only infect a buried plant if roots are present in the buried portions of the plant. The signal for nematodes to move into the deposited substrate is the development of plant roots (De Rooij-van der Goes *et al.* 1998). It takes at least 4 to 6 weeks under warm weather conditions for new roots to develop on a buried culm. Thus, not only is the plant safe from nematodes during that period but the burial deposit will also not be of any benefit to the plant because the resources can not be utilized. However, both parasitic and saprophytic fungi moved up into the burial deposit before the formation of roots (De Rooij-van der Goes *et al.* 1998). Second, as soon as new roots were produced by the plant they were immediately infected by nematodes (De Rooij-van der Goes *et al.* 1995b) that had moved into the deposited sand. Apparently, nematodes can move quite fast into the soil. For example, *Meloidogyne incognita* was able to move 7.5 cm within 1–2 days (Robinson 1994) while *Heterodera rostochiensis* moved 2–3 cm in 16 hrs (Wallace 1960). Thus, escape from nematodes, if any, following a burial episode is very brief indeed. Third, soil contains a complex system of both useful and harmful soil organisms which have strong positive and negative interactions among them within the soil core and especially in the rhizosphere (Killham 1994). For example, the possible useful role of arbuscular mycorrhizal (AM) fungi (Chapter 9) can not be ignored. Since Van der Putten *et al.* (1988) applied fertilizer to both treated and control plants it would have dele-teriously affected the colonization of roots by AM fungi in the control pots.

7.2.6 Decline of mycorrhizal fungi

The colonization ability of mycorrhizal fungi declines because of a decrease in root develop-ment (Boudreau and Houle 2001). According to Little and Maun (1996) the best way to evalu-ate the escape hypothesis would be to examine the impact of all three soil factors, burial, plant parasitic nematodes and mycorrhizal fungi on plant growth, simultaneously. They grew 48 seedlings of *A. breviligulata* in sand steri-lized with 1 Mrad of gamma radiation to kill nematodes and fungi and then experimentally tested the effects of eight treatments: control, plant parasitic nematodes (N), mycorrhizal fungi (M), burial (B) and their reciprocal inter-actions, M + N, M + B, N + B and M + N + B on the growth of seedlings. When they were 2 weeks old, the root zones of 24 seedlings were inoculated with AM fungi by placing sweet corn roots colonized by mycorrhizal taxa from sand dunes. The remaining 24 seedlings received corn roots with no mycorrhizal fungi. The endoparasitic nematode inoculates of two plant parasitic nematode species of the genera, *Pratylenchus* and *Heterodera*, were extracted into a distilled water medium from the roots of glasshouse pot cultures of *A. breviligulata*. Twenty-four plants were then inoculated with 20 mL of nematode suspension by inserting a needle fitted with a syringe and delivering it into the root zone. Control plants were injected with 20 mL of distilled water. The plants were allowed to grow for another 3 weeks to allow the establishment of nematodes before the bur-ial treatment was imposed. For burial, 5 cm deep and 10 cm diameter plastic drainage pipes were placed around each plant and then filled with gamma irradiated sterile sand.

Mycorrhizal fungi started to increase leaf area per plant at 8 weeks of age and continued to increase it until 14 weeks when other burial treatments—(B, B + M, B + M + N)—caught

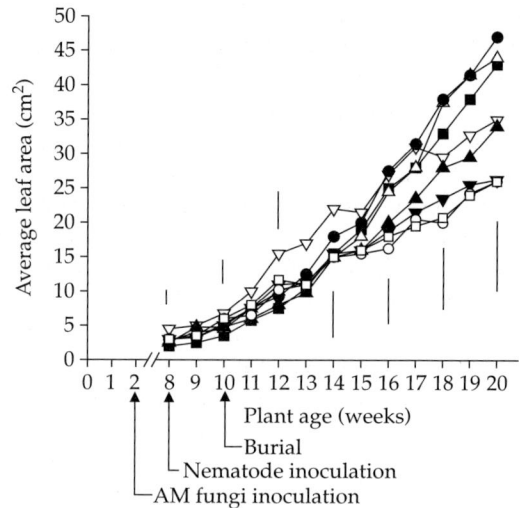

Figure 7.4 Leaf area of *Ammophila breviligulata* seedlings exposed to (○) control (C), (▽) AM fungi (M), (□) nematodes (N), (△) burial (B), (▼) M + N, (■) M + B, (▲) B + N, and (●) M + N + B. Plant age at which the three treatments were imposed is indicated by arrows (adapted from Little and Maun 1996).

up and then surpassed the M alone treatment (Fig. 7.4). Presence of nematodes did not sig-nificantly lower the leaf area per plant than control but their presence in the substrate lowered the positive influence of both burial (B + N) and mycorrhizal fungi (M + N). The negative impact of nematodes on leaf area per plant was removed only in the presence of both mycorrhiza and burial (M + N + B). The dry weight of roots was significantly lower in plants infected with nematodes and then bur-ied with sand (N + B). The biomass data indi-cated that burial alone (B) and mycorrhizal fungi (M) alone increased plant biomass while plant parasitic nematodes (N) alone decreased biomass. In two factor combinations (B + N) and (M + N), nematodes decreased the biomass of plants irrespective of the positive influence of burial and mycorrhizal fungi, however, it was not significantly different than control. Only in the three-way interaction (M + N + B) were the injurious effects of nematodes allevi-ated. When a plant is buried it must use energy stored in its roots and rhizomes to emerge above the sand surface. Similarly, mycorrhizal fungi depend on the host for energy to survive and grow in the soil. Also, the nematodes are completely parasitic on the plant. In other words all three treatments require the plant to spend energy. However, in return for photosynthates, the mycorrhizal fungi provide the plant with phosphorous, other nutrients and many other benefits (see Chapter 9). The expenditure of energy in survival after a burial episode is also temporary because burial increases the soil volume thus permitting the expansion of AM hyphae and plant roots into the burial deposit. Thus, the escape hypothesis can not explain the improvement of plant vigour. Four arguments will be provided in favour of this statement:

1. It was clearly evident that plant parasitic nematodes were injurious to the growth of plants and burial by itself did not alleviate the effect of nematodes. In fact, the lowest weight of roots per plant was found in buried plants that had been infected with nematodes primarily because both treatments required the expenditure of energy to withstand their impact.
2. The highest leaf area per plant and total dry weight were obtained only when plants containing both nematodes and AM fungi were buried in sand. Nevertheless, the presence of mycorrhizal fungi was essential because the injurious effects of nematodes were mitigated by burial only when the sand contained mycorrhizal fungi. Thus, any enhancement in vigour was more likely due to AM fungi than to burial.
3. Burial in sand and mycorrhizal fungi both had a positive effect on the growth of plants, probably because of increased soil volume and expansion of root-absorbing surface by AM fungi. A healthier plant and a greater root system would allow the plant to withstand more root pathogens.
4. The soil is a complex system with both useful and harmful soil organisms which have strong interactions among them. Patho-complexes containing plant parasitic nematodes and pathogenic fungi are probably acting together rather than each acting on its own. However, there is evidence that AM fungi mitigate the impact of pathogenic fungi (Newsham *et al.* 1994; 1995) and plant parasitic nematodes. Moreover, fungivorous nematodes will also reduce the impact of patho-complexes.

7.2.7 Genetic differentiation

The debilitation of populations may be due to a genetic differentiation in the species in older populations (Laing 1967; Gray 1985) because of changes in selective pressures such a decline in sand deposition, decreased soil salinity, lower pH and little or no salt spray. Imbert and Houle (2000) tested this possibility by examining ecophysiological differences among *Leymus mollis* populations across a subarctic dune system. They measured net carbon assimilation rate of plants at different distances from the beach to the older dunes. They found that *Leymus mollis* ramets on older ridges generally had lower net CO_2 uptake and water use efficiency than young dunes. They then collected ramets from different distances and grew them under controlled conditions in a greenhouse and repeated the measurements of net CO_2 uptake and water use efficiency. They found no differences between populations collected at different distances from the shoreline and suggested that field differences were caused by environmental rather than genetic differences between populations.

7.2.8 Other factors

There is little evidence for some of the other proposed hypotheses, ageing process, accumulation of toxic substances, decrease in pH of substrate and a decrease in aeration of roots and rhizomes. The ageing process (Wallén 1980) can not be a factor because older shoots are regularly replaced by younger ones in debilitated populations of *A. breviligulata* (Eldred and Maun 1982). The average vegetative growth rate of a population of *A. breviligulata* was 0.4 ± 0.3 year^{-1}. Moreover, since moribund populations rejuvenate following the application of fertilizer or burial in sand, the problem must be related to environmental factors rather than intrinsic ageing process. Experiments by Laing (1954) clearly showed that there were no apparent toxic effects (Tansley 1953) related to accumulation of leaf litter of *A. breviligulata* (auto-allelopathy) on the growth of seedlings. According to Salisbury (1952) a decrease in pH may lead to the deterioration of plants. However, our observations show that *A. breviligulata* can grow in a variety of sandy soils with widely different pH ranges (Maun and Baye 1989). For example, the plants grow vigorously in highly silicious sand with a rather low pH of 4.8 to 5.6 along the Atlantic coast of Maine. In contrast, along other parts of the Atlantic coast the pH may range between 6.0 and 7.5 and along Lake Huron from 8.0 to 8.5. Similarly, measurements of oxygen diffusion rates of sand in robust and debilitated populations of *A. breviligulata* did not show any differences in the aeration of roots and rhizomes.

7.3 Stimulation response

As shown in Chapter 6, all studies agree on two aspects of burial in sand dune systems. First, burial has a strong positive influence on growth and flowering of plants and second, debilitated populations of foredune plant species can be rejuvenated by sand deposition. What are the possible causes of stimulation in growth by burial? The 'enhancement in plant vigour' has been debated for over a century (Eldred and Maun 1982) but no consensus has been reached about the possible reasons for this phenomenon. Researchers have been trying to attribute enhanced vigour to single factors. In fact, plant performance is controlled by interactions between resources and abiotic conditions of the habitat (Houle 1997).

There is mounting evidence that a buried plant is influenced by multiple factors. Eldred and Maun (1982) used a multivariate approach to the problem and suggested that no single factor is responsible for the enhancement of plant vigour. Instead several variables may combine to produce growth stimulation. I will call it the *'multifactor hypothesis'* in which five major reasons may be advanced for enhanced vigour in plants following burial: (i) increased soil volume, (ii) increased soil resources, (iii) increased activity of mycorrhizal fungi, (iv) a physiological hormonal response by the plant (reactive growth according to Danin 1996) and (v) higher capacity of sink (Fig. 7.5).

7.3.1 Increased soil volume

Burial creates more space which becomes available to the plant for the expansion of the root absorbing surface. However, the plant will have to first emerge above the sand surface by utilizing its stored energy reserves and then initiate new roots. It usually takes a few weeks for the production of new roots. Nevertheless, the increased soil volume of moist sand would eventually allow increased development of new roots.

7.3.2 Increase in soil resources

The sand deposit contains small amounts of nutrients that become available to plants. For

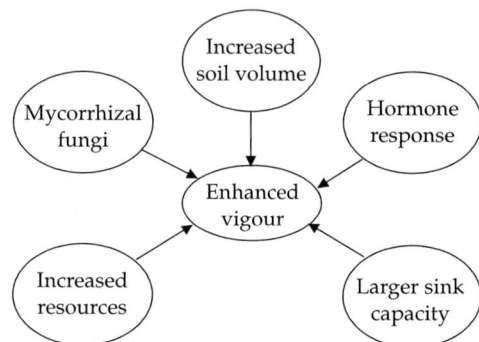

Figure 7.5 A 'multifactor hypothesis' to explain the stimulation response of sand dune species following burial in sand (modified from Maun 2004).

example, it contains apatite that provides a source of phosphorous for the plant.

7.3.3 Growth of mycorrhizal fungi

As shown above, following burial by sand the roots must grow into the deposit before they can exploit the meagre resources within. However, the ubiquitous mycorrhizal fungi immediately grow from the plant roots into the fresh sand and expand the total absorbing surface of the plant.

7.3.4 Physiological hormonal response

Upon sand accretion, a plant must emerge from burial or face local extirpation. The plant reacts strongly to the burial stress and directs all of its energy towards emergence above the new sand surface. If the amount of burial is below the threshold of tolerance the plant eventually emerges, probably because of an etiolation response. The response is a physiological change which is most probably mediated by hormonal production by the plant (Baye 1990).

7.3.5 Higher sink capacity

Plants under burial conditions are robust and have a well-developed underground rhizome network. The higher net CO_2 uptake and water use efficiency closer to the beach may most probably be due to higher sink capacity of these plants as shown by Imbert and Houle (2000) in a population of *Leymus mollis* across a subarctic sand dune system.

7.4 Summary

The so called *Ammophila* problem signifies a deterioration of plant vigour following cessation of moderate burial of plants by sand. Foredune plants are regularly exposed to sand accretion by wind and have evolved adaptations not only to withstand burial but to benefit from it by exhibiting better growth

and reproduction. In fact these species require regular burial to maintain vigorous growth. As soon as burial ceases they start to lose vigour and debilitated stands of dune-forming species become a common site. Several possible hypotheses have been proposed by various authors to explain this debilitation. First, burial provides nutrients and in its absence this input is cut off; however, the hypothesis has been questioned because burial by leached sand with no nutrients also improved plant growth. Second, with increased stability of the habitat and no burial, new species of plants are able to establish, thus increasing competition for space and meagre resources. This hypothesis also has drawbacks because the removal of competitive species from such stands did not improve plant growth. Third, cessation of burial reduces root production and old roots are not replaced by young roots. Moreover, the inherent trait of plants to grow upwards brings the growing point and adventitious roots into the dry soil surface area where they are desiccated. Fourth, in the absence of burial, pathogenic fungi and nematodes proliferate and debilitate populations of plants. Burial allows plants a brief respite from these pathogens. This hypothesis has also been challenged because the harmful soil organisms can move very fast into the new sand deposit and respite, if any, is very brief indeed. Finally, as a plant ages, its root system becomes less effective in attracting useful organisms such as mycorrhizal fungi. Burial in sand provides an opportunity for mycorrhizas to explore greater volumes of sand and release nutrients to the plant. Mycorrhizas also mitigate the impact of pathogenic fungi and plant parasitic nematodes. Other proposed hypotheses include genetic differentiation, ageing process, accumulation of toxins, decrease in soil pH and aeration; however, there is no evidence in support of these suggestions.

The stimulation response of plants to burial may be explained by a multivariate approach to the problem. A multifactor hypothesis consisting of five major factors may be advanced for the enhancement of vigour. Burial by sand increases the volume of soil into which the buried plants can expand. This soil contains nutrients, however small, that can be utilized especially by aid from the activity of mycorrhizal fungi whose hyphae proliferate and move into the deposited layer of sand. The plant also has a physiological hormonal response as a reaction to the burial episode and increases its growth to overcome the episode. Finally, the increased plant growth enhances the sink capacity of the plant, thus allowing it to proliferate even more.

CHAPTER 8

Salt spray and soil salinity

8.1 Introduction

Salt spray is an important abiotic stress that affects plant and other biotic communities in the vicinity of sea coasts. Salt stress refers only to excess of ions in the environment, but along sea coasts it specifically involves increased amounts of Na^+ and Cl^- ions. Anyone who has visited a sea coast on windy days has experienced the landward movement of salt in the form of salt spray. Salisbury (1805) reported details of a marine storm in England following gale force winds from the east for one week. He writes: 'On the 14th of January 1803, I observed an east window of my house, which had been cleaned a few days before, covered on the outside with an apparent hoar frost.' Chemical analysis showed that it was salt from salt spray that had been deposited on plants, buildings and other objects. In spring of that year he made two observations: (i) plant taxa showed differential tolerance to salt spray and (ii) injury was more pronounced on the windward than the leeward sides of plants.

Salt spray acts as a strong environmental stress and populations of biotic organisms have evolved traits that allow them to tolerate the effects of salt. The salt crystals also act as condensation nuclei in the air and damage plants by abrasion during wind storms. However, salt spray may also be beneficial because it improves plant growth by providing some essential nutrients. Many researchers have examined the role of salt spray on survival, distribution and growth form of plant species. In this chapter the effects of salt spray and soil salinity on seed germination, seedling survival and plant growth will be examined. Symptoms of injury to plants, mechanisms of

salt tolerance and comparisons between plant species native to the coastal dunes will also be discussed.

8.2 Formation of salt spray

Three factors—wind speed, wave amplitude and wind direction—influence the formation of salt spray. Waves of high amplitude produce four basic types of breakers: surging, collapsing, plunging and spilling (see Chapter 1, Fig. 1.3). The steepest waves with high turbulence create spilling breakers with a bore accompanied by large quantities of small foam bubbles. Shattering of small foam bubbles does not eject droplets of water into the air but foam itself may be blown a few metres inland. Bubbles are not created at a wind velocity of <5 m s^{-1} but when the winds exceed 6.5 m s^{-1} large bubbles are formed (Boyce 1954). When large bubbles burst very small droplets of seawater are produced and ejected into the air stream. However, if the crater of a large bubble is filled with liquid a water jet is ejected that breaks into a few large droplets (Knelman et al. 1954). Woodcock et al. (1953) took high-speed photographs of different stages in the collapse of bubbles and formation of droplets (Fig. 8.1) and showed that approximately five

Figure 8.1 Schematic drawing of stages in the collapse of bubbles, production of jets and release of droplets (after Woodcock *et al.* 1953).

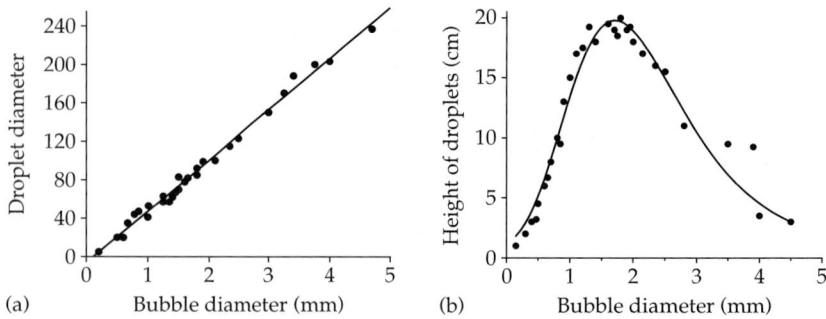

(a)   Bubble diameter (mm)

(b)   Bubble diameter (mm)

Figure 8.2 (a) Relationship between bubble diameter and droplet diameter ejected from bubbles of various diameters at 21°C and 83% relative humidity and (b) bubble diameter and maximum height to which droplets are ejected (from Boyce 1954).

droplets may be ejected by each large bubble to heights of 0.5–15 cm.

Boyce (1954) showed that droplet diameter ranged from about 25–200 μm and was positively correlated with bubble diameter (Fig. 8.2a) and bubbles of 1 to 2 mm diameter ejected droplets to the greatest heights of up to 20 cm (Fig. 8.2b). Once in the air stream the inland transportation of a droplet depended on wind velocity. During its transport, the water may evaporate depending on wind speed, relative humidity and size of droplets. Evaporation reduces the weight of droplets and creates salt nuclei or crystals usually ranging in mass from 10^{-8} to 10^{-14} g or less (Knelman et al. 1954). Reduction in droplet size decreases the rate of settlement of salt crystals and at any given wind velocity salt spray will be carried farther inland under low rather than high humidity conditions.

8.3 Salt spray deposition

The fallout of salt nuclei and droplets from the air stream follows the basic principles of physics of blowing wind carrying particles of different sizes over irregular surfaces and encountering obstacles of different sizes, densities and morphology. At wind speeds of <6.0 m s^{-1}, spray intensities and salt deposition were very low; however, there was an abrupt increase in salt deposition when wind velocity increased from 6.0 to 7.5 m s^{-1} or higher

Table 8.1 Average salt spray deposition (mg salt dm^{-2} hr^{-1}) on cheesecloth traps installed at Long Beach, North Carolina (from Boyce 1954)

Wind speed (m sec^{-1})	Distance from the mean tide line (m)		
	20	95	270
3.0	1.9	0.3	0.2
5.5	2.3	0.5	0.2
6.0	3.4	0.9	0.3
7.5	6.2	4.4	1.5
8.0	6.8	4.2	1.6
9.5	7.1	4.2	1.8
11.0	8.4	4.7	2.2

(Table 8.1). Salt spray is measured by using stationary or swivel type of aerial salt traps with wind vanes and collection surfaces perpendicular to the shoreline and facing onshore winds. Collectors installed on these traps are usually made of cheesecloth or filter paper.

Barbour (1978) measured salt spray deposition at Point Reyes National Seashore along the Pacific coast of California and found that traps set at 60 m from the mean tide line, received 4 to 5 times as much spray as those located at 80 m from the mean tide line (Fig. 8.3). Other studies also showed a similar regular decrease (Table 8.1) in salt spray deposition with increasing distance from the mean tide line (Boyce 1954; Oosting and Billings 1942; Gooding 1947). Salt spray may travel long distances inland at high-energy coasts with strong

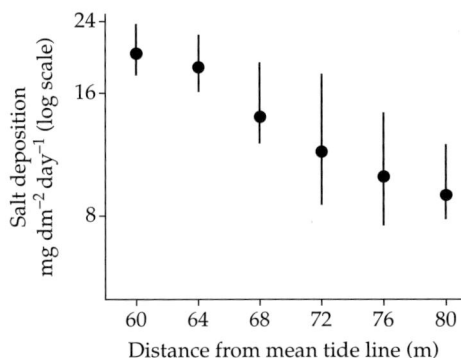

Figure 8.3 Average salt spray (mg salt dm^{-2} day^{-1}) for six rows of eight traps each installed at various distances from the mean tide line at Point Reyes National Seashore along the Pacific coast of California (from Barbour 1978).

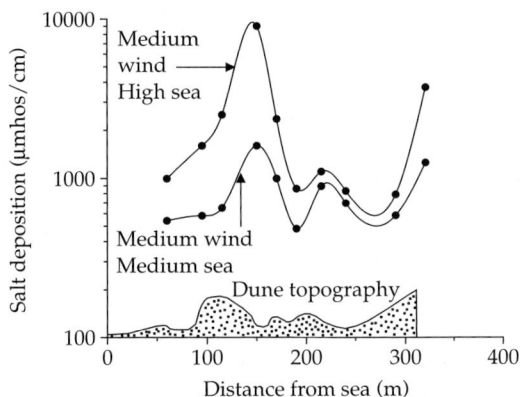

Figure 8.4 Relationship between salt deposition (µmhos cm^{-1}) and distance from the sea at two wave amplitudes on a sand dune with irregular topography (from Randall 1970).

winds. For example, in the Aberystwyth area of Wales even though a significantly greater amount of salt (114 µg cm^{-2} NaCl) of marine origin was deposited at 1.1 km, a substantial amount (77 µg cm^{-2} NaCl) was received at 6.3 km and beyond (Edwards and Claxton 1964). There was also a highly significant correlation (r = 0.57, P < 0.001) between weekly run-of-wind and total weekly salt deposition; traps with the highest 15-week average had a deposition of 23.2 mg dm^{-2} day^{-1} compared to 5.4 mg dm^{-2} day^{-1} for traps with the lowest average (Barbour 1978).

Since there is a marked relief in sand dunes, salt deposition varies at different elevations above the sand surface. At Bogue Bank North Carolina, Oosting and Billings (1942) found that 40–50% of the total salt spray was deposited on the foredune, 20–30% on the top of the first dune ridge, 10–20% on its lee slope and 1–9% in the slack; however, salt deposition increased again to 10–25% on the top of the second dune ridge. Salt deposits increased with an increase in wave amplitude and a change in topography (Randall 1970; Fig. 8.4). The slacks had significantly lower salt deposition than dune ridges and ground traps received only 17% of that deposited in vertical traps placed 85 cm above ground. Wind velocity has a major impact on salt deposition at

various heights. At wind speeds of 4.3 km hr^{-1}, the relative proportion of salt received on traps set at 22, 14 and 0 cm heights was 1.0, 0.13 and 0.13, respectively while an increase in wind speed to 8.5 km hr^{-1} changed the relative proportions at the same heights to 1.0, 0.28 and 0.15 (Barbour 1978). In the Azores salt rains have been frequently recorded at altitudes as high as 200–500 m (Sjögren 1993).

The amount of salt deposition depends not only on the size of plant parts such as leaf area index, but also on the position, orientation and surface properties of plant surfaces as well as airflow characters of wind. Small thin leaves of *Spartina patens* (a halophyte) and needles of pine trees were more efficient collectors of smaller salt droplets than large flat leaves of plant species (Boyce 1954). Leaf orientation to the wind also affected salt accumulation. However, interactions between a great variety of leaf sizes and orientation did not permit simple correlation to be made between leaf traits and salt spray tolerance. The leaf area index of plant communities on coastal sand dunes may vary from approximately 4–10, hence the amount of salt spray intercepted by plant canopies would obviously be much higher than bare ground. Deposition on leaf surfaces was also lower than the amount measured in salt traps. Leaching of horizontal leaves of

Ambrosia chamissonis located at 5 cm above the sand surface yielded 0.2 mg dm^{-2} day^{-1} of salt compared to 5.5 mg dm^{-2} day^{-1} on surrounding traps (Barbour 1978). In contrast, vertical leaves of *Elymus mollis* at 16 cm above the sand surface averaged about 0.4 mg of salt dm^{-2} day^{-1} compared to 22.5 mg dm^{-2} day^{-1} on traps of similar height. Salt deposition also increased immediately behind obstacles because of the eddying effect. The windward sides acted as efficient collectors and reduced the leeward spray intensity. After a gale force wind storm lasting less than 24 hrs, plant samples collected from the windward sides of a hawthorn hedgerow about 1.6 km from the sea contained a significantly ($P < 0.01$) higher amount of salt (404.8 μg NaCl cm^{-2}) than the leeward side (93.2 μg NaCl cm^{-2}) (Edwards and Claxton 1964). Usually, the ratio of deposition on windward to leeward sides of a plant was about 4:1.

8.4 Composition of salt spray

The overall average concentration of dissolved salts in the open ocean is approximately $3.5 \pm 0.2\%$ by weight of salts, which is about 300 times greater than in river water (Brown *et al.* 1989). The variation in salinity of seawater is dependent almost entirely on the extent of precipitation and evaporation, flow of rivers into oceans and mixing and circulation of surface and deep waters. In estuaries, the concentration of salts is lower (~2.5%) because of dilution by river water that has a salinity of approximately 0.012%. In contrast, high concentrations of ≥4.0% salt may exist in shallow bays with little or no input of freshwater, little or no internal mixing and limited exchange with the open ocean. Inland seas or lakes may have very high salt concentrations on their shores and in water because they have no outlets and salts accumulate as the water evaporates. For example, the salt content of the Dead Sea is about 80% while it is approximately 28% in Salt Lake, Utah.

Eleven major elements make up about 99.9% of the dissolved constituents in seawater (Table 8.2) and the proportion of anions

Table 8.2 Average concentrations of principal ions in seawater, per mill (‰ by weight) or in parts per thousand (from Brown *et al.* 1989)

Ion	‰ by weight	
Chloride, Cl^-	18.980	
Sulphate, SO_4^{2-}	2.649	
Bicarbonate, HCO_3^-	0.140	Negative ions
Bromide, Br^-	0.065	(Anions)
Borate, $H_2BO_3^-$	0.026	Total = 21.861 ‰
Fluoride, F^-	0.001	
Sodium, Na^+	10.556	
Magnesium, Mg^{++}	1.272	Positive ions
Calcium, Ca^{++}	0.400	(Cations)
Potassium, K^+	0.380	Total = 12.621 ‰
Strontium, Sr^{++}	0.013	

‰ = parts per thousand

(2.186% by weight) is almost twice that of cations (Brown *et al.* 1989). The main reason is the very high concentration of chlorides that are derived partly from the weathering of rocks but mainly by gas released from volcanoes. Most of the other elements in seawater originate from the weathering of earth's crust. Thus, seawater has greater concentrations of salts that are more soluble in water. For example, the three most abundant elements in the earth's crust, silicon (Si) (28.2%), aluminium (Al) (8.2%) and iron (Fe) (5.6%) are not found in ocean water because of their extremely low solubility. They are represented in the oceans only in the solid phase as components of sand, clay and colloidal particles. In contrast the concentrations of calcium (Ca)$^{++}$ (4.2%), Na^+ (2.4%), K^+ (2.4%), Mg^{++} (2.0%), titanium (Ti) (0.6%), manganese (Mn) (0.1%) and P (0.1%) in the earth's crust is relatively low, yet, because of their high solubility these elements are over-represented in the oceans as dissolved salts. An important concept in oceanography states that in spite of the variability of seawater composition in different areas of the ocean, the ratio of any major ion to the total

Table 8.3 Abundance of essential nutrient elements in seawater (from Brown *et al.* 1989)

		Concentration (mg L^{-1}/or ppm)	Probable form
Essential elements			
Chlorine	Cl	1.95×10^4	*Cl$^-$
Magnesium	Mg	1.29×10^3	*Mg^{2+}
Sulphur	S	9.05×10^2	*SO$_4^{2-}$, NaSO$_4^-$
Calcium	Ca	4.12×10^2	*Ca^{2+}
Potassium	K	3.80×10^2	*K$^+$
Carbon	C	28	HCO$_3^-$, CO$_3^{2-}$, *CO$_2$
Nitrogen	N	11.5	N$_2$ gas, *NO$_3^+$, *NH$_4^+$
Boron	B	4.4	B(OH)$_3$, B(OH)$_4^-$, H$_2$BO$_3^-$
Phosphorous	P	6×10^{-2}	*HPO$_4^-$, PO$_4^{3-}$, H$_2$PO$_4^-$
Molybdenum	Mo	1×10^{-2}	*MoO$_4^{2-}$
Iron	Fe	2×10^{-3}	Fe(OH)$_2^+$, Fe(OH)$_4^-$
Nickel	Ni	1.7×10^{-3}	*Ni^{2+}
Zinc	Zn	5×10^{-4}	ZnOH$^+$, *Zn^{2+}, ZnCO$_3$
Manganese	Mn	2×10^{-4}	*Mn^{2+}, MnCl$^+$
Copper	Cu	1×10^{-4}	CuCO$_3$, Cu$^-$OH$^+$
Non-essential elements			
Sodium	Na	1.077×10^4	Na$^+$
Silicon	Si	2	Si(OH)$_4$

* Available form for land plants.

or to any other major ion remains relatively constant (Brown *et al.* 1989).

There is ion separation when bubbles burst and salt spray droplets are ejected into the atmosphere (Bloch *et al.* 1966). Concentration of selected ions in seawater and water collected in rain gages on the coast showed that even though the concentrations of Na$^+$ > Mg^{++} > Ca^{++} > K$^+$ followed the same sequence as seawater, their relative concentrations had changed (Clayton 1972). During transport as salt spray, generally the chloride content decreased more rapidly than cations.

8.4.1 Beneficial effects of salt spray

Although seawater is a weak nutrient solution that contains all the essential nutrients required for the growth of terrestrial plants, the relative proportion of elements is physiologically unbalanced for plant growth because the major constituents of oceanic spray are salts of Na$^+$, Ca^{++}, K$^+$, Mg^{++}, Cl$^-$ and S with only minute quantities of all other essential elements (Table 8.3).

Farmers along the Venezuelan coast believe that 'a good rain after the dry season is better than a thousand waterings'. Apparently, an *alisio* (salting) wind blows steadily from the ocean during the dry season. The airborne salt spray particles dry quickly leaving brine suspended in the air, it forms a thick haze that may reduce visibility to almost zero. This brine contains gypsum (CaSO$_4$), Cl$^-$, cations of Na$^+$, Ca^{++}, K$^+$, Mg^{++} and traces of other elements (Zuloaga 1966). Brine dissolves during the rainy season, thus enriching the soil with some essential nutrients. Measurements by Art *et al.* (1974) in an *Ilex opaca, Sassafras albidum* and *Amelanchier canadensis* dominated forest on a barrier island off Long Island, New York, (0.3 km from the ocean) showed that the quantities of K^+, Na^+, Ca^{++} and Mg^{++} deposited in aerosol salt spray were 0.73, 14.1, 0.98 and 1.9 g m^{-2} year^{-1}, respectively. This meteorological input was sufficient to compensate for the low rates of weathering and allowed the maintenance of the same rates of production as the inland temperate forest ecosystems. Etherington (1967) showed that in a South Wales coastal grassland calcium carbonates were leached from the surface to lower layers in the soil, however, there was no similar reduction in K^+ levels in upper soil horizons because leaching loss was balanced by the annual increment of K^+ from the salt spray. He also suggested that an adequate supply of micronutrients was deposited in coastal dune systems as salt spray.

Coastal foredunes are, however, very deficient in nitrogen. Salt spray alone will not provide sufficient nitrogen because the concentration of N in seawater is only 11.5 ppm of which 9 ppm is dissolved N_2 gas which can not be utilized by plants (Table 8.3). The amount of useable forms, NO_3^- and NH_4^+ is only 2.5 ppm in total, of which the average concentration of nitrate is about 0.5 ppm. The useable forms originate primarily from organic sources on land that are transported to the oceans by rivers and streams. Wilson (1959) argued that the thin upper layer of the ocean is minerally enriched and contains greater concentrations of organic N, NH_4^+, K^+ and dead remains of partially decomposed planktonic organisms. Larger quantities of nitrogen were contained in bursting bubbles close to the seashore than the open ocean. For example, sea foam contained 25% solid material consisting of bacteria, diatoms and fragments of phyto- and zooplanktons. Its total N content was 235 ppm. Similarly, snow along the windward coast of New Zealand (surrounded by

thousands of kilometres of ocean water) generally contained 5 to 25 times more organic N than ocean water and it is believed that N probably originated from foam blown inland by wind.

8.5 Soil salinity

Along sea coasts increase in soil salinity may occur by deposition of salt spray, inundation of beach by seawater and by light rain that washes the salt from plant surfaces to the soil. However, the amount of salt retained in the soil depends on adsorption by clay and silt colloids (Randall 1970). Since sandy soils of sea coasts contain little or no silt and clay, the salt is readily leached out of the rhizosphere by soaking rainstorms. Measurements recorded over the last 100 years by various authors along coasts around the world are presented in Table 8.4. In three locations along the Atlantic coast the soluble salt content of coastal sandy soils from the surface to 90 cm depth ranged from 0.003–0.009%. The values decreased with distance from 50–150 m inland on the beach. Along the Pacific coast the concentration of water soluble salts in soil (0–30 cm depth) was 0.02–0.19 at Bodega Head (Barbour and De Jong 1977), 0.09–0.13% in soil (at 0–60 cm depth) at Long Beach, California (Kearney 1904), and 0.008–0.28% on the driftline (0–10 cm depth) of 34 beaches (Barbour *et al.* 1976). This difference in soil salt content between the Atlantic and Pacific locations was ascribed to differences in quantity and distribution of rainfall. The mean annual rainfall at Los Angeles, California is only 39 cm compared to 130 cm, at Norfolk, Virginia. Generally soil salt content did not change with depth in regions with high rainfall; however, in hot arid regions with low rainfall such as along the Mediterranean in Egypt, the high salt concentrations were found only in the top few cm layers and decreased rapidly with depth (Ayyad 1973).

The chloride content in sand samples was also very low (Table 8.5). At Long Beach,

Table 8.4 Concentration of soluble salt (%) in soil samples collected from different depths and distances recorded at five locations by different authors over the last 100 years

Distance (m)	Depth (cm)	Salt (%)	Location and authority
2	20	0.005	Fremantle, Australia (Olsson-Seffer 1909)
5	35	0.004	Fremantle, Australia (Olsson-Seffer 1909)
7	30	0.004	Fremantle, Australia (Olsson-Seffer 1909)
10	30	0.009	Fremantle, Australia (Olsson-Seffer 1909)
15	40	0.006	Fremantle, Australia (Olsson-Seffer 1909)
50	15	0.013	Barbados, West Indies (Gooding 1947)
100	15	0.0082	Barbados, West Indies (Gooding 1947)
150	15	0.0064	Barbados, West Indies (Gooding 1947)
—	0–30	0.003–0.006	Woods Hole, Massachusetts (Kearney 1904)
—	40–60	0.004	Woods Hole, Massachusetts (Kearney 1904)
—	0–30	0.003–0.004	Norfolk, Virginia (Kearney 1904)
—	40–60	0.003–0.009	Norfolk, Virginia (Kearney 1904)
—	70–90	0.003–0.007	Norfolk, Virginia (Kearney 1904)
—	0–30	0.13 ± 0.005	Long Beach, California (Kearney 1904)
—	40–60	0.09 ± 0.020	Long Beach, California (Kearney 1904)
Driftline*	10	0.008–0.280	34 beaches, Pacific coast, USA (Barbour et al. 1976)
Driftline*	30	0.020–0.190	Bodega Bay, California (Barbour and De Jong 1977)

* Leading edge of vegetation.

Table 8.5 Concentration of chloride (%) in soil samples collected at two depths and different distances from Long Beach, North Carolina, after 4 days of south-west winds with no rain, after an afternoon rain and 3 days after a rain (Boyce 1954)

Habitat	Depth (cm)	No rain	Rain	3 days after rain
Top of foredune	1	0.0043	0.0008	0.0041
	15	0.0008	0.0007	0.0008
Lee of foredune	1	0.0035	0.0007	0.0036
	15	0.0008	0.0008	0.0008
75 m	1	0.0034	0.0008	0.0032
	15	0.0008	0.0008	0.0007
100 m	1	0.0035	0.0007	0.0030
	15	0.0008	0.0008	0.0008

North Carolina samples collected from the soil surface (1 cm depth) at four locations, 75 and 100 m from the high tide mark and on the top and lee of the foredune contained 0.0034 to 0.0043% chloride (w/oven dry weight) of sand after 4 days of no rain (Boyce 1954). However, following an afternoon rain, these values dropped to 0.0008% because of leaching from the soil. The chloride content in the surface 1 cm layer increased again to the same level three days after the rainfall, undoubtedly because of salt spray deposition. At 15 cm depth, the values remained 0.0008%. Even the driftline zone directly exposed to oceanic spray and occasional inundation is not consistently saline (Olsson-Seffer 1909; De Jong 1979). In fact, the soils of coastal strands may not be saline; because even the highest concentration of water-soluble salts did not exceed the 0.02 to 0.2% found in cultivated agricultural soils. Thus, plant species occupying coastal strands, although resistant to salt spray and drought, can not be classified as halophytes or xerophytes (Kearney 1904; De Jong 1979).

8.6 Plant response to salt

Foredune species respond to salt stress in a number of ways. Some species escape stress by completing their growth during less stressful conditions when the frequency of storms is low, while others make metabolic adjustments to either avoid or tolerate the impact of salt. Nevertheless, during periods of high salt stress individuals of populations of some plants do succumb to the impact.

8.6.1 Morphological symptoms of injury

Boyce (1954) sprayed several foredune species, *Phytolacca americana*, *Erigeron canadensis*, *Iva imbricata*, *Physalis maritima* and *Croton punctatus*, three times per day with four solutions containing total salt concentrations equivalent to that of seawater but with varying ion ratios and recorded symptoms of injury. Necrosis first began on leaf tips, progressed to upper margins of leaves and eventually developed into an inverted V-shaped pattern. Injury was manifested first in young rather than old leaves of plants. On species with deep leaf lobes such as *Quercus* or *Vitis* species, injury appeared first on the tip of each lobe. Injury from high concentrations of chloride ions manifested itself long before that of other ions and caused extensive tip and margin injury, necrosis of leaves and finally death of coastal plants. Chloride ions accumulated in leaf tips at higher concentrations than other parts of leaves and irrespective of point of entry into the plant, they were translocated preferentially to the tips of young leaves. On compound leaves, the tip of the terminal leaflet was affected first and then it spread to the tips of all other leaflets. Several samples of

necrotic leaves contained as much as 5.6 g of chlorine per 100 g of oven dry weight. In ferns, *Osmunda regalis* and *O. cinnamomea*, chloride burning appeared first in leaflets in the acropetal end of the rachis and then proceeded basipetally down the rachis and peripherally along the leaflets (Martin 1959). Following gale force winds on the Finland coast leaf tips of *Pinus sylvestris* trees became necrotic, necrosis proceeded towards the base of needles, a dark brown band developed between the necrotic and green basal part of the leaf, and finally the needles fell off the branches (Pyykkö 1977). Similar observations were made on *Picea mariana* trees along the coast at Basin Head, Prince Edward Island, Canada. Needles on branches exposed to the full force of wind on the seaward side were killed and shed by the trees in summer each year. Apparently, abscission is a typical manifestation of salt spray injury along sea coasts. In temperate regions salt spray storms with gale force winds occur primarily in the autumn and winter months. Normal growth of both herbaceous and woody species is resumed in spring and summer and the new leaves on the windward side are killed again by salt spray during the following winter. Along the Cape Cod coast, salt spray kills twigs of *Quercus velutina*, *Q. ilicifolia* and *Prunus maritima* after the abscission of leaves (Boyce 1954).

Trees and shrubs on sea coasts exhibit asymmetric growth forms that may be caused by the killing action of salt spray rather than the high wind velocities (Wells and Shunk 1938). Boyce (1954) explained that during periods of low salt spray twigs develop normally but when spray intensity increases, the seaward twigs are killed while those on the lee side survive and continue to grow in the landward direction. Radial growth finally covers scars of dead twigs on the seaward side and the stem curves into a wind-trained form. The trees are also dwarf because salt spray kills the terminal buds of branches and the leader.

Small lacerations, splits and scratches produced by mechanical bending of leaves by high wind velocities serve as points of entry of salt into the leaves. Wetting properties of leaves, cuticle thickness, hairiness and leaf structure also affect leaf penetration. Pyykkö (1977) showed that following a wind storm salt crystals of various sizes were embedded on the surfaces of injured leaves of *Pinus sylvestris* and after entry into the leaf salt killed the chloroplast, the number of resin ducts decreased, sclerenchyma was poorly developed and the number of vascular bundles increased from the normal two to three. Salt also had indirect effects on the future growth of trees. For instance, if more than 80% of the needles were killed on a tree, the dormant buds, although undamaged, failed to open in the spring. The stems of *Pinus sylvestris* showed a decrease in thickness of annual rings owing to a decrease in the number of cell layers (Pyykkö 1977). The annual rings on the seaward side of trees were smaller than those on the leeward side and the length of terminal shoots, needles and the number of terminal buds and short shoots on the seaward branches was reduced to approximately half that of the leeward branches. New leaves contributed to the formation of late wood in the main stem but the growth of secondary xylem was reduced, probably because of the loss of older leaves.

8.6.2 Effects on seed germination

Substrate salinity can act as a major selective force in seed germination and seedling emergence. In spring seeds of annual and perennial species may be exposed to differing levels of soil salinity because of salt spray deposition and periodic inundation by seawater during winter months. The mid-beach is of course inundated more frequently than the upper beach. *Cakile maritima* occurs on sandy beaches close to the high tide mark. Barbour (1970a) tested seed germination and root growth of *C. maritima* under different concentrations of NaCl and found that total percentage germination was not significantly different from control at 0.01 and 0.1% salt solutions. However,

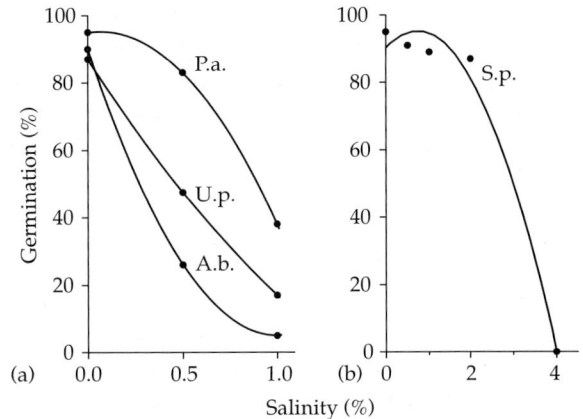

Figure 8.5 (a) Effect of different levels of salinity on percentage seed germination of dominant grasses: *Ammophila breviligulata* (A.b.), *Uniola paniculata* (U.p.), and *Panicum amarulum* (P.a.) collected from sand dunes and (b) *Spartina patens* (S.p.) from low moist sand flats along the North Carolina coast (Seneca 1969).

strong inhibition occurred at 1, 2 and 3% salt solutions. Similarly, there was a linear decrease in percentage germination of *Ammophila breviligulata*, *Uniola paniculata*, *Panicum amarulum* and *Spartina patens*, collected from the Outer Banks of North Carolina, with an increase in concentration of NaCl from 0 to 0.5, 1.0, 2.0 and 4% weight over volume (w/v) (Seneca 1969). Seeds of the former three species failed to germinate in 1% NaCl (Fig. 8.5). Germination of the halophyte (*S. patens*) remained relatively stable up to a concentration of 2% but no germination was observed at 4% NaCl. The maximum tolerance limits for seed germination of *A. breviligulata* and *U. paniculata* were between 1–1.5%, for *P. amarulum* 1.5–2.0% and for *S. patens* approximately 4%.

Woodell (1985) tested the effects of salinity on seed germination of 29 plant species from driftline, shingle, dune and marsh habitats in England. Seeds were placed in Petri dishes containing filter papers and 5 mL of distilled water (0 mM NaCl), half-strength seawater (300 mM NaCl), full-strength seawater (600 mM NaCl) or 1.5 strength seawater (900 mM NaCl). Based on the germination response, the species were grouped into three categories. Type 1 consisted of 12 species: seven from driftline, four from dune and one (*Aster tripolium*) from marsh habitats in which germination was inversely proportional to salinity. Seeds of Type 2 species experienced strong inhibition in 0.5 or higher strength seawater. Of the nine species in this group, three were from dune, two from marsh, three from shingle and one from driftline habitats. Ten species (seven from marsh and three from driftline habitats) were classified as Type 3 and their response to salinity was variable; marsh species showed reasonably good germination in half-strength seawater but species from driftline habitats did not germinate in any of the salinity treatments. Germination for four species is shown in Fig. 8.6. The data clearly indicated that the species of driftline, dune and marsh habitats differed markedly in their

germination response to seawater. Many strand plants do indeed use sea waves as important dispersal avenues for their propagules to a new habitat when seeds or fruits are exposed to full-strength seawater (Chapter 3). Controlled studies clearly showed that germination was not affected by immersion because seeds of all species germinated after being recovered from the seawater solutions and washed in distilled water (Seneca 1969; Barbour 1970a; Woodell 1985). In fact, seashore species including halophytes do not require salt solution for seed germination (Darwin 1857; Woodell 1985; Ungar 1978). Seeds of some species regularly exposed to immersion in seawater may, however, exhibit the phenomenon of 'salt stimulation' or enhanced rate of germination (Woodell 1985).

The studies just described were all conducted under controlled conditions and do not fully

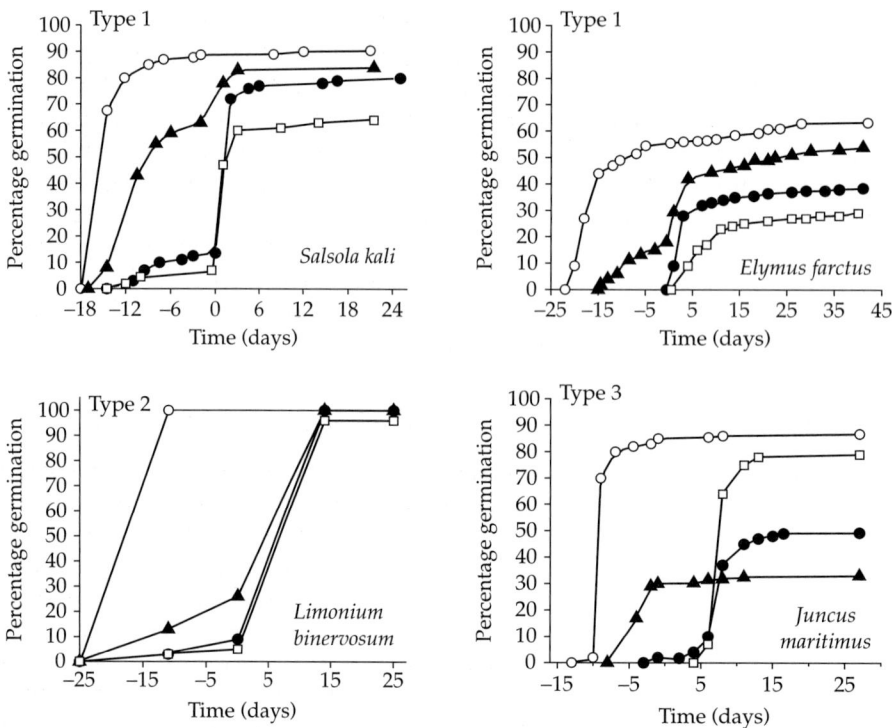

Figure 8.6 Effect of different concentrations of seawater on seed germination of four species of driftline, dunes, shingle and marsh habitats in England. Day 0 is the day of transfer of seeds to freshwater and negative days are before transfer: (○) freshwater, (▲) half-strength seawater, (●) full-strength seawater, (□) 1.5-strength seawater (Woodell 1985).

reveal what happens to propagules under natural salinity conditions along seashores when they are cast onshore following dispersal by seawater. Nevertheless, there is strong evidence that soil salinity of coastal strands is not deleterious to seed germination of beach and driftline species, because as explained earlier the salinity of soil on the strand is only 0.003 to 0.13% (Table 8.4) and well below the critical inhibitory level of about 1% for seed germination of susceptible species. This is orders of magnitude less than that required to kill highly resistant species (Table 8.6).

8.6.3 Physiological effects on seedlings

Seneca (1972) tested the effects of soil salinity on the growth of 6–10-month-old seedlings of four sand dune species: *Ammophila breviligulata*, *Panicum amarulum*, *Spartina patens* and *Uniola paniculata*. An inverse linear relationship existed between growth and increased salinity in three species, *A. breviligulata*, *Spartina patens* and *Uniola paniculata*. However, *P. amarulum* showed a curvilinear relationship with first an increase and then a decline in growth. *Ammophila breviligulata* and *Uniola paniculata* seedlings showed moderate growth

Table 8.6 Maximum soil salinity at which the following plant species were killed (Olsson-Seffer 1909)

Species name	Salinity (%)
Argentina anserina	1.9
Aster tripolium	2.6
Atriplex hastata	3.1
Cakile maritima	2.9
Crambe maritima	2.5
Elymus arenarius	2.6
Erythraea vulgaris	1.9
Glaux maritima	2.7
Juncus gerardi	2.2
Matricaria inodora maritima	2.3
Plantago maritima	2.8
Sonchus arvensis maritima	2.6
Triglochin maritima	2.1

at salinities of 0.5 and 1% or less but barely survived at soil salinity of 2% and showed typical symptoms of injury: twisting and rolling of leaves, yellowing and necrosis of tips and then the whole leaf. *Panicum amarulum* seedlings performed much better at 2% soil salinity but all three species died under 4% soil salinity treatment. *Spartina patens* seedlings were very resistant and some seedlings survived 4.0% concentrations of NaCl for four weeks. The descending order of seedling tolerance to soil salinity was *S. patens* > *P. amarulum* > *U. paniculata* > *A. breviligulata*, an order that correlated well with positions of seedlings on the strand. Barbour (1970a) tested the effects of salinity on seedling growth of *Cakile maritima* by planting them in trays filled with sterile sand up to a depth of 5 cm and watering them from below with tap water (0.1% salinity), 1/12 (0.8%) and 1/3 (1.5%) strength seawater. Only a slight decline in growth of seedlings occurred at 0.8% salinity, however, a significant decline in biomass, number of leaves and leaf size occurred at 1.5% salinity (Table 8.7). Seedlings survived in 1.5% salinity treatment but lost turgor and developed necrosis on cotyledon tips. In another experiment the root growth of *C. maritima* seedlings in Petri dishes was less affected by salinity than seed germination in 0.01, 0.1 and 1.0% salt solutions, however, a significant decline occurred at 2 and 3% solutions (Barbour 1970a).

8.6.4 Physiological effects on adult plants

Plant species differed widely in their tolerance to salt spray. A list of plant species considered as highly resistant, resistant, susceptible and very susceptible is presented in Table 8.8. Martin (1959) tested the tolerance of 50 species of herbaceous and woody plants to salt spray on sand dunes along the seashore at Island Beach State Park. He thoroughly wetted plant leaves of each species with undiluted seawater each day for 15 consecutive days using a fine mist sprayer. Highly resistant species, graminoids or succulents, on the beach strands along the

Table 8.7 Effect of different levels of salinity on the growth of *Cakile maritima* seedlings over a 2-week period in a growth chamber maintained at 21°C and 1500 ft-c light intensity (Barbour 1970a)

Salinity (%)	Dry weight (mg)	Number of leaf pairs	Length of first leaf pair (mm)
0.1	23	1.4	10.7
0.8	25	1.3	10.5
1.5	13	0.4	0.2

Table 8.8 Salt spray resistance of major plant species, based on burning of foliage (%) by 15 consecutive daily sprays of undiluted seawater, growing on the strand along the Atlantic coast of North Carolina (from Martin 1959). A partial list of species from other authors is presented in the second half of the table

Foliar damage (0%) highly resistant
Ammophila breviligulata, Artemisia stelleriana, Cakile edentula, Carex kobomugi, Iva frutescens, Solidago sempervirens, Spartina alterniflora, Spartina patens

Foliar damage (5–40%) resistant
Arctostaphylos uva-ursi, Convolvulus sepium, Ilex glabra, Juniperus virginiana, Panicum virgatum, Pinus rigida, Robinia pseudoacacia, Vaccinium macrocarpa

Foliar damage (75–90%) susceptible
Amelanchier canadensis, Kalmia angustifolia, Prunus serotina

Foliar damage (100%) very susceptible
Acer rubrum, Euphorbia polygonifolia, Hudsonia tomentosa, Hypericum virginicum, Parthenocissus quinquefolia, Phytolacca americana, Prunus maritima, Quercus ilicifolia, Smilax rotundifolia

Other authors*
Highly resistant: *A. arenaria,*[6] *Baccharis halimifolia,*[7] *Croton punctatus,*[3] *Desmoschoenus spiralis,*[6] *Elymus farctus,*[6] *Elytrigia junceiformis,*[5] *E. pungens,*[5] *Physalis viscosa,*[3] *Spartina patens,*[3] *Spinifex sericeus,*[2] *Uniola paniculata*[4]
Resistant: *Chloris petrea,*[3] *Elytrigia repens,*[5] *Ilex vomitoria,*[7] *Iva frutescence,*[3] *I. imbricata,*[3] *Myrica cerifera,*[7] *Quercus virginiana,*[7] *Solidago mexicana*[5]
Susceptible: *Andropogon scoparius,*[4] *Ceratodon purpureus*[1]
Very susceptible: *Aulacomnium palustre,*[1] *Chenopodium ambrosoides,*[3] *Cynodon dactylon,*[3] *Heterotheca subaxillaris,*[3] *Leptilon canadense,*[3] *Pinus taeda,*[7] *Polytrichum commune,*[1] *Strophostyles helvola*[3]

1, Boerner and Forman (1975); 2, Maze and Whalley (1992a); 3, Oosting (1945); 4, Oosting and Billings (1942); 5, Rozema *et al.* (1983); 6, Sykes and Wilson (1989); 7, Wells and Shunk (1938).

North Atlantic exhibited no foliar damage after 15 days of daily spraying and up to 15 days after the last treatment. The second group of species, woody shrubs, was resistant and suffered 5–40% damage to their leaves but showed little change 15 days after the discontinuation of treatments. The third group consisting of three species was susceptible and suffered 40–75% foliar damage by the end of treatments but this damage increased to 75–90% within 15 days of the last spray. Another group of species was very susceptible and suffered 50–100% foliar damage by the end of treatments or 100% damage 15 days after the last treatment. A small group of woody shrubs, *Robinia pseudoacacia, Viburnum dentatum, Rhus radicans, Sassafras albidum, Rosa rugosa* and *R. virginiana*, showed 100% burning of foliage after 15 days of spraying but plants recovered later and produced new leaves. Sykes and Wilson (1988, 1989) exposed one batch of 29 species (20 native, 9 introduced) of New Zealand foredunes and

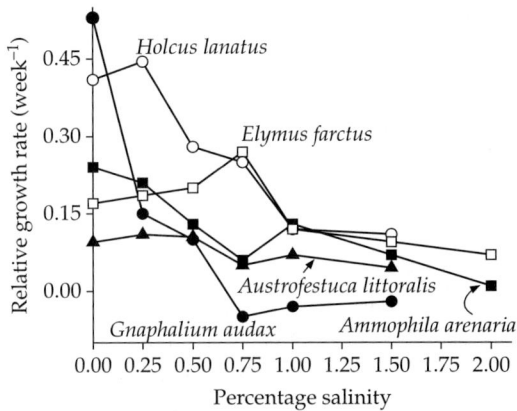

Figure 8.7 Effect of percentage of salinity on the relative growth rate per week of five plant species based on the mean weight of four plants at each harvest: (○) *Holcus lanatus*, (□) *Elymus farctus*, (■) *Ammophila arenaria*, (▲) *Austrofestuca littoralis*, (●) *Gnaphalium audax* (Sykes and Wilson 1989).

slacks to root salinities of 0, 0.25, 0.50, 0.75 and 1% for all species, 1.5% (20 species) and 2% (9 species) in hydroponics and another batch of the same species growing in pots to salt spray under greenhouse conditions. In general, there was an inverse relationship between salinity and RGR, although species exhibited considerable differences to salinity tolerance (Fig. 8.7). Thirteen species were classified as glycophytes and the most tolerant species was *E. farctus*, which was considered a facultative halophyte. Almost half of the species showed less tolerance to salinity and salt spray than wheat (*Triticum aestivum*), which served as the glycophytic control in their study. Among the *Elytrigia* species growing at different sites, *E. pungens* (salt marshes) was the most resistant, *E. junceiformis* (foredunes) intermediate and *E. repens* (inland sites) the least resistant to high soil salinity, however, *E. junceiformis* was more resistant to salt spray than both *E. repens* and *E. pungens* (Rozema *et al.* 1983).

Seawater spray reduced plant size, seed production and seed mass of the annual *Triplasis purpurea* (Cheplick and Demetri 1999). Maze and Whalley (1992a) showed that the total dry weight of *Spinifex sericeus* plants sprayed with

seawater or watered with seawater once every two weeks and then flushed the next day was not significantly different from control. Coastal dune mosses, *Aulacomnium palustre*, *Polytrichum commune* and *Ceratodon purpureus* sprayed daily with full-strength seawater for 33 days showed that none of the populations were tolerant and did not recover within 15 days of the last spray (Boerner and Forman 1975).

Barbour (1978) found no correlation between salt spray and total coverage of plant species growing on the dune strand along the Pacific coast of California at Bodega Head. However, there were indications that percentage cover of plants declined above a certain critical salt spray load, suggesting that plant distribution may be affected under some conditions. Below the critical level of salt other factors may be more limiting to growth.

8.7 Mechanisms of resistance

Several traits of avoidance and tolerance are prevalent in strand species.

1. *Hypertrophy:* Hypertrophy, the abnormal enlargement of cells, is a common occurrence in some annual and perennial species of seashores by which they rely on high ion uptake to maintain cell turgor under conditions of low water potential. These plants also accumulate salt in their vacuoles and keep the concentration of Na^+ and Cl^- in the cytoplasm at low levels. Hypertrophy is caused primarily by chloride ions because mesophyll cells containing high concentrations of chloride ions are 2 to 3 times larger than those containing Na^+, K^+ and SO_4^- ions. However; the degree of tolerance of a plant species to chloride is not a general phenomenon related to the degree of succulence because species of the same genera may respond differently and hypertrophy occurs only in dicots. Leaf thickness of dune populations of *Solidago sempervirens* was significantly greater than those in dune hollow, and intermediate in Bayshore habitats primarily because of local environmental effects on phenotypic expression rather than genetic differences (Cartica and Quinn 1980). Similarly, the leaves on the seaward side of *Iva imbricata* shrubs were more thick and succulent than on the leeward side because of slow movement of chloride to that side (Boyce 1954). When *Cakile edentula* var. *edentula* plants were sprayed with 0, 20 and 90 mg dm^{-2} day for four weeks the leaf thickness increased from 462 µm to 558 and 820 µm, respectively. Other species exhibiting succulence along sea coasts are *Amaranthus pumilus, Cakile* species, *Convolvulus sepium, Honkenya peploides, Iva imbricata, Physalis maritima, Salsola kali, Solidago sempervirens* and many others (Boyce 1951).

2. *Annual habit:* Some annual species although susceptible to injury from salt spray thrive in this habitat because they complete their short life cycles between storms, are able to survive in protected habitats, have higher relative growth rates, reproduce prolifically and have high phenotypic plasticity. For instance, *Erigeron* species, although the most intolerant of species to chloride salts, was the most abundant species in the salt spray zone (Oosting 1945), probably because of more favourable conditions other than their tolerance to intracellular chlorides.

3. *Prostrate growth habit:* Several species of sea coasts such as *Baccharis pilularis, Euphorbia polygonifolia, Hudsonia tomentosa, Oenothera humifusa, Parthenocissus quinquefolia* and *Rhus radicans*, although they rate low to extremely low in salt resistance, occur regularly in the high salt spray zone. They grow at an elevation where the amount of salt spray is below their level of tolerance and is only a fraction of what is received by a species with an erectile canopy (Randall 1970).

4. *Reduced uptake:* Some grass species such as *Elytrigia junceiformis, Agropyron junceum, Ammophila arenaria, Elymus arenarius* and, *Spinifex sericeus* respond to high salt spray concentrations by limiting the influx of ions into the leaves because of low wetting properties of cuticular surfaces, through beading especially during rain and then rolling the seawater droplets off the plant leaves (Boyce 1954; Rozema *et al.* 1983; Maze and Whalley 1992a). The presence of sclerenchyma surrounding the parenchyma in grasses may also reduce the amount of chlorides reaching the parenchyma cells.

5. *Loss of salt from roots:* Some species move salt from the shoots to the roots and then leach it into the soil. Clayton (1972) applied ^{22}Na to the leaves of *Baccharis pilularis* plants and found that 12.1% of the applied Na was exuded into the soil. Similarly, a small proportion of ^{22}Na applied to leaves of *Spinifex sericeus* was extruded from the roots into the Hoagland solution (Maze and Whalley 1992a).

6. *Shedding of old leaves:* Salt resistance is acquired by sequestering of salt into old leaves and then shedding them. For example, *Rhus radicans* can be as tall as 60 cm above the sand surface; however, it survives in this habitat by losing leaves following a heavy salt spray episode and then recuperates rapidly by new growth from lateral buds.

7. *Other mechanisms:* The rate of entry of salts into the leaves may also be prevented by thick epidermis, sunken stomata, thick cuticle, thick layers of epicuticular wax, hairy leaves, closure of stomata and changes in leaf position as in *Ipomoea pes-caprae*.

8.8 Salt stress and evolution

A critical examination of the possible impact of salt spray and soil salinity on shoreline species indicates that plants are largely unaffected by the natural levels of salt deposition. All species have some capacity to withstand salts encountered along sea coasts. Some species such as *Leymus arenarius* (Greipsson and Davy 1996), *Spartina patens, Suaeda maritima* and many others that grow in high salinity substrates close to the sea coast have developed the ability to tolerate salt and are classified as halophytes. Some halophytes such as mangroves (*Rhizophora mangle*) are salt regulators by not absorbing salt and excreting it from

Table 8.9 Percentage germination of *Ammophila breviligulata* (Seneca and Cooper 1971) and *Lathyrus maritimus* (Little and Maun 1993) seeds from lacustrine and maritime populations in filter papers containing different concentrations of NaCl solutions

NaCl concentration (%)	*Ammophila breviligulata*		*Lathyrus maritimus*	
	Lacustrine %	Maritime %	Lacustrine %	Maritime %
0	86	92	85.3	95.3 ns
0.25	68	90	—	—
0.5	12	78	84.0	80.0 ns
0.75	0.4	42	—	—
1.00	2	33	73.3	76.6 ns
1.25	0	4	—	—
1.5	0	0	26.7	65.3 **
3.0	—	—	0	0

** $P = 0.006$; ns, not significant.

roots, while others such as *Tamarix articulata* excrete salts, specifically, Na^+, Cl^- and HCO_3^- ions, through salt glands on their leaves.

As mentioned earlier, species of coastal strand are often salt-tolerant non-halophytes, probably because levels of salt in the environment are relatively low, salt spray events and seawater inundation are episodic and seasonal, and salt is leached away by rain through the porous sand. Thus the selection pressure for tolerance to substrate salinity is weak and the usual means of coping with salt stress for strand species along sea coasts is avoidance.

Over time salt tolerance may be lost from populations no longer exposed to salinity. Populations of *A. breviligulata*, *Lathyrus maritimus* and *Cakile edentula* on the shores of the Laurentian Great Lakes have been separated from the maritime populations since the last glaciation about 13 000 years ago. The differences in salinity tolerance between freshwater and maritime populations were tested by several authors. Studies by Seneca and Cooper (1971) showed that seeds from lacustrine populations of *Ammophila breviligulata* from Lake Michigan had lost salt tolerance and showed significantly lower germination at 0.5% salinity after 40 days than seeds from the maritime

populations along the North Carolina coast (Table 8.9). After 30 days in Petri plates, no significant differences in seed germination were found between seeds of lacustrine (Great Lakes) and maritime (Atlantic coast) varieties of *Lathyrus maritimus* at 0, 0.5 and 1% salinity; however, the maritime variety showed significantly ($P = 0.006$) higher germination than the lacustrine variety at 1.5% NaCl solution (Little and Maun 1993). In contrast, the freshwater variety *Cakile edentula* var. *lacustris* from Lake Huron shores germinated as well as variety *edentula* of the Pacific coast in 0, 0.01, 0.1 and 1% salt solutions (Boyd and Barbour 1986).

The seed germination of three populations of *Leymus arenarius* located at different distances from the sea showed that seeds from littoral populations temporarily inundated by seawater produced significantly higher germination in salt solutions containing 300 mM NaCl than populations located at <5 or >5 km from the sea coast (Greipsson *et al.* 1997). The inland populations lost their ability to germinate under saline conditions within a few generations, suggesting that they have adapted to low salinity in just a few decades, probably because only a few genes are involved in salt tolerance.

The growth responses of plant species exposed to different salinity levels gave variable results. Eight-week-old seedlings of lacustrine and maritime varieties of *Lathyrus maritimus* sprayed with 0, 1.5 and 3.0% NaCl solutions did not show any differences in visible damage rating index between the two varieties (Little and Maun 1993). The salt water inundation of seedlings using the same concentrations of salt showed a significant decrease in leaf area, stem and root dry weights but there were no differences between the two varieties. Similarly, increasing salinity reduced the dry mass of roots and shoots, leaf area, leaf length and number of leaves of both coastal and inland populations of *Leymus arenarius* (Greipsson and Davy 1996). However, the coastal population consistently produced more total dry biomass under high salinity conditions, primarily because of increase in root biomass and an increase in the number of tillers at 200–400 mM NaCl. The lacustrine populations of *Ammophila breviligulata* from Lake Michigan were as tolerant of salinity as the maritime populations of the Atlantic coast at North Carolina (Seneca and Cooper 1971). In contrast, Boyd and Barbour (1986) found a difference in salt spray tolerance between lacustrine and coastal varieties of *Cakile edentula* after 4 weeks of twice-daily sprays with full-strength seawater. The reproductive output of marine variety was unaffected while there was a marked depression in the lacustrine variety.

The populations of some species on the foredunes extend from the seashore to the base of the first dune ridge. Cheplick and White (2002) tested subpopulation differences, if any, in salt spray tolerance of the annual *Triplasis purpurea* but did not find any evidence of differentiation because reproductive fitness was not negatively impacted and plastic response acted as a viable buffer against selective elimination of less fit genotypes.

Salt tolerance will be a desirable trait for plant species occupying the foredunes where they are regularly exposed to salt spray and occasional seawater inundation. After their separation from the Atlantic and growing in a freshwater strand along the Great Lakes, salt spray and salinity are no longer environmental stresses. In evolutionary terms the removal of a significant stress should allow the survival of inferior genotypes in the population that would not normally be able to survive (Mayr 1977). Therefore, we will expect that the foredune species would not lose the salt-tolerance trait immediately and as long as salt tolerance does not decrease their fitness, the trait will not be lost from the population. And if the trait serves some other useful purpose in the environment or has a pleiotropic effect for another trait the salt tolerance trait will be retained mainly because it will increase fitness in the freshwater environment. Nevertheless, considering the relatively short period of time since their separation the results presented here do show an ecological divergence. Some divergence in seed germination, growth and reproduction of some species has occurred since their separation that may possibly have been caused by other selective forces such as more continental climate, wind velocities and biotic factors in the lacustrine environment. It is also possible that the concentration of seawater was lower at the time of separation of the two populations. It thus appears that derived populations may retain salt adaptations for thousands of years, possibly through plastic responses which prevent selective elimination of less fit genotypes.

8.9 Summary

Along seashores salt spray acts as a strong environmental stress on plants and other biotic organisms of coastal sand dunes. High waves also inundate part of the seashore, thus increasing the salinity of the soil. Salt spray is formed by the bursting of bubbles that eject droplets of seawater into the air which are carried inland by wind. Because of a marked relief in sand dunes, salt deposition varies at different elevations above the sand surface and different distances from the sea coast. Usually dune ridges receive higher amounts of salt spray than low-lying areas such as the slacks. Seawater is a weak nutrient solution that contains salts of Na^+, Ca^{++}, K^+, Mg^{++}, Cl^- and S with only minute quantities of all other essential elements.

Symptoms of salt spray injury to plants manifest themselves as necrosis that begins on leaf tips, progresses to upper margins of leaves and eventually develops into an inverted V-shaped pattern. Trees and shrubs on sea coasts exhibit asymmetric growth forms because the seaward twigs are killed by salt spray while leeward twigs receive very low salt spray and grow normally. Seed germination of strand species was not affected by immersion in seawater, because seeds of all species germinated after being recovered from the seawater solutions and washed in distilled water. Germination does not require salt solution although some species may exhibit the phenomenon of 'salt stimulation' after being exposed to salt water. On average the soluble salt content of coastal sandy

soils is too low (0.003–0.28%) to influence the growth of plant species because all species are capable of withstanding up to about 1% salinity in the substrate. They have been classified as salt-resistant non-halophytes; however, they differ in their tolerance levels to salt spray from highly resistant to very susceptible.

Some strand species respond to salt spray by completing their growth during periods of low salt spray storms or by making metabolic adjustments to either avoid or tolerate salt stress. Some species accumulate excess ions in the vacuoles and keep concentrations of Na^+ and Cl^- at low levels in the cytoplasm, while others avoid excess salts by limiting their entry into leaves because of thick epidermis, sunken stomata, thick cuticle, thick layers of epicuticular wax, hairy leaves, closure of stomata, and changes in leaf position. Some other species have higher relative growth rates, reproduce prolifically and have high phenotypic plasticity. Development of salt tolerance is energetically expensive and strand species have primarily evolved avoidance strategies rather than salt tolerance.

CHAPTER 9

Mycorrhizal fungi

9.1 Introduction

Mycorrhizal fungi (mycobionts) form a ubiquitous mutualistic symbiotic association with the roots of higher plants (phytobionts) in coastal sand dunes worldwide. These obligate biotrophs perform vital functions in the survival, establishment and growth of plants by playing an active role in nutrient cycling. As such they serve as a crucial link between plants, fungi and soil at the soil–root interface (Rillig and Allen 1999). Mycorrhizas occur in a wide variety of habitats and ecosystems including aquatic habitats, cold or hot deserts, temperate and tropical coastal dunes, tropical rainforests, saline soils, volcanic tephra soils, prairies and coral substrates (Klironomos and Kendrick 1993). Simon *et al.* (1993) sequenced ribosomal DNA genes from 12 species of arbuscular mycorrhizal (AM) fungi and confirmed that mycorrhizas (fungal roots) fall into three families. He estimated that they originated about 353–462 million years ago and were instrumental in facilitating the colonization of ancient plants on land. Further evidence was provided by Remy *et al.* (1994) who discovered arbuscules in an early Devonian land plant, *Aglaophyton major,* and concluded that mycorrhizal fungi were already established on land >400 million years ago. Thus the nutrient transfer mechanism of AM fungi was already in existence before the origin of roots. Plant roots probably evolved from rhizomes and AM fungi served as an important evolutionary step in the acquisition of water and mineral nutrients (Brundrett 2002).

Over evolutionary time the divergence among these fungi has accompanied the radiation of land plants, and about 200 species of AM fungi have been recognized (Klironomos and Kendrick 1993) that exist in association with about 300 000 plant species in 90% of families (Smith and Read 1997), indicating that AM fungi are capable of colonizing many host species. Approximately 150 of the described mycorrhizal species may occur in sand dunes (Koske *et al.* 2004).

Most host–fungus associations are beneficial to both the plant and the fungus and are thus regarded as mutualistic (++); however, the widespread use of the term mutualism (mutual benefit) for mycorrhizal interactions has been questioned because all associations are not beneficial to both the plant and fungus (Brundrett 2004). For example, some species such as Indian pipe (*Monotropa uniflora)* are heterotrophic and are unable to provide any carbon to the fungus (Allen 1991): Brundrett (2004) called this exploitative mutualism (+−). In other cases the fungus exploits the plant host. In fact, the current view is that the association between plants and mycorrhizal fungi ranges from balanced mutualism to parasitism depending on the genotype of plant species and fungi as well as prevailing environmental conditions (Jones and Smith 2004).

In general mycorrhizal fungi depend entirely on plants for carbon compounds that may constitute up to 10% of the energy expenditure by the plant. In return, the plant is compensated by the supply of essential nutrients and water: the benefits to the plant outweigh the increased energy expenditure. The transfer of organic compounds from the plant to the mycorrhizal fungi also enriches the soil ecosystem because some compounds of carbon eventually become available to other fungi and organisms in the below-ground food chain.

Even though mycorrhizal fungi have been known for a long time, detailed understanding of their function has begun only recently. Research has focused on nutrient acquisition, development of mycorrhizal connections, taxonomy and physiology of both ectomycorrhizal and arbuscular mycorrhizal (AM) fungi (Klironomos and Kendrick 1993). In this chapter a general description and possible role of mycorrhizal fungi in coastal sand dunes will be elucidated.

9.2 Classification and colonization process

The material in this section has been assembled from Smith and Read (1997), Mukerji *et al.* (2000), Peterson *et al.* (2004) and Brundrett (2002, 2004). Mycorrhizas have been recognized into two broad subdivisions, ectomycorrhizas and endomycorrhizas. Endomycorrhizas have been further classified into arbuscular mycorrhizas, ectendomycorrhizas, ericoid, arbutoid, monotropoid and orchid mycorrhizas (Peterson *et al.* 2004). In each case the root cells are invaded by fungal hyphae but the actual development of intracellular hyphae may differ considerably.

9.2.1 Arbuscular mycorrhizas

The most common and widespread types of mycorrhizal fungi in sand dunes are arbuscular mycorrhizas which form an association with more than 80% of plant species and develop inter- and intracellular hyphae in cortical cells (Peterson *et al.* 2004). The AM fungi as classified by Morton and Benny (1990) are presented in Table 9.1. The most prevalent AM genera in sand dunes are *Acaulospora*, *Gigaspora*, *Glomus* and *Scutellospora*. With the exceptions of some annuals and biennials on the driftline of coastal dunes all herbaceous and woody angiosperms, some gymnosperms

Table 9.1 Classification of arbuscular mycorrhizal fungi found in sand dune systems (Morton and Benny 1990)

Phylum:	Glomeromycota (Zygomycetes)	
Order:	Glomales	
Suborder:	a. Glominae	
Families:	Glomaceae	Genera: *Glomus*, *Sclerocystis*
	Acaulosporaceae	Genera: *Acaulospora*, *Entrophospora*
Suborder:	b. Gigasporinae	
Family:	Gigasporaceae	Genera: *Gigas*, *Scuttelospora*

(trees and shrubs) and pteridophytes of foredunes and fixed dunes have AM associations. Mycobionts produce asexual spores singly or aggregated into sporocarps. Spore size is about 50–600 μm in diameter and the spore wall may be about 30 μm in thickness (Kendrick 1992). During colonization the fungal hyphae encounter the root, and produce attachment structures called appressoria that develop a penetration peg. The peg then enters the epidermis or cells of root cortex (but never meristematic or endodermal cells) and forms finely branched structures called arbuscules by repeated dichotomous branching of intracellular hyphae (Fig. 9.1). Arbuscules are the major sites of nutrient and photosynthate exchange between the plant and fungus, they are short-lived and break down within about 4–15 days (Kendrick 1992). There are two main types of AM fungi (Peterson *et al.* 2004), the Arum-type (Fig. 9.1a) and the Paris-type (Fig. 9.1b). In the Paris-type extensive hyphal coiling occurs within the cortex while in the Arum-type a hypha frequently forms a coil before entering the intercellular space system of cortex; however, the growth of hyphae between cortical cells is linear (Peterson and Massicotte 2004). Many, though, not all AM fungi also produce vesicles—spherical swellings at the end of intercellular or intracellular hyphae that contain lipids and cytoplasm (Fig. 9.1). They apparently serve as storage

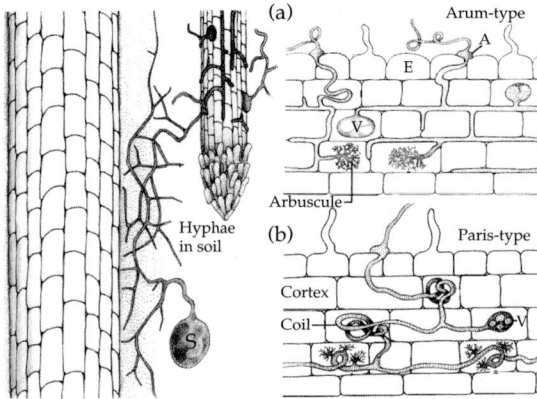

Figure 9.1 (a) Arum-type arbuscular mycorrhizal association. Hyphae penetrate the epidermal cells (E) from appressoria (A) and enter the cortex and the intercellular space. Arbuscules and vesicles (V) (depending on the species) form within inner cortical cells. (b) Paris-type arbuscular mycorrhizal association. Early events are similar to arum-type but extensive hyphal coiling occurs within the cortex, note the arbusculate coils. Root surface is colonized either by germinating spores or from previously colonized root (from Peterson *et al.* 2004 with permission).

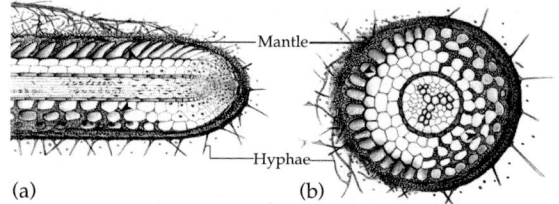

Figure 9.2 (a) Diagram of longitudinal section of root showing Hartig net (arrowheads), mantle and extra-radical mycelium of ectomycorrhizal fungi. The upper half of the diagram shows that the Hartig net is only confined to the epidermal cells in angiosperms. In conifers the Hartig net forms around both epidermal and cortical cells. (b) Diagram of a transverse section of a root showing Hartig net (arrowheads), mantle and extra-radical mycelium of ectomycorrhizal fungi in angiosperms (left half of the section), and in conifers (right half of the section) (from Peterson *et al.* 2004 with permission).

organs (Hussey and Roncadori 1982) but may also germinate and form hyphae. Because of their widespread distribution and association with primitive groups of plants, AM fungi are thought to precede ectomycorrhizas in an evolutionary sense (Brundrett 2002).

9.2.2 Ectomycorrhizal fungi

The majority of ectomycorrhizal fungi belong to families in the Basidiomycetes but a few are placed in the Ascomycetes. However, the genus *Endogone* belonging to Zygomycetes can also form ectomycorrhizas (Smith and Read 1997). Approximately 5500 species of fungi are known to form ectomycorrhizas, of which about 80% produce above-ground reproductive structures while others have their structures below-ground (Peterson *et al.* 2004). About 140 genera of dominant woody tree species, belonging to families Betulaceae, Pinaceae, Fagaceae and Salicaceae, along with a few shrub and herbaceous species, form

this association. In ectomycorrhizal fungi the hyphae first penetrate the root cortex, grow in the intercellular spaces and form a network of fungal hyphae around epidermal and cortical cells called the Hartig net, they then establish a sheath or dense mass of interwoven hyphae called the mantle around the root and finally with hyphae aggregated into rhizomorphs grow into the soil (Fig. 9.2). The Hartig net is involved in exchange of nutrients whereby the hyphae pass mineral nutrients and water to the root cells and receive sugars from the host (Peterson *et al.* 2004). The hyphae expand slowly between cells of the cortex and reach the endodermis but do not enter it. Ectomycorrhizal fungi can be grown axenically in the lab and produce morphologically distinct macroscopic fruiting bodies. In contrast, a major barrier to the study of AM fungi has been the inability to culture them *in vitro*.

9.2.3 Ectendomycorrhizas

These are similar to ectomycorrhizas in that they have a Hartig net and a mantle but form intracellular hyphae in epidermal and cortical cells just like endomycorrhizas. Nevertheless, even though they penetrate the cortical cells, they do not form arbuscules or vesicles. This mutualism mainly occurs in two conifer genera, *Pinus* and *Larix*.

9.2.4 Septate endomycorrhizas

(Arbutoid, Monotropoid, Ericoid, Orchid): Except for the Ericoid group that belongs to either Basidiomycetes or Ascomycetes, all others are classified as Basidiomycetes.

9.2.5 Ericoid mycorrhizas (*Pezizella, Oidiodendron* spp.)

They colonize plants in the families Ericaceae, Empetraceae and Epacridaceae with the exception of small number of ericaceous genera such as *Arbutus* and *Arctostaphylos*. Some ericaceous shrubs such as *Vaccinium* do occur in foredunes and stable dunes.

9.2.6 Arbutoid mycorrhizas

These fungi are restricted to plant taxa in the Ericales. They are similar in some ways to ect-endomycorrhizas in that they produce a mantle and a Hartig net but in addition produce extensive intracellular hyphal coils in the root epidermis. In sand dunes, host genera include *Arbutus*, *Arctostaphylos* and *Pyrola*. The most common plant colonized by arbutoid fungi in the foredunes along the Great Lakes is *Arctostaphylos uva-ursi*, which establishes first on the south slope of the first dune ridge and later becomes a major component of the dune vegetation in the tall grass prairie component in the transition zone between foredunes and forested areas (Maun 1993).

9.2.7 Monotropoid mycorrhizas

They colonize plants strictly in the Monotropoideae (Ericaceae). *Monotropa* plants are heterotrophic and depend completely on carbon from the fungus that forms connections with roots of neighbouring autotrophic plants (Allen 1991).

9.2.8 Orchid mycorrhizas

As the name suggests, these fungi form a mutualistic relationship only within the family Orchidaceae. The fungal species form complex intracellular hyphal coils called 'pelotons' (Peterson *et al.* 2004). The mycobionts (e.g. *Rhizoctonia*, *Ceratobasidium*) are required for the establishment of orchid seedlings and to provide carbon from other plants to the seedling in its initial stages of growth when it does not have any chlorophyll. Since the orchid appears to provide nothing to the fungus during early stages of development, the association may be described as exploitative mutualism. Later the hyphae do colonize the parenchyma cells and the nutrient exchange occurs. The only orchid species in the dune complex of the Great Lakes is *Epipactis helleborine*, but *Cypripedium calceolus* is also found in the transition zone between grassy and woody vegetation (A Maun personal observations).

9.3 Spore germination and host association in the dune complex

In temperate latitudes spores over-winter as dormant propagules and require cold stratification for high germination (Gemma and Koske 1988). In spring when the temperatures rise above 15–20°C, the germination of spores and growth of hyphae usually coincides with the initiation of root and shoot growth of plants. Spores produce germ tubes up to 10 cm or more in length that are attracted to the plant roots by volatile root exudates (Koske 1982). The degree of root colonization is influenced by the mycorrhizal inoculum potential (MIP). There is a zone of molecular communication between the fungus and plant roots in the rhizosphere and/or rhizoplane. Root exudates may trigger germination of spores, stimulate hyphal growth and facilitate entry of the fungus into the root system. On the root itself, the AM fungi are not evenly distributed; the root cap and the zone of root elongation have few endophytes while 92% of root hairs may be penetrated (Nicolson 1959).

9.4 Species richness of arbuscular mycorrhizal fungi

Mycorrhizal species richness in sand dunes varies according to the host plant species and geographical location. Arbuscular mycorrhizal (AM) fungi are diverse taxonomically, physiologically and morphologically and can colonize a diverse group of herbaceous and woody plants. Nonetheless, several species of AM fungi can associate with a single host. For example, in Dutch, Portuguese and European sand dunes several *Glomus*, *Scutellospora* and *Acaulospora* species have been found in association with *Ammophila arenaria* (Kowalchuk *et al.* 2002; Rodríguez-Echeverría and Freitas 2006). Koske and Halvorson (1981) found six species of AM fungi in association with *Ammophila breviligulata*, *Lathyrus japonicus*, *Myrica pennsylvanica* and *Solidago sempervirens*. Of those, *Gigaspora gigantea* and *Acaulospora scrobiculata* were the most abundant. The most common species in Wisconsin dunes was *Glomus etunicatum* (Koske and Tews 1987). On a latitudinal gradient along the Atlantic coast of USA some species were abundant at the cooler northern end of the range while others were common in the warmer south. At the southern

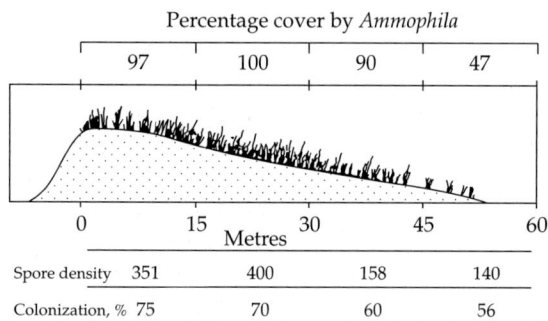

Percentage cover by *Ammophila*

| 97 | 100 | 90 | 47 |

| | | | |

| 0 | 15 | 30 | 45 | 60 |

Metres

| Spore density | 351 | 400 | 158 | 140 |
| Colonization, % | 75 | 70 | 60 | 56 |

Figure 9.3 Profile of a barrier dune at Moonstone beach, Rhode Island, showing percentage cover by *Ammophila breviligulata* and spore density 100 g^{-1} of soil and percentage root length colonized by vesicular-arbuscular (VA) fungi (from Koske and Polson 1984 with permission).

location, 6–14 species of AM fungi were associated with *Uniola paniculata* (Koske 1987). On coastal dunes of India 41 species belonging to 6 genera of AM fungi were associated with roots of *Ipomoea pes-caprae* (Beena *et al.* 2000a). Plants of high vigour have greater AM colonization potential than low vigour populations. For example, in a population of *Ammophila breviligulata* extending from the beach to 60 m inland the spore density and percentage colonization decreased with a decrease in vigour of plants (Fig. 9.3).

A partial list of species of AM fungi identified in temperate, subtropical and tropical sand dune systems around the world is presented in Table 9.2 (Dalpé 1989; Beena 2000). Much work is being done around the world on the taxonomy of mycorrhizal fungi and new species are being added to the list regularly.

The species specificity in each region is related to interactions between soil mesofauna (Koehler *et al.* 1995), plant species, soil properties and climatic conditions. For example, dunes of northern Atlantic USA contain a different AM fungal species complement than the southern Atlantic, and there was a significant positive correlation (r = 0.98, P < 0.01) in species richness as one moved south along the Atlantic coast (Koske 1987). Similarly, in the tropical and subtropical coastal sand dune systems the mycorrhizal fungi communities are different from temperate regions (Table 9.2).

On the foredunes, monospecific stands of some colonizing pioneer plant species such as *Ammophila arenaria* in Europe, *A. breviligulata* in North America and *Ipomoea pes-caprae* in the subtropics are abundant, hence root system of single species may be the only hosts available for colonization by AM fungi. The major drawback of several AM species colonizing a single host species (a common occurrence) is interspecific competition among them. Competitive differences arise because of the differential ability of AM species to penetrate and colonize the roots of the same host plant (Wilson and Tommerup 1992). The rank and abundance graph of AM species showed that

Table 9.2 A partial list of arbuscular mycorrhizal species reported from soil collections and plant species of sand dunes around the world (assembled from lists compiled by Dalpé (1989), Beena (2000), Sridhar and Beena (2001)

Glomaceae (36 species)

Glomus aggregatum (NA, Ja, Fl, Haw, Can, Ind), *G. albidum* (Can, Ind), *G. caledonium* (Can, NA), *G. claroides* (Can, Ind), *G. clarum* (NA, Ind), *G. constrictum* (NA, Br, Can, Eu, Haw, Ind), *G. convolutum* (Ind), *G. corymbiforme* (Eu, Po), *G. deserticola* (Fl), *G. dimorphicum* (Ind), *G. etunicatum* (NA, Br), *G. fasciculatum* (NA, Can, Eu, Aus, Ind), *G. fecundisporum* (NA, Ind), *G. fistulosum* (Ind), *G. gibbosum* (Eu, Po), *G. globiferum* (NA, Fl, Eu), *G. halon* (Eu), *G. intraradices* (Haw, Can, Ind), *G. lacteum* (Haw, Ind), *G. macrocarpum* (Can, Sing), *G. microaggregatum* (NA, Can, Haw, Ind) *G. microcarpum* (NA, Ind, Sing), *G. minutum* (Po), *G. monosporum* (NA, Ind), *G. mosseae* (Eu, Can, Ind), *G. nanolumen* (Haw), *G. occultum* (NA), *G. pallidum* (Ind), *G. pansihalos* (NA), *G. pubescens* (NA, Ind), *G. pustulatum* (NA, Can, Ind), *G. reticulatum* (Ind), *G. spurcum* (Haw), *G. tortuosum* (NA, Ja, Ind), *G. trimurales* (NA), *Sclerocystis rubiformis* (Can).

Acaulosporaceae (15 species)

Acaulospora elegans (NA), *A. denticulata* (Ind), *A. dilatata* (Can), *A. gadanskensis* (Eu), *A. koskei* (Eu), *A. lacunosa* (NA), *A. laevis* (NA, Eu, Aus), *A. mellea* (NA), *A. morrowiae* (Fl), *A. nicolsonii* (Can, Ind), *A. polonica* (Eu), *A. scrobiculata* (NA, Aus, Fl, Br, Haw), *A. spinosa* (Can, Ind), *A. taiwanica* (Ind), *A. trappei* (NA).

Gigasporaceae (26 species)

Gigaspora albida (NA, Fl, Br), *G. calospora* (NA), *G. decipens* (Ind), *G. gregaria* (Ind), *G. gigantea* (NA, Aus, Fl, Ind), *G. globiferum* (Can), *G. margarita* (NA, Ind), *G. rosea* (NA, Sing), *Scutellospora arenicola* (NA, Ind), *S. calospora* (NA, Eu, Aus, Ind), *S. coralloidea* (NA, Br), *S. dipapillosa* (NA, Fl), *S. erythropa* (NA, Fl, Ind), *S. fulgida* (NA, Fl), *S. gilmorei* (NA), *S. gregaria* (NA, Ja, Br, Haw, Ind), *S. hawaiiensis* (Haw), *S. heterogama* (Ita, NA), *S. minuta* (Cu), *S. nigra* (Ind), *S. pellucida* (Fl, Ind), *S. persica* (NA, Eu), *S. reticulata* (NA), *S. scutata* (Br), *S. verrucosa* (NA, Fl), *S. weresubiae* (NA, Fl, Br).

NA, North America; Eu, Europe; Ja, Japan; Aus, Australia; Fl, Florida; Br, Brazil; Can, Canada; Cu, Cuba; Haw, Hawaii; Ind, India; Ita, Italy; Sing, Singapore; Po, Poland; NZ, New Zealand.

the most abundant species were allocating more resources to sporulation than others, but no conclusions could be drawn about resource acquisition (Stürmer and Bellei 1994). In stable dunes diversity of plant species is substantially higher and roots of many more species become available for colonization. However, there is also an increase in the diversity of AM species with definite spatial and seasonal heterogeneity (Pringle and Bever 2002). Some proliferate in spring while others become more abundant in summer or autumn. In the Poaceae, there was a close association between AM fungi and the metabolic pathway of the grass species; the C_3 plants were facultatively mycotrophic while C_4 were obligate (Hetrick *et al.* 1990). Fungal diversity influences plant biodiversity, productivity and ecosystem functioning (Van der Heijden *et al.* 1998; Klironomos *et al.* 2000). Similarly, increase in plant species richness increases below-ground fungal diversity. For example, when *Oryzopsis*

hymenoides was growing in association with other grasses in an inland sand dune system there was a significant increase in mycorrhizal colonization compared to when it was growing alone (Al-Agely and Reeves 1995).

9.5 Dispersal mechanisms

Sand originating from the seabed and deposited by waves on sea coasts does not have any AM fungal inoculum. However, high waves may erode a beach and transport AM spores from there to another beach by longshore currents (Koske and Gemma 1990). Once on a beach, wind disperses spores along with sand from the lower to the upper beach and dunes. The fungi also co-disperse with vegetative fragments of rhizomes, stolons and ramets of plants in seawater or freshwater from one beach to another (Koske and Gemma 1990). Within habitat, dispersal is facilitated by birds, rabbits, rodents, ants and insects by their nesting, excretion and burrowing activities (Koehler *et al.* 1995). Ectomycorrhizal fungi produce large quantities of spores in epigeous fruiting bodies that are dispersed by wind and other vectors. Hypogeous fruiting bodies are also eaten by animals, passed through the gut unaltered and transported to new areas.

9.6 Spore density and sporulation period

The spore density of AM fungi in coastal sand dunes is low and varies in different geographic regions depending on climate, plant communities, sand movement and environmental stresses. In temperate dunes spore densities usually range between 0 to 3.4 spores g^{-1} of soil (Koske *et al.* 2004). Similar values were also reported from tropical dunes. In a stand of *Ipomoea pes-caprae* on coastal sand dunes of south India the mean number of spores of all AM species was 57 per 100 g of soil (Beena *et al.* 2000a). The most abundant AM species was *Glomus mosseae* followed by *G. albidum*, *G. lacteum* and *Gigaspora margarita*: the most frequent species in 60 samples was *Glomus albidum* (62%) followed by *G. mosseae* (53%). Subtropical regions usually contain higher number of spores g^{-1} of soil than temperate regions (Sridhar and Beena 2001). Along a Brazilian island coast, the most frequent species on all dates were *Gigaspora albida*, *Scutellospora weresubiae* and *Acaulospora scrobiculata*, with about 0–69 spores per 100g of soil for all species except *A. scrobiculata* that had 60–247 spores per 100 g soil (Stürmer and Bellei 1994). The number of spores in 10 g of soil on a coast in New South Wales, Australia, ranged from 0–110 (Koske 1975). Koske and Polson (1984) showed an increase in spore density and species richness of AM fungi with an increase in distance from the foredune to the second dune ridge (Table 9.3). In a pioneer dune population of *Uniola paniculata* in northeastern Florida the total spore density showed a negative bimodal distribution with distance

Table 9.3 The density of AM spores per 100 g of dry sandy soil and average species richness of AM fungi from root zones of plants in the foredune, first dune ridge and second dune ridge on the southern coast of New South Wales, Australia (from Koske and Polson 1984, with permission)

	Foredune	I dune ridge	II dune ridge
Distance (m)	0	20	47
Spore density 100 g^{-1}	26	51	122
Richness of AM species	1.5	2.1	2.4

from water indicating that spores were not randomly distributed (Sylvia 1986). Bare areas contained few spores with large aggregations in the vicinity of root systems. The actual density was 0–677 per 100 g of sand and the most abundant AM fungi were two species each of *Acaulospora* and *Gigaspora* and three of *Glomus*.

Spore density may fluctuate over time because the sporulation rate of different species of AM fungi varies seasonally (Stürmer and Bellei 1994; Giovannetti 1985). In Rhode Island dunes, spore formation of *Gigaspora gigantea* occurred primarily in late summer and autumn and their density reached a peak in December (Gemma and Koske 1988). Plant species richness, the host species and the age of plants also have an influence on spore abundance. Spore formation is intricately related to the phenology (periodic phenomenon) of plant species. In temperate North America, spores are usually formed in autumn when temperatures start to decline and plants start to become dormant. Drought usually induces sporulation as well. For example, in the Mediterranean region of Portugal drought and heat in July increased sporulation of AM fungi associated with *Ammophila arenaria* (Rodríguez-Echeverría *et al.* 2004). Similarly, on the south coast of India spore density was highest during summer (58–133 per 100 g) because of drought and high temperatures, and lowest (10–35 per 100 g of sand) during the post-monsoon season when plants were growing vigorously

(Beena *et al.* 2000a). Different AM species may exhibit different seasonal trends in spore production. For example, on a Massachusetts dune highest spore abundance of *Acaulospora scrobiculata* and *Gigaspora gigantea* was found in October, *Scutellospora calospora* in August and *S. pellucida* and *S. persica* in October and May (Gemma *et al.* 1989). In contrast, along the Brazilian coast spore abundance of each species showed little variation throughout the season (Stürmer and Bellei 1994).

9.7 Effect of soil biota and physical environment

Burrowing animals such as gophers, moles, rabbits, insects and soil mesofauna break up roots and disrupt connections with fungal mycelia (Koehler *et al.* 1995). Other soil animals such as phytopathogenic nematodes and springtails reduce root connections by consuming mycorrhizal fungal hyphae. However, this may be beneficial to the plant because the fungi provide an alternate source of energy to these potentially injurious organisms.

Mycorrhizal development is strongly influenced by soil conditions and the quantity of available nutrients. For example, very high levels of phosphorous and nitrogen in the soil may strongly suppress the development of connections between the host and mycobiont. During primary succession, however, sandy substrate has extremely low concentrations of N and P, so neither of these elements will limit mycorrhizal root association in coastal dune systems.

Mycorrhizal fungi decreased with soil depth because of a decline in the number of functional plant roots (Al-Agely and Reeves 1995), lower temperatures, higher moisture and lower oxygen levels. Similarly, spore abundance declined with soil depth. The majority of AM fungal spores were associated with rhizomes where the sand was almost always moist (Sridhar and Beena 2001). In temperate regions AM fungi over-winter in the soil as spores, but mycelia in the soil may also survive

over winter and remain infective in the spring (Kendrick 1992). Connections to plant roots or root fragments are not necessary for survival during winter.

Soil pH is an important factor in spore germination and AM development. A decrease in soil pH increases the concentration of H^+, Al^{+++}, Mn^{++} and decreases concentrations of basic cations. Different species of mycobionts have variable pH optima for spore germination, root colonization and growth (Hayman and Tavares 1985) and for most AM species it ranges between 5 and 8. However, AM fungi have been recorded in soils with pH as high as 9.2 and as low as 2.7 (Gupta and Mukerji 2000). Species composition of AM fungal flora was altered by high aluminium concentrations at low pH (Wang *et al.* 1985). Dunes with low pH have a different AM fungal community than those with higher pH. Very high soil pH also strongly decreases species of mycorrhizal fungi (Al-Agely and Reeves 1995).

AM fungi are also negatively affected by very low or very high soil moisture content. For example, under low soil moisture conditions growth of AM fungi is reduced, probably because of declining photosynthate production by plants. Reduced fungal activity decreases the diffusion of nutrients into the root, thus limiting its growth. High soil moisture limits AM fungal colonization by reducing the availability of oxygen (Al-Agely and Reeves 1995). Drying of soil has a significant impact on percentage colonization by AM fungi. However, different species responded differently to drought. For example, drying reduced the percentage colonization of *Glomus mosseae* and *G. etunicatum* by 37 and 11% respectively, had no effect on *G. intraradices* and increased colonization by *Acaulospora denticulata* and *Scutellospora calospora* (Klironomos *et al.* 2001). In Baja, California, the root colonization started near the soil surface with the rainy season and as the soil dried the colonization shifted to greater depths that were moist (Sigüenza *et al.* 1996).

Colonization by AM fungi is also strongly affected by a change in seasons. In Baja, California, Mexico the highest percentage colonization occurred in June for most of the six species in fixed dunes (Sigüenza *et al.* 1996). In sand dunes of Tentsmuir Point, Scotland, root colonization by *Glomus fasciculatum* showed two peaks, one in June–July and the second in September–November (Nicolson and Johnston 1979), while in the Netherlands the highest AM colonization of *Agrostis stolonifera* and *Calamagrostis epigeios* occurred during summer (Ernst *et al.* 1984). Root and rhizome samples from three plant species—*Helichrysum stoechas*, *Ammophila arenaria* and *Eryngium maritimum*—on a stable sand dune in Tuscany, Italy showed that colonization was low in summer, relatively constant in autumn and high from January to the end of May (Giovannetti 1985). In Argentina the root colonization of C_4 plants by AM fungi was high from summer to autumn and C_3 plants in spring and winter (Lugo *et al.* 2003). The quantity of fungal structures such as arbuscules, entry points and coils were more common under high temperatures, intermediate soil moisture and good solar radiation, while vesicles were more abundant under dry conditions. In the subtropics, AM fungi colonization was most abundant in early spring and in the post-monsoon period under luxuriant plant growth conditions (Beena 2000). However, AM colonization did not vary seasonally on a tropical sand dune along the Gulf of Mexico (Corkidi and Rincón 1997a).

Some plants such as *Artemisia campestris* may have allelopathic effects on mycorrhizal colonization. Aqueous extracts of roots and shoots of *A. campestris* significantly reduced colonization of *Agropyron psammophilum* and *Elymus canadensis* roots by AM fungi (Yun and Maun 1997).

Do elevated CO_2 levels have an impact on the incidence of arbuscular mycorrhizas? According to Remy *et al.* (1994) AM fungi evolved at a time when CO_2 levels were considerably higher than at present. Hence, proliferation of AM fungi would be predicted in a CO_2-enriched atmosphere. Díaz *et al.* (1993) observed a negative impact of elevated CO_2 on two non-mycorrhizal species but *Calluna vulgaris*, a strongly mycorrhizal plant of sand dunes, benefited from elevated CO_2 because mycorrhizal fungi provided a carbon sink for assimilates. Thus, shifts in plant and microbial community composition should be anticipated with an elevation of CO_2 concentration. Elevated CO_2 will probably change the plant community first and then the mycorrhizal species composition (Staddon and Fitter 1998). Rillig and Allen (1999) argued that the contribution of AM fungi to elevated CO_2 levels must be viewed not only at the individual plant level but also at the plant population, community and ecosystem levels, because AM fungi influence the system by causing: (i) homeostatic adjustment of the individual host plant, (ii) variable responses of individuals within a population, (iii) variable responses of species within a plant community, (iv) variable responses of different plant communities within an ecosystem and (v) acting as a carbon sink in the soil.

9.8 Effects of disturbance

Natural disturbance caused by drought, salinity, freezing and thawing of soil, grazing by animals, loss of vegetative cover, sand accretion and/or erosion of substrate and parasitism by microfauna may destroy hyphal networks and thereby alter fungal communities (Smith and Read 1997). Similarly, severe artificial disturbance of dunes by human trampling and removal of organic debris significantly reduced richness of AM fungal species, growth of hyphae, species colonization, spore production and spore density (Beena *et al.* 2000b). For example, in the moderately disturbed habitats dominated by *Ipomoea pescaprae* the spore density was 0.44 g^{-1} of sand, AM species richness was 3.2 species sample^{-1} and fungal colonization was 58%. In contrast, in the severely disturbed habitats, spore density was 0.05 g^{-1} of sand, species richness was 0.6 per sample and colonization was 49%. A number of animal species disturb sandy substrate and create bare areas in coastal sand dunes, thus exposing them to erosion and disrupting AM fungi. These patches may be recolonized by a new complement of plant species and AM fungi (Allen 1991). Apparently, severe disturbance lowered N and P levels as well as organic matter content of sandy substrate below the optimum required for the growth of plant roots and AM fungi (Beena *et al.* 2000b), thus reverting the plant community to an earlier stage in succession.

9.9 Benefits of AM fungi to sand dune plant species

By far the most common mycorrhizas in the sand dune complex are arbuscular mycorrhizas (AM). AM fungi are of vital importance for the colonization and proliferation of pioneer plant species of sandy beaches (Koske and Polson 1984). However, plant species differ significantly in their responses to mycorrhizal fungi. The below-ground plant mutualism by AM fungi also influences the above-ground mutualism of plants (Wolfe *et al.* 2005) by increasing pollinator visits and percentage seed set of *Chamerion angustifolium* over that of non-mycorrhizal plants. The response of plants to mycorrhizal fungi depends on three main factors: (i) compatibility of the host root with the fungus and establishment of connections with plant cells, (ii) development of extramatrical mycelium by AM that extends into a large volume of soil, and (iii) the ability of the host to accept absorbed nutrients and release photosynthates to the fungus. Dune plants benefit greatly from AM fungi by improved establishment, greater biomass accumulation, faster colonization of bare areas, improved water relations, large increase in relative growth rate, leaf area, total biomass and increased seed output (Corkidi and Rincón 1997b). In fact, some dune species will not establish

or survive in dune systems without mycorrhizal fungi. For instance, after 3 months of growth in a greenhouse *A. breviligulata* plants containing AM fungi showed 78% survival compared to only 20% (P < 0.001) in control (Koske and Polson 1984). Some of the major benefits of mycorrhizal fungi to plants are explained below.

9.9.1 Exploitation of soil

The association of mycorrhizal fungi with roots allows the plant to exploit a larger volume of soil (Hayman and Mosse 1972). The mycelial network in the soil essentially functions as an extension of the root system and may be even more effective than roots in absorbing nutrients. Hyphae can grow several centimetres from the roots into the soil, thus enlarging the absorptive surface. Each gram of sand in a Florida foredune dominated by the foredune grass, *Uniola paniculata*, contained about 12 m of all fungal hyphae g^{-1} of sand, and 592 m of hyphae cm^{-1} of colonized root (Sylvia 1986). When plants are buried by sand in foredunes the AM hyphae can also extend upwards into the new substrate long before the plant produces new roots in the recently deposited sand (Maun 2004).

9.9.2 Transfer of organic nitrogen to plant roots

The quantity of nitrogen in coastal foredunes is extremely low yet no plant species show symptoms of nitrogen deficiency. As shown in Chapter 2, nitrogen available to plants normally originates from organic detritus cast on shore, input from rainfall, fixation by leguminous plants and small amounts from salt spray. Do AM fungi contribute N to the plant by extracting it from the soil? AM fungi can absorb inorganic forms of nitrogen but can not mineralize organically bound nitrogen (Ames *et al.* 1984). However, ectomycorrhizal fungi have the ability to utilize organic forms of nitrogen and provide it to the plant

(Read 1989). The preferred mineral source of nitrogen by ectomycorrhizal fungi is ammonium, whose absorption leads to the extrusion of H^+ ions which acidifies the substrate thus leading to increased proteolytic activity of the fungal associate and release of nucleotides, nucleic acids, peptides and free amino acids into the soil (Read 1989). Abuzinadah and Read (1989a) showed that ectomycorrhizal fungi (*Hebeloma crustuliniforme, Amanita muscaria, Paxillus involutus*) utilized peptides and transferred nitrogen from these peptides to *Betula pendula*. Ericoid mycorrhizas also transfer amino acids and peptides to ericaceous host plants (Read 1989). Along with the transfer of nitrogen, experimental studies in a dune slack confirmed that ectomycorrhizal fungi associated with the root system of silver birch (*Betula pendula*) contributed up to 9% of the carbon from heterotrophic assimilation (Abuzinadah and Read 1989b). Some species of ectomycorrhizal fungi such as *Laccaria bicolor* acted as a predator of springtails in soil and transferred animal-origin nitrogen to the plant host (Klironomos and Hart 2001). Since ectomycorrhizal fungi do not occur on the strand, the pioneer herbaceous plants will not be able to benefit from them. Nevertheless, woody vegetation replacing the grassy foredune vegetation in dune slacks, and in later stages of primary succession in coastal dunes, have endo-, ecto- or ectendomycorrhizal fungi associated with their roots that will enhance nitrogen uptake.

9.9.3 Supply of phosphorous and other nutrients

The colonization of plant roots by arbuscular mycorrhizal fungi greatly increases the uptake of phosphorous, nitrogen, calcium, potassium and zinc (Gupta *et al.* 2000). Phosphorous is immobile in the soil, and external hyphae are extremely efficient in extracting and mobilizing it through biological weathering from a larger volume of soil beyond the depletion zone of roots of mycorrhizal plants. Arbuscular mycorrhizal fungi produce extracellular enzymes such as alkaline phosphatase and hydrolyse insoluble inorganic and organic phosphates, thus facilitating the uptake of phosphorous (Allen *et al.* 1981). George and colleagues (1995) quantified the uptake of nitrogen and phosphorous by AM fungi by separating the growing zones of roots and hyphae and found that many but not all fungal isolates absorbed phosphate, NH_4^+ and NO_3^- and AM fungi absorb these nutrients primarily from soluble rather than from insoluble nutrient sources. Coastal sand dunes in the tropics usually develop rather quickly into rainforest because of high rainfall, relative humidity and temperatures. These conditions also are highly conducive to rapid biodegradation and release of nutrients. Mycorrhizal fungi play a very significant role in absorbing these nutrients and transferring them to plant roots before they are lost from the soil by leaching.

9.9.4 Soil aggregation

The sand deposited by waves on a beach consists of single grains without any adhesion between them. Formation of sand aggregates occurs by (i) microbial secretion of binding agents, (ii) adhesion of sand particles into micro-aggregates and (iii) physical association of micro-aggregates and sand into larger aggregates. Mycorrhizal fungi significantly enhance the aggregation of sand particles as the weight of soil adhering to the roots was three times greater in mycorrhizal than non-mycorrhizal plants (Sutton and Sheppard 1976). Arbuscular mycorrhizal fungi played a significant role in producing water-stable aggregates by secreting copious amounts of insoluble glue-like glycoproteins called glomalin into the soil (Wright and Upadhyaya 1998; Piotrowski *et al.* 2004). In addition, AM fungi produce a massive mycelial network of hyphae that envelope sand particles (Koske *et al.* 1975). They estimated that 5–9% of the sand in the top 30 cm of soil was composed of spherical sand aggregates of about 1 cm in diameter. Thus, glomalin production by AM fungi, by-products of other micro-organisms, soil organic matter content and weaving by hyphae, algae, microflora and plant roots (Fig. 9.4) increase the aggregation of sand particles (Koehler *et al.* 1995). These aggregates are a source of major carbon input into dune habitats with low organic matter such as the coastal dune systems (Allen 1991).

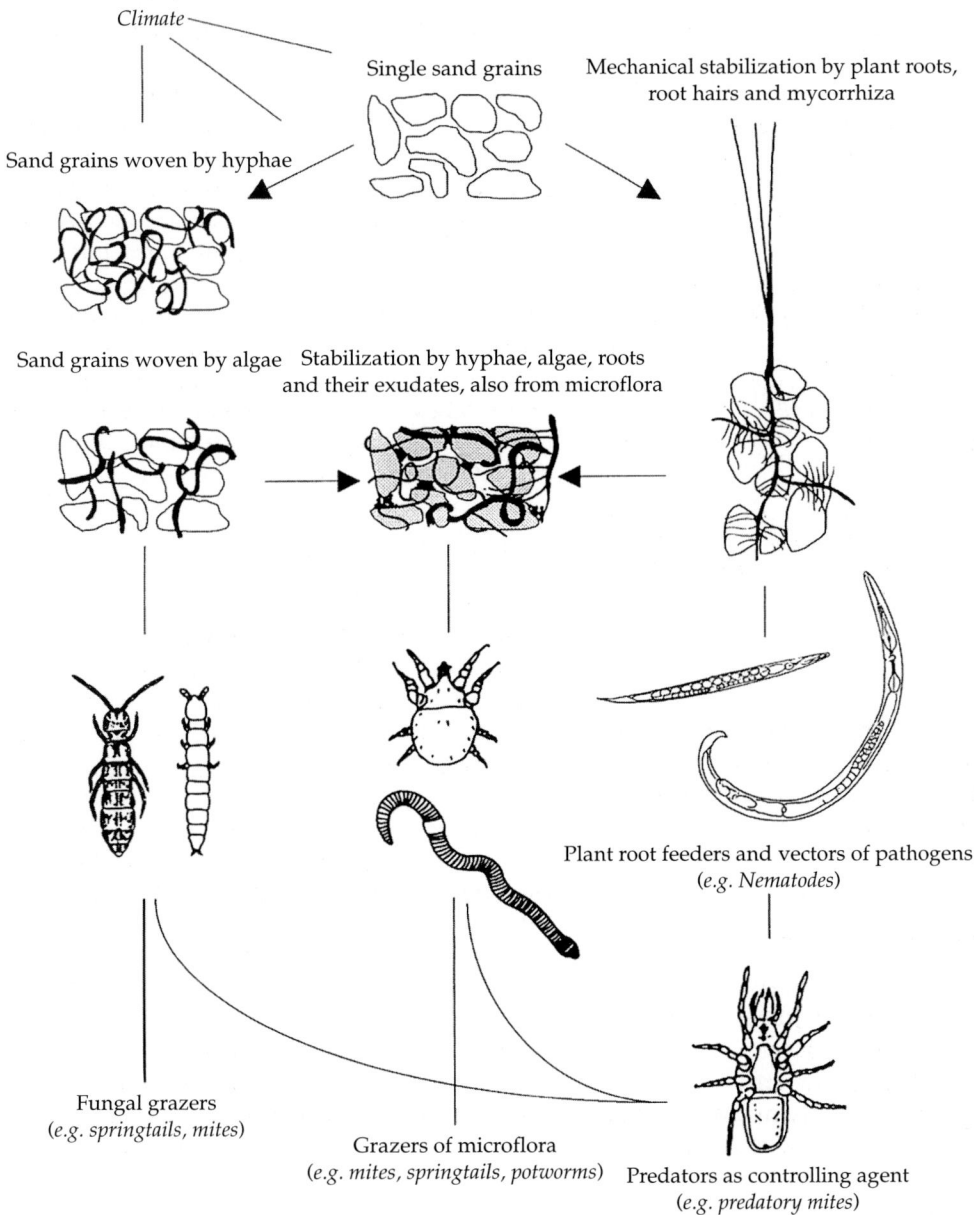

Figure 9.4 Ecological interactions between mycorrhizal fungi, algae, plant roots, herbivores, pathogens and insects leading to the formation of soil aggregates and sand stabilization in coastal dunes (from Koehler *et al.* 1995 with permission).

9.9.5 Relief from water stress

Mycorrhizal fungi increase leaf water potential, enhance transpiration rate and lower stomatal resistance. For instance, the stomata of mycorrhizal plants of *Bouteloua gracilis*, a dominant species of short grass prairies, remained open for longer periods of time than control under high light conditions but conserved water by closing stomata under low light conditions (cloudy days) when the photosynthetic activity was reduced (Allen *et al.* 1981). Mycorrhizal plants showed 35% more transpiration than non-mycorrhizal plants over a 16-hrs period (Faber *et al.* 1991). The fungal hyphae increase the hydraulic activity of roots and transport water and nutrients to the plant roots (Allen 1991). Since there are hundreds of fungal entry points per centimetre of root surface and surface area of fungal hyphae per gram of soil is very large, the total amount of water transferred to the plant is substantial. Other indirect evidence of relief from water stress is suggested by a delay in the date of flowering, reduced leaf mortality and improved leaf retention by mycorrhizal plants.

9.9.6 Protection of roots from pathogens

The soil rhizosphere acts as a zone of interaction between the plant root and soil organisms. Micro-organisms in this zone include disease-causing fungi, decomposer fungi, AM fungi, nematodes, mites, Collembola—springtails, bacteria, protozoa and large soil fauna—all of which have strong animal–animal and plant–animal interactions (Koehler *et al.* 1995). For example, some species are fungivorous, some consume plant roots while others are carnivorous. Grazing of mycorrhizal mycelium by fungivores limits growth of external mycelium and severs connections between the external and internal mycelium of the root with negative consequences for the host. The internal mycelium becomes parasitic until the external mycelium is regenerated and re-established. There is evidence that colonization by AM

fungi reduces the negative impact of pathogenic fungi by providing a physical barrier to root infection, secretion of antibiotics and utilizing surplus carbohydrates. Newsham *et al.* (1995) inoculated seedlings of *Vulpia ciliata*, an annual grass of maritime dunes of Europe and some parts of Africa and India, with a factorial combination of AM fungi and root pathogen (*Fusarium oxysporium*) before transferring them into a natural phosphorous-rich open grassland population from which both fungi had been removed. They clearly showed that AM fungi protected plant roots from pathogenic fungi. Similarly, arbuscular mycorrhizal fungi limited the activity of plant parasitic nematodes, ameliorated their detrimental effects and enhanced growth and vigour of plants buried by sand (Little and Maun 1996, 1997). In the case of endoparasitic nematodes (*Pratylenchus brachyurus*) that migrate within the cortex, AM fungi reduced their numbers by competing for space in the cortex (Hussey and Ronacadori 1982). The external mycelium of some *Glomus* species reduced the relative abundance of Collembola (Insecta) in the soil. In young coastal dune soils, mycorrhizas were a favourite food of Protura that would have a deleterious effect on dune vegetation (Koehler *et al.* 1995).

9.9.7 Transfer between plants through mycorrhizal fungal hyphae

There is evidence for the transfer of nutrients between plants by the grafting of roots, with fungi acting both as facilitators and conduits for the movement of nutrients and photosynthates. Radioactive tracer studies using mycorrhizal (source) and non-mycorrhizal (sink) plants of *Festuca ovina* and *Plantago lanceolata* showed that mycorrhizal and non-mycorrhizal plants established connections readily and nutrients were directly transferred through hyphal connections from source to sink plants (Whittingham and Read 1982; Francis and Read 1984). Miller and Allen (1992) called the improved well-being of its neighbouring plant by mycorrhizal fungi 'direct facilitation'. Its significance to a non-legume plant such as *Ammophila* may be considerable in sand dunes if it makes contact with a legume such as beach pea (*Lathyrus japonicus*). Mycorrhizal transfer may also increase the survival and longevity of plants in stressed microhabitats. For example, a plant growing close to the beach and inundated by salt water will benefit by its connection with another host through extra radical mycelium (Read *et al.* 1985). Ectomycorrhizal fungi also establish hyphae connections between and within plant species. Finlay and Read (1986) fed ^{32}P to ectomycorrhizal mycelial network interconnecting plants of *Pinus contorta* and *P. sylvestris* tree roots and traced its movement to other parts. They found that ^{32}P was readily absorbed by the mycelium and then transferred to sinks on other parts of the mycelium, the roots and the tree, suggesting that the mycelial network was a functional extension of the root system. In a dune slack ectomycorrhizal fungi of silver birch (*Betula pendula*) contributed up to 9% of the carbon from heterotrophic assimilation (Abuzinadah and Read 1989b).

9.9.8 Salinity tolerance

As mentioned earlier, the spores of AM fungi are dispersed in seawater by longshore currents, but are they capable of germination following immersion in seawater? Koske *et al.* (1996) mixed spores of *Gigaspora gigantea* with sand, sealed about 600 mL of mixture in each of eight polyethylene mesh bags and then suspended them in seawater (salinity 32‰) from a dock. Bags were retrieved after 1, 2, 3, 4, 5, 10, 15 and 20 days of immersion in seawater. The germination of spores in an incubator showed that immersion of spores in seawater for 5 days did not reduce their germination. However, even though there was a significant linear decrease in germination of spores with an increase in days of immersion, 61% of the spores still germinated after 20 days of immersion (Fig. 9.5).

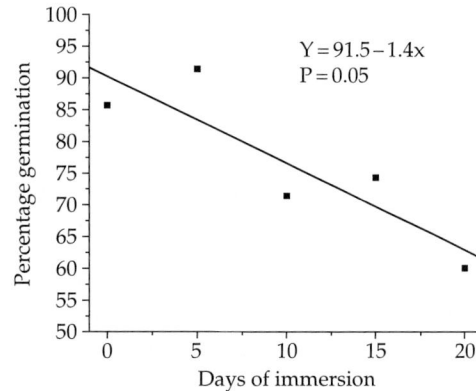

Figure 9.5 Germination of *Gigaspora gigantea* spores following immersion in seawater for a different number of days (modified from Koske *et al.* 1996 with permission).

Are plants associated with mycorrhizal fungi more resistant to salinity than non-mycorrhizal plants? Along sea coasts, *Strophostyles helvola* occurs in the mid-beach area where it is regularly inundated by seawater. Tsang and Maun (1999) periodically inundated the soil containing mycorrhizal and non-mycorrhizal plants of *S. helvola* with 0, 1, 2 and 3% sea salt solutions. There was a significant increase in chlorophyll content, shoot dry weight (Fig. 9.6) and number of root nodules (Fig. 9.7) in mycorrhizal plants exposed to salt solutions compared to control. Thus the negative impact of salt was apparently mitigated by association with AM fungi. Higher salt concentrations promoted hyphal colonization and number of nodules but decreased distribution of arbuscules. This translated into an overall benefit to the plant.

9.10 Role in sand dune succession

Do mycorrhizal fungi play a role in colonization, growth and succession of plant species in sand dunes? This question does not have an easy answer. Succession in sand dunes is essentially a change in plant species over time. With time, the soil and aerial physical environment is modified by plants, meso- and

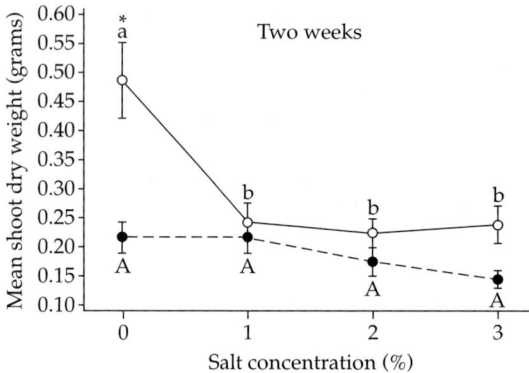

Figure 9.6 Effect of four different concentrations of salt on mean ± 1 standard error (SE) of shoot dry weight of mycorrhizal (○) and non-mycorrhizal (●) plants of *Strophostyles helvola*. Means with the same letter within each mycorrhizal treatment are not significantly different at P < 0.05. * denotes significant difference between mycorrhizal and non-mycorrhizal plants at each salt treatment as influenced by in the soil (from Tsang and Maun 1999 with permission).

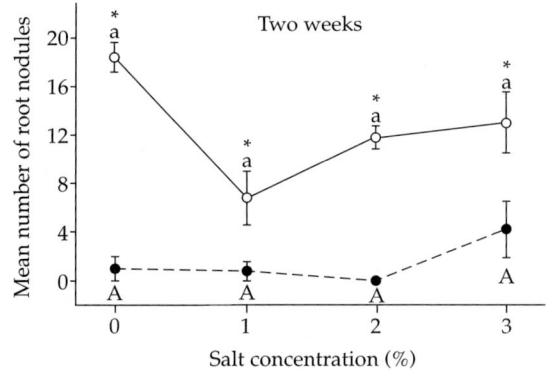

Figure 9.7 Effect of four different concentrations of salt on mean ± 1 SE number of root nodules of mycorrhizal (○) and non-mycorrhizal (●) plants of *Strophostyles helvola*. Means with the same letter within each mycorrhizal treatment are not significantly different at P < 0.05. * denotes significant difference between mycorrhizal and non-mycorrhizal plants at each salt treatment as influenced by in the soil (from Tsang and Maun 1999 with permission).

macrofauna, organic matter accumulation and many other factors (see Chapter 12). There are numerous positive, neutral or negative inter-actions and feedbacks (Wardle *et al.* 2004) within the plant community. The AM fungi play a vital role in succession of plant species in coastal sand dunes irrespective of whether a plant species is non-mycorrhizal, facultative or obligate mycorrhizal in its requirements of mycobiont. The incidence of AM colonization varies with dune succession (Nicolson 1960). The role of mycorrhiza in plant succession has been examined from two viewpoints: (i) the influence of AM fungi on plant succession and (ii) influence of plant communities and soil development on fungal succession (Koske *et al.* 2004). The change in plant species over time is well documented but no attempts have been made in natural sand dune seral commu-nities to determine the sequence or change, if any, of mycorrhizal species at different stages of dune succession. In artificial plantings of *Ammophila breviligulata*, however, there was an increase in AM species richness, spore density and soil inoculation potential over seven years (Koske and Gemma 1997). However, the oft-

repeated statement that plant succession over long periods of time is regulated in part by mycorrhizal fungi (Gemma and Koske 1997) is poorly supported. Mycorrhizal fungi pro-vide immense benefits, but only in concert with a myriad of other factors that change in the dune habitats over time. As detailed earl-ier, their major contribution would be aid in aggregation of sand particles, improvement in soil structure, provision of nutrients and other benefits, but these contributions would not constitute successional regulation of plant species. Perhaps the best term to describe the benefit will be direct 'facilitation'.

Worldwide most pioneer annual, biennial and ruderal dune species on the driftline and mid-beach areas are non-mycotrophic (Table 9.4). These species may actively reject AM fungi. For example, *Salsola kali* produces an incompatible reaction with the fungus. Even when AM fungi penetrate the roots, the invaded root tissue becomes necrotic and dies within one or two days (Allen *et al.* 1989). Many of these species belong to non-mycorrhizal families, Brassicaceae, Caryophyllaceae and Chenopodiaceae. In general, they depend

Table 9.4 Possible mycorrhizal types associated with plant species at different stages during the succession of plant species in coastal sand dune systems

Stages in succession, plant species and mycorrhizal types associated with them

Driftline (annuals, biennials, ruderals): Non-mycotrophic
Cakile edentula, Corispermum hyssopifolium, Equisetum arvense (Perumal 1994); *Salsola kali* (Allen *et al.* 1989); *Abronia maritima* (Sigüenza *et al.* 1996); *C. maritima, Beta maritima, Salsola kali, Atriplex hastata, Honkenya peploides* (Read 1989); *Aira praecox, Erophila verna, Cardamine hirsuta* (Ernst *et al.* 1984).

Foredunes, mobile and early fixed dunes: Arbuscular mycorrhizal
Ammophila breviligulata, A. arenaria, Agropyron psammophilum, Elymus canadensis, Panicum virgatum, Melilotus alba, Oenothera biennis, Tussilago farfara, Calamovilfa longifolia, Andropogon scoparius, Cirsium pitcheri (Perumal 1994); *Abronia umbellata, Camissonia californica, Haplopappus venetus, Helianthus niveus, Lotus distichus, L. bryantii* (Sigüenza *et al.* 1996); *Senecio jacobea, Eryngium maritimum, Festuca rubra, Ononis repens, Euphorbia paralias, Viola tricolor* (Read 1989); *Rosa hybrida* (Augé *et al.* 1986); *Solidago sempervirens, Lathyrus japonicus, Myrica pensylvanica, Deschampsia flexuosa* (Koske and Halvorson 1981); *Ipomoea pes-caprae, Canavalia rosea, Sporobolus virginicus* (Beena 2000); *Palafoxia lindenii, Trachypogon gouinii, Chamaecrista chamaecristoides* (Corkidi and Rincón 1997a); *Agrostis stolonifera, Calamagrostis epigeios* (Ernst *et al.* 1984).

Fixed dunes (grasses, shrubs and some trees): Arbuscular mycorrhizal or obligate
Ammophila breviligulata, A. arenaria, Agropyron psammophilum, Elymus canadensis, Panicum virgatum, Calamovilfa longifolia, Andropogon scoparius, A. gerardii, Liatris spp., *Sorghastrum nutans, Koeleria cristata, Prunus pumila, Salix interior, Cornus stolonifera, Juniperus virginiana, J. communis, Populus balsamifera, P. deltoides, Thuja occidentalis.*

Slacks (grasses and shrubby trees): Arbuscular mycorrhizal or ectomycorrhizal fungi
Salix repens, Betula pendula.

Dune heath: Arbuscular, arbutoid or ericoid mycorrhizal fungi
Arctostaphylos uva-ursi, Calluna vulgaris.

Dune woods: Ectomycorrhizal, ectendomycorrhizal, orchid or monotropoid mycorrhizal fungi
Quercus velutina, Q. rubra, Q. muehlenbergii, Q. prinoides, Q. alba, Pinus resinosa, Pinus strobus, Vitis riparia, Hamamelis virginiana, Prunus virginiana, P. serotina, Amelanchier stolonifera, Ceanothus americanus, J. virginiana, Rhus aromatica, Fagus grandifolia, Acer spp.

primarily on nutrient-rich detritus washed up on the beach and aerosol spray from the ocean. However, a few annuals and biennials of the mid-beach and driftline areas are mycorrhizal. Additionally, the non-mycorrhizal status of some plant species needs to be carefully re-examined (Allen and Allen 1990).

The perennial dune-building grasses, vines, shrubs and trees in the incipient mobile foredunes above the high tide line are facultatively mycorrhizal or mycorrhizal (Table 9.4). The percentage colonization by AM fungi of *Ammophila arenaria* (Nicolson 1960) and *A. breviligulata* (Koske and Polson 1984) was low in this zone because of the mobility of the substrate and continuous burial and erosion (Table 9.5). Burial, however, creates only a

temporary decline until new roots develop in the newly deposited sand and fungal hyphae expand upwards from lower layers (Perumal and Maun 1999). Nevertheless, as the substrate becomes more stable with increasing distance from the beach, mycorrhizal activity that promotes establishment, root formation, production of tillers and inflorescences increases sharply (Gemma and Koske 1997). However, significant hyphal networks are only established after about six years. In artificial dune rehabilitation programmes this process can be hastened by inoculation of planting stock.

What role do mycorrhizal fungi play during later stages of dune succession? Several studies indicate that further stability of the substrate and increased diversity of vegetation

Table 9.5 Percentage of colonization and external mycelium in *Ammophila arenaria* in mobile foredunes and fixed dunes at Southport, England (modified from Nicolson 1960)

	Colonization (%)	External mycelium (%)
Mobile foredune	1	0.5
Fixed yellow dune		
(Site 1)	23	19
(Site 2)	22	35
Older fixed dune	13	10

with an increase in dune age were positively correlated with soil spore density, AM colonization and AM species richness. Some plant species that have an obligate requirement for colonization by mycorrhizal fungi would be able to establish in this community only if the required fungi are present or if the inoculum abundance is above the critical level. The required species must have a mechanism to survive at the site until the obligate host establishes in the community. The AM species may persist as spores in the sand, or may be a part of a residual fungal community adapted to live with early successional plant species. For example, do all of the 41 species of AM fungi associated with the root system of *Ipomoea pes-caprae* have a mutualistic positive relationship with the host (Beena *et al.* 2000a)? Probably some contribute significantly to the welfare of the host while others do not. Thus an AM fungus may not require a specific host, even though some plant species may require a mycorrhiza belonging to a particular group: nevertheless many AM species do associate themselves with a single host (Allen 1991).

The fungal plant association may change from a facultative to non-obligate mycorrhizal type in foredunes to obligate type in later stages of sand dune succession (Allen and Allen 1990). At the same time, there is a change from the arbuscular mycorrhizal fungi in foredunes to the ectomycorrhizal, arbutoid or ericoid mycorrhizal type in the later stages

of the sand dune chronosequence (Allen and Allen 1990; Read 1989). This is associated with the transition from predominately herbaceous to predominately woody vegetation.

Some pioneer tree species such as *Populus balsamifera, P. deltoides, Juniperus virginiana, J. communis, Prunus pumila, P. virginiana, Salix interior, S. amygdaloides, Cornus stolonifera* and *Thuja occidentalis* are first facultatively mycorrhizal and then are colonized by arbuscular mycorrhizal fungi (Table 9.4). *Salix* and *Populus* species can be colonized by both AM and ectomycorrhizal fungi. In the later stages of chronosequence, the woody species of *Quercus, Pinus, Fagus* and others have a similar complement of mycorrhizal fungi as in other forest ecosystems. In a tropical sand dune system along the Gulf of Mexico, 97% of the root samples collected from 37 plant species along a successional gradient from mobile to stable dunes were mycorrhizal (Corkidi and Rincón 1997a). Over the successional sequence in artificially planted populations of *Ammophila breviligulata* there was an increase in root colonization and the mycorrhizal inoculation potential (MIP) of the soil (Koske and Gemma 1997). The presence of mycorrhizal fungi and their importance to the overall health, maintenance and succession of plant communities at different locations is well documented, but we do not know the actual composition of mycorrhizal species at different stages of successional sequence.

9.11 Summary

Plants emerged from water about 400 million years ago. The successful colonization of plants on land following emergence from water was made possible by an intimate alliance with fungi that had emerged from water earlier and were well adapted to explore the soil for essential nutrients. However, they desperately needed a source of carbon for their energy requirements. The availability of plants on land and release of carbon and amino acids from their bases attracted fungi and provided them with an opportunity to develop both harmful and mutually beneficial relationships. Evidence of this relationship is found in fossils of Devonian plants that contain well-preserved fungal structures similar to those of present day healthy plant species. The beneficial relationship with fungi named as mycorrhiza (fungus root) not only provides plants with much needed nutrients and water, but also protects them from pathogens and increases their tolerance to water stress and salinity.

Two main kinds of mycorrhizal fungi, endotrophic and ectotrophic, have been recognized. They colonize more than 90% of plant species. Endomycorrhizas enter the root cortex and contain both septate and aseptate fungi, while ectomycorrhizas contain almost all septate fungi and form a Hartig net and mantle around the root and between the cortical cells but do not penetrate the cells. Endomycorrhizas are very common and are found in association with approximately 300 000 herbaceous and woody plant species. Ectomycorrhizas colonize more than 2000 species of plants that include the most important woody plants belonging to the families, Betulaceae, Pinaceae, Fagaceae and Myrtaceae. The most common mycorrhizal fungi in the sand dunes complex are arbuscular mycorrhizas. More than 100 species of AM fungi have been isolated from coastal sand dune systems. The fungi are diverse taxonomically and many species are able to colonize the same host.

In temperate latitudes the spores survive over winter as dormant propagules and undergo the required cold stratification to break the dormancy. They germinate in spring and their germ tubes are attracted to plant roots by volatile root exudates and an intimate connection between the root and fungus is established. In order to be of mutual benefit to both the plant and fungus the host roots must be compatible with the fungus, the fungus must be able to exploit a large volume of soil and there must be an effective mechanism for the exchange of absorbed soil nutrients to the host roots and carbon compounds to the fungus.

Spore density of AM fungi in dune soils ranges from 0–11 spores per gram of soil, fluctuates over time and is intricately connected to the phenology of plants. Many animals disrupt host–fungus connections by their burrowing activities. Soil organisms such as nematodes and springtails consume AM fungal hyphae thus reducing their populations. AM fungi are also negatively affected by adverse environmental variables, e.g. low or high temperatures, moisture, or pH, low oxygen levels and drought conditions. Soil disturbance also has a major impact on hyphal networks, spore density, species richness and colonization rates. Dispersal of mycorrhizal fungi occurs by diverse means. Major agents of dispersal are wind, water, animals, and fragments of rhizomes and stolons. The spores can disperse in seawater without losing viability. Animals disperse spores by their nesting and burrowing activities and ingesting but not digesting spores.

Major benefits of mycorrhizal fungi to the host are more effective exploitation of soil nutrients, transfer of organic nitrogen, phosphorous and other nutrients, soil aggregation by hyphae and glomalin production, relief from water stress, protection from pathogenic fungi and other harmful organisms, increased tolerance to salinity and salt spray and transfer of carbon from one plant to another through mycorrhizal fungal hyphae connections.

Plant succession in coastal sand dunes is a process of change over time in plant species composition that is mediated by a change in stability of substrate, organic matter, nutrient status, and micro-organism complement of the soil. Mycorrhizal fungi play a vital role in facilitating the establishment, growth and reproduction of plant species. The pioneer annuals that establish on the beach and driftlines of sea coasts are non-mycotrophic and obtain their nutrients from nutrient-rich detritus deposited by high waves in spring.

The upper beach and foredunes are occupied by perennial dune-building grasses such as *Ammophila arenaria* and *A. breviligulata* in temperate regions and *Ipomoea pes-caprae* and *Spinifex* species in the subtropics and tropics. These species are either facultative, mycorrhizal, or mycotrophic, and are responsible for stabilizing the substrate. Greater stability and soil development allows more plant species to establish with a corresponding increase in fungal density and diversity. In later stages of succession, graminoids and shrubs of open highlight dunes are replaced by woody species such as oaks and pines that are obligately mycorrhizal in their requirements for establishment, growth and reproduction. The mycorrhizal type also changes from primarily AM to ectomycorrhizal fungi; however, the actual change in fungal species composition has not yet been determined.

CHAPTER 10

Animal–plant interactions

10.1 Introduction

Population dynamics of plant species of coastal sand dunes is influenced directly, both above and below the soil surface, by a wide variety of organisms. Plants serve as sources of carbon and pathogens including viruses, insects, bacteria, fungi, birds, and mammals of various kinds. Some enhance plant performance while others have deleterious effects. Positive interactions include pollination of flowers by useful insects in return for nectar and pollen, nutrient acquisition from soil by mycorrhizal fungi in exchange for carbon and acquiring nitrogen (N) from N-fixing bacteria.

In the history of co-evolution between plants and organisms over one hundred million years plants have developed many mechanisms to defend themselves from pathogens. Morphology may be altered by producing epicuticular waxes, developing trichomes over leaves, producing tough leaves with deposition of celluloses, lignin, suberin and callose, developing thorns on stems and branches or producing secondary plant metabolites that retard development, intoxicate or kill herbivorous insects. Herbivory may induce a plant to produce chemicals that signal to advertise the presence of insects feeding on them and attract parasites to reduce their numbers. Phenological escape is also employed, such as delay of leaf expansion during periods of insect abundance. Some indirect mechanisms of plant defence involve the use of insects such as ants for protection from other phytophagous insects. However, the predators have also evolved the ability to break down the defence mechanisms of the plant. For example, they

may use phytochemicals for their own defence or as olfactory clues for feeding. In this chapter a brief account of organisms of the coastal dune communities, including species of the intertidal zone, scavengers of the sea coast, reptiles, birds, insects, mammals and their possible interactions with terrestrial vegetation is presented.

10.2 Life in the intertidal zone

For biological organisms of the seashore the intertidal zone is the most important for food and shelter. The sand-dwelling species of the seashore must be able to contend with four limiting factors: (i) rush of water from the approaching or receding high tide and pounding breakers, (ii) low salinity of the top surface of sand (iii) desiccation of surface by high winds and sunshine and (iv) extreme changes in temperature of topsoil. The species composition of beach communities depends on the energy contained in the waves, beach type (reflective or dissipative) and physical characters of the sandy substrate. McLachlan (1990) showed that the species diversity, relative abundance, biomass and individual biomass of organisms on sandy beaches was closely related to physical environmental variables: wave height, sand texture, beach slope, sand moisture and availability of food. From the morphodynamic viewpoint dissipative beaches with flat slopes, wide surf zones, fine sand, heavy wave action and longer swash periods support richer fauna than reflective beaches with steep slopes, no surf zone, coarse sand, low wave energy and low swash period. For example, on reflective beaches waves with 10 sec periods will have a swash every 10 sec while the most dissipative beaches will have a swash every 40–60 sec. The longer the swashes the greater are the opportunities for organisms to exploit the beach and feed themselves. According to McLachlan (1990) the fauna not only exhibited a zonation pattern between species but also within populations of single species according to their environmental tolerances. On the Moolach beach along the Oregon coast he recognized four zones: the drying zone (area from the driftline to the dunes), retention zone (mid-littoral area below the driftline), resurgence zone (centred around the low tide line) and saturation zone (constitutes the low tide swash zone and glassy layer). The boundary between these zones is not sharp, and the day-to-day positions may change according to tidal cycles, inundation and exposure times, rainfall, groundwater flow and faunal responses. The intertidal region provides two main habitats for the fauna: the interstitial spaces between and on sand particles provide a home for the meiofauna, while the surface and upper layers of sand are primarily occupied by macrofauna.

10.2.1 Meiofauna

Large numbers of microscopic organisms such as algae, diatoms, bacteria, fungi, protozoa and metazoa living in interstitial places between sand particles or on sand grains include cyanobacteria such as *Beggiatoa*, *Oscillatoria* and others that produce mucilaginous mats and fix nitrogen. Many bacteria are important in decomposition of organic matter aerobically. For decomposition, the bacteria require oxygen which if not available is acquired indirectly from nitrate (NO_3), sulphate (SO_4), carbon dioxide (CO_2) or bicarbonate (HCO_3). Primary production in the sediments is usually from diatoms, such as *Nitzschia* and *Hantzschia*, and from dinoflagellates. Many of these species are motile and may undergo vertical migrations mediated by day length, wetting and mechanical disturbance by tidal surges. Diatoms

tend to dominate in finer sediments while harpacticoid copepods are more characteristic of coarser sediments (McLachlan *et al.* 1981). Usually as the sand grains become finer the microbial diversity increases. Meiofauna consume a variety of food items such as bacteria, protozoa, diatoms, detritus and dissolved organics. Heterotrophic flagellates and ciliates (*Trachelocera* and *Ramanella*) occur in mud and sand, where they feed on bacteria, diatoms and dead animals.

10.2.2 Macrofauna

Since sandy seashores do not provide a firm base for attachment they do not support seaweeds, barnacles, hydroids, sponges, polyzoans or sea squirts. The force of pounding breakers is another factor limiting survival in this habitat. However, three principal groups—polychaetes, crustaceans and molluscs—typically occupy this habitat. According to McLachlan (1983) the most exposed beaches are dominated by crustaceans, the most sheltered by polychaetes and the intermediates by molluscs (Fig. 10.1). Crustaceans are more abundant in tropical beaches or more exposed beaches while molluscs are more abundant in temperate or less exposed beaches. Polychaetes burrow deeper in sand where salinity is higher, temperature is lower and sand is always

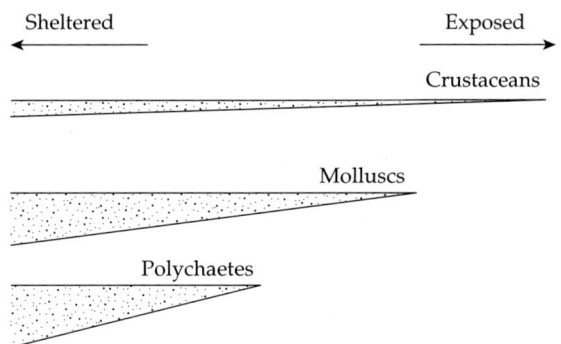

Figure 10.1 Effect of exposure gradient on the occurrence of major groups of macrofauna on sandy beaches (modified from McLachlan 1990).

wet. At high tide the barnacles, mussels and tube worms of rocky shores consume suspended material largely consisting of phytoplankton. Tube worms, e.g. *Lanice conchilega* and *Pectinaria belgica*, feed on organic matter deposited on the sand surface when the tide is in. Carnivorous worms such as *Nephtys hombergi*, *N. californiensis* and *Glycera* spp. consume other organisms that burrow in sand while lug worm (*Arenicola marina*) and ribbon, blood, segmented, ciliated and peanut ingest sand and utilize organic matter contained in it.

Among the Crustaceans (shrimp, barnacles, isopods, amphipods, protozoans, crabs, mysids, lobsters) the most common copepods on sandy shores are isopods, amphipods and decapods. On sandy beaches the most common amphipods are the sand hoppers, *Talitrus saltator* and *Orchestia agilis*, that occur in immense numbers along the driftline where they burrow in sand under plants and detritus. When disturbed they may jump several decimetres, the mass arising like a cloud from the sand surface.

Sandy beaches in the tropics are also home to many species of semi-terrestrial crabs. Several species belonging to the genera—*Ocypode*, *Uca*, *Scopimera*, *Dotilla*, *Coenobita* and *Virgo*—are adapted to live an amphibious lifestyle (Fishelson 1983) and go into the ocean only to spawn. Some in the south Pacific live on the seashore and climb coconut trees. Ghost crabs (*Ocypode albicans*, *O. arenaria*) primarily consume carrion washed onshore by high waves and storms and are also predators of mole crabs and coquina clams.

While the molluscs, clams, mussels, qua hogs, sea slugs, squids, octopuses, scallops, chitons and snails are successful in the sea, only clams, mussels and qua hogs are exposed when the tide goes out. Cockles (*Cardium* spp.) and clams are essentially filter feeders while many other bivalves such as *Tellina tenuis*, *Donax serra*, *D. vittatus* and *Gari depressa* are deposit feeders. *Donax* species are predominantly tropical in their distribution and only a few species occur in temperate regions (Ansell

1983). Sand burrowing snails are most common along tropical shores. In temperate regions, some species of *Natica* (*N. alderi* in UK) and *Polinices* burrow into sand using their powerful feet. They are carnivorous and may feed on small bivalves such as *Tellina* spp.

The organisms of the intertidal zone provide food to shorebirds and four principle classes of avifauna have been recognized in the intertidal zone of South Africa on the basis of feeding behaviour: the African black oystercatcher, curlew sand piper and whimbrel are *deep probers*; the grey plover and knot are *medium probers*; sanderling are *shallow probers*; and the turnstone, ringed plover, white-fronted plover and cape wagtail are *surface peckers* (Hockey *et al.* 1983). Gulls, oystercatchers and sanderlings prey on *Donax* species (Ansell 1983). Many of the migratory shorebirds depend on molluscs to replenish energy resources during travel to their southern destinations.

Carrion of different types such as dead bodies of fish, birds and insects wash up on the beach driftline, especially during spring and autumn. The scavengers include raccoons, opossums, turkey vulture (*Cathartes aura*), bald eagle (*Haliaeetus leucocephalus*), grackles, herring and ring-billed gulls (*Larus argentatus* and *L. delawarensis*), crabs, hyena, foxes, jackals, many other animals all around the world's sea coasts and also many invertebrates such as blow flies and rove beetles.

10.3 Reptiles

Reptiles including turtles, geckos, lizards, iguanas, skinks, chameleons and snakes are found in sand dunes along sea coasts. Except for turtles, the impact of other species on plants is minimal. Seven species of sea turtles in oceans of the world spend their lives at sea but use coastal sandy beaches as nesting sites (Johnson and Barbour 1990; Bustard 1973). Their diet is slightly variable. For example, the green turtle (*Chelonia mydas*) is primarily vegetarian consisting of algae and angiosperms, the loggerhead turtle (*Caretta caretta*) is carnivorous and feeds on crabs and other crustaceans; hawksbill turtles (*Eretmochelys imbricata*) are omnivorous and as adults consume molluscs, crustaceans and algae, and the leatherback turtle (*Dermochelys coriacea*) feeds on shellfish, sea urchins, squid, crustaceans, fish, algae and floating sea weeds. The favourite food of kemps ridley (*Lepidochelys kempi*) and olive ridley turtles (*Lepidochelys olivacea*) is crabs, snails, clams and various marine plants. Since they live in water their impact on terrestrial vegetation is minimal. Only the female turtles disrupt beach vegetation by their digging activities which are associated with nesting onshore, in the course of which they disturb roots and rhizomes of plants.

10.4 Birds and their impact on dunes and vegetation

Many bird species, including herring gull (*Larus argentatus*), ring-billed gull (*L. delawarensis*), eastern kingbird (*Tyrannus tyrannus*), bank swallow (*Riparia riparia*), barn swallow (*Hirundo rustica*), brown thrasher (*Toxostoma rufum*) and mourning warbler (*Oporornis philadelphia* Wilson) nest in the open dunes. Other breeding birds include many species of terns (Family Laridae); common (*Sterna hirundo*), sandwich (*S. sandvicensis*), arctic (*S. paradisaea*), roseate (*S. dougallii*), sooty (*S. fuscata*), least (*S. antillarum*), royal (*S. maxima*), Caspian (*S. caspia*) and gull-billed (*S. nilotica*). Sooty, least and royal terns are birds of the tropics and subtropics while roseate tern has a worldwide distribution. Gulls generally nest on rocky ledges or inaccessible places but may also nest on dredge spoils, open tundra, gravel bars and sandy beaches, with easy access to food. The black-headed gull (*Larus ridibundus*) nests in the grasses on sand dunes. Many species of plovers—ringed (*Charadrius hiaticula*), semi-palmated (*C. semipalmatus*), piping (*C. melodus*), snowy (*C. alexandrinus*), Wilson's (*C. wilsonia*), Eurasian golden (*Pluvialis apricaria*) and American golden (*P. dominica*)—use

thickets, slacks and sand dunes of coastal beaches for nesting and raising their young. The eggs of shore birds are eaten by raccoons, weasels, birds of prey, foxes and other predators (Cameron 1998).

Birds have both beneficial and deleterious impacts on vegetation. One of the most useful influences is habitat enrichment by guano (Gillham 1956). They also incorporate remains of prey, feathers, eggshells and carcasses into the soil near their nests. Guano adds three macronutrients—N, P and K—but has less water-holding capacity than organic matter and decreases soil pH which is neutralized by salt spray. Ellis (2005) reviewed studies detailing the impact of colonial nesting seabirds on plant biomass, richness of plant species and plant community composition. The plant biomass increased 11.8-fold in wet years but in dry years it decreased, probably because large amounts of guano created toxic conditions for plant growth. Plant species richness varied according to density in areas influenced by sea birds. Under high density richness decreased but under lower density conditions birds introduced seeds of alien species, thus increasing species richness.

Negative impacts include trampling of vegetation, uprooting of plants while burrowing, leaf removal and reducing photosynthetic activity because of coverage with guano. Continuous disturbance of the substrate by seabirds reduces seedling establishment and increases damage to adult plants. Usually woody species are replaced by perennial and biennial herbs, thus returning the plant communities to an earlier stage of plant succession. Burrowing and excavation of tunnels by bank (*Riparia riparia*) and rough-winged (*Stelgidopteryx ruficollis*) swallows increase soil erosion, decrease water retention and decrease native woody and herbaceous perennials. Breeding populations nest in colonies and destroy vegetation in local areas. The eider duck (*Somateria mollissima*) builds nests among grasses on dunes close to the shoreline while shelducks (*Tadorna tadorna*) nest in rabbit burrows along the sea coasts in many areas of the dunes. Their impact on vegetation is usually minimal.

10.5 Insects and their impact on vegetation

Shelford (1977) found many species of invertebrates: the white tiger beetle (*Cicindela lepida*), bronze tiger beetle (*C. scutellaris*), sand spider (*Trochosa cinerea*), burrowing spider (*Geolycosa pikei*), white grasshopper (*Trimerotropis maritime*), long-horned locust (*Psinidia fenestralis*), sand locusts (*Ageneotettix* spp., *Spharagemon* spp.), migratory locusts (*Melanoplus* spp.), digger wasps (*Microbembex monodonta*) and ants (*Lasius* spp.) in the open dune, grassland-heath vegetation and dry forest in the Chicago region (Shelford 1977). Some ant species tend aphids by storing and guarding their eggs over winter, hatching them and then placing them on preferred food plants. Ant lions (*Cryptoleon* spp.) are common in coastal sand dunes. They build conical pits in sand and wait at the bottom for an unsuspecting prey to fall into it.

The mesophytic forest also contains numerous species of invertebrates: most characteristic (Shelford 1977) are the green tiger beetle (*Cicindela sexguttata*), millipedes of genera *Fontaria* and *Spirobolus*, centipedes of genera *Lithobius* and *Geophilus*, betsy beetles (*Passalus* spp.), woodroaches (*Blattidae* spp.) and wood snails (*Cepaea* spp.).

A tally of insects grazing on *Cakile edentula* showed that the plants were palatable to a large number of both generalist and specialist species. Cutworms (*Euxoa* spp.; Family Noctuidae; Order Lepidoptera) fed on roots and cut off stems either at or below the sand surface. In one study about 9% of the seedlings were lost in 1979 (Payne and Maun 1984). Cabbage butterfly (*Pieris rapae*; Family Pieridae; Order Lepidoptera) larvae consumed leaves of seedlings and killed them. Banded-wing grasshoppers (*Trimeroptropis verraculatus*; Family Acrididae; Order Orthoptera), chewed large parts of leaves thus decreasing the leaf

area and photosynthesis. Flea beetles (*Phyllotreta cruciferae* and *P. striolata*; Family Chrysomelidae; Order Coleoptera) were also very serious pests of this species. The insects rasped small pits in the foliage, chlorophyll was lost and seedlings died. The diamond-back moth (*Acrolepia xylostella*; Family Plutellidae; Order Lepidoptera) caused minor damage by eating from the underside of leaves and producing many small holes. Snout beetles (*Hypera postica* and *H. nigrirostris*) were particularly abundant on cold windy days and fed on cotyledons of young seedlings in early spring. Aphids (*Hyadaphis erysimis*; Family Aphididae; Order Homoptera) infested about 37% of *Cakile* plants on the beach in July of each year and caused curling of leaves and stunting of stems. Ladybird beetles did exert some control on aphid populations but their numbers were small.

Many insects have been reported to damage *Ammophila breviligulata* (Maun and Baye 1989). Aphids infected the apical meristems and a beetle (*Strigoderma arboricola*) entered the open florets and consumed both the male and female parts of the flower and developing seeds. A soft scale insect (*Eriococcus caroliniae*), was a serious pest that consumed leaves and sheaths of plants. Caterpillars of the family Danaidae consumed all parts of *A. breviligulata* including the glumes, lemma and palea. Other insects of minor importance consisted of a gall midge (*Mayetiola ammophilae*), a moth (*Chenoriodes arenella*), a bug (*Phyllophaga rugosa*), banded-wing grasshopper (*T. verraculatus*) and nematodes. In *Lithospermum caroliniense* the insect damage was primarily caused by larvae of *Ethmia longimaculella* that constructed webs around the cymes of stems and consumed leaves and flowers inside the webs (Westelaken and Maun 1985a). Approximately 8–17% of the flowers per plant on the first dune ridge at the Pinery were consumed. Tent caterpillars (*Malacosoma americanum*) are a common sight on *Prunus virginiana* in spring of each year and may completely defoliate shrubs.

10.6 Fungi and their impact on vegetation

Many fungi infect sand dune vegetation. *Ammophila breviligulata* is susceptible to fungi such as Marasmius blight (*Marasmiellius mesosporus*). Ergot (*Claviceps purpurea*), a fungus of the family Poaceae, replaced seeds in about 2% of the florets on a panicle (Krajnyk and Maun 1981b, 1982). Other fungi of minor importance consist of rusts (*Puccinia* spp.) and a smut (*Ustillago striiformis*). *Prunus virginiana* is also infested by black knot fungus (*Apiosporina morbosa*) that eventually kills branches of the shrub. Seedlings are highly susceptible to diseases. In a population of *Cakile edentula*, damping off fungi (*Pythium* spp.) were the first to infect a certain proportion of seedlings soon after their emergence and caused chlorosis of leaves, blackening of hypocotyl and taproot and eventually killed them. As the summer temperatures increased, the plants were infested by *Albugo candida*, an obligate parasite of the family Cruciferae. This rust produced white blisters on leaves which ruptured and released spores, thus spreading the disease to other seedlings, but it did not kill them. The fungus also caused root rot and infected seeds in the seed bank. Deeply buried seeds are particularly vulnerable to fungal infection because of their high moisture environment.

10.7 Mammalian herbivores and their impact on vegetation

Common mammalian herbivores found in dune habitats are the chipmunk (*Tamias striatus*), white-footed mouse (*Peromyscus leucopus noveboracensis*), deer mouse (*Peromyscus maniculatus bairdii*), meadow vole (*Microtus pennsylvanicus*), red squirrel (*Tamiasciurus hudsonicus*), grey and black squirrels (*Sciurus carolinensis*), rabbits and white-tailed deer (*Odocoileus virginianus*).

They have a severe impact on plant communities. Grazing of dunes by cattle, horses and sheep has had obvious consequences of destruction of vegetation and destabilization of the environment. Each species has its own food preferences. For example, horses and cattle prefer grasses, deer prefer woody plants and rabbits prefer shrubs, grasses and forbs. For each class of herbivore, plants may be classified from very palatable to unpalatable. Up to a certain level, herbivory is beneficial to the plant community because regular removal of leaves does not allow the build-up of old leaves and prevents plant senescence. However, when the population of grazing animals exceeds the carrying capacity of the plant community, palatable species are decimated and the ecological balance between plants is upset. Unpalatable species increase in coverage because of reduced competition and alien species invade the ecosystem and alter species composition.

Small native mammals (*Peromyscus* spp.) consume seedlings of *Cakile maritima* (Boyd 1988) and seeds of *Uniola paniculata* (Wagner 1964). A large proportion (8–29%) of *L. caroliniense* seedlings on Lake Michigan foredunes was killed by the tunnelling activity of moles who severed developing roots (Weller 1989). The impact of two major naturally occurring herbivores is presented below.

10.7.1 Rabbits

Gillham (1963) showed that rabbits are important agents of dispersal of some species on beaches of some South African islands and have a significant impact on dune vegetation because of their habit of selective grazing of palatable species. Prior to the increase in populations of rabbits a *Zygophyllum morgsana* climax community was maintained and used by cormorants for building nests in its canopy. Since *Z. morgsana* shrubs were palatable to rabbits its populations declined while that of unpalatable species, *Helichrysum ericaefolium*, increased. The altered habitat allowed *Oxalis pes-caprae*, *Urtica urens*, *Australina*, *Anthericum* and other species to invade the plant community. The plant community was so greatly altered that it could no longer be used by cormorants for nesting: thus over-grazing created a new climax community maintained only by heavy gazing pressure from rabbits.

The removal of grazing pressure also alters dune communities. Ranwell (1960) documented changes in sand dune vegetation of Newborough Warren, Anglesey, after the decimation of rabbit populations by myxomatosis in 1954. The component plant species either increased or decreased in frequency. Some plant species—*Geranium molle, Cerastium tetrandrum, Veronica polita, Plantago maritima, Leontodon leysseri, Carex pulicaris, Viola tricolor* ssp. *curtisii* and *Rhodobryum rosseum*—disappeared from the community. New species also established (Table 10.1). A later study by Hodgkin (1984) indicated that release of dune habitat from grazing pressure led to a significant expansion of *Crataegus monogyna* populations into the dunes. Heavy grazing pressure from rabbits had evidently been responsible for the suppression of woody plant species in the plant community. Many other tree and shrub species such as *Prunus spinosa, Ulex europaeus, Betula pendula, Calluna vulgaris* and *Hippophae rhamnoides* also established later in the plant community.

Similarly, the grazing of *Spinifex sericeus* sand dunes by agile wallabies (*Macropus agilis*) in southern Queensland affected the structure and composition of grassy vegetation in foredunes and adjacent woodlands (Ramsey and Wilson 1997). When the ground cover by grasses, sedges and forbs was low, grazing of *Spinifex* increased. However, the impact was more evident on foredunes with low *Spinifex* abundance. Grazing pressure clearly depressed performance of *Spinifex*.

Other consequences of release from grazing pressure included enrichment of soil nutrient levels. For example, Hodgkin (1984) showed that beneath the newly established *Crataegus monogyna* plants in Wales, nitrogen and phosphorous content was enhanced which allowed subsequent invasion of dunes by weedy species. Similarly, rabbit grazing of *Schoenus* tussock grasses in the Netherlands released nitrogen and facilitated invasion by N-responsive late successional species (Grootjans *et al.* 1998).

10.7.2 White-tailed deer

White-tailed deer (*Odocoileus virginianus*) were a component of the southern forest region of Ontario, Canada, prior to the settlement by Europeans, but their population size was then regulated by predation by wolves, mountain lions and other carnivores, disease, cold weather and competition for food with other large herbivores such as elk and moose. The

Table 10.1 List of plant species that increased, decreased or invaded the plant community released from grazing pressure by rabbits following myxomatosis in 1954. The species records were made in 1958, four years after the release of dunes from grazing pressure (from Ranwell 1960, courtesy, *Journal of Ecology*)

Increasers	Decreasers	Invaders
Carex arenaria	*Agrostis tenuis*	*Cardamine pratense*
C. flacca	*Luzula campestris*	*Cynosurus cristatus*
Festuca rubra	*Poa pratensis*	*Ononis repens*
Salix repens	*Prunella vulgaris*	*Vicia angustifolia*
Thymus drucei	*Trifolium repens*	*Eurhynchium praelongum*
Hylocomium splendens	*Veronica chamaedrys*	*Mnium undulatum*
Pseudoscleropodium purum	*Bryum* spp.	
Rhytidiadelphus triquetrus	*Climacium dendroides*	

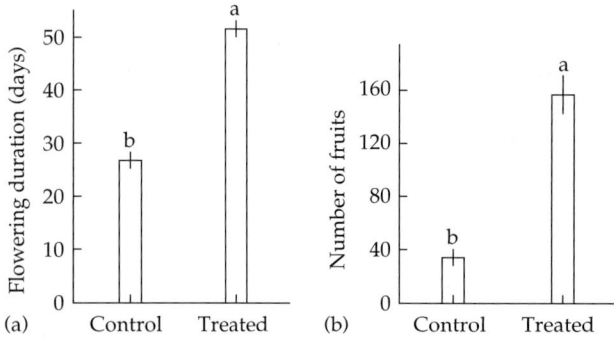

Figure 10.2 (a) Flowering duration and (b) number of fruits per plant of *Cakile edentula* in response to the continuous removal of fruits as they were formed on a plant under greenhouse conditions (after Gedge and Maun 1992).

conversion of land for agriculture and urbanization has removed most natural controls and populations of white-tailed deer have exploded. Estimates based on faecal counts of animals in the Pinery sand dunes revealed a population of 880 animals, which is more than four times the natural carrying capacity of the dune system. The evidence of over-grazing was first evident on the first dune ridge from the decimation of *Prunus pumila* and killing of *Rhus aromatica, Ceanothus americanus* and young saplings of all oak species, *Quercus velutina, Q. rubra* and *Q. prinoides* in the dune complex. In addition the animals ate oak acorns and oak seedlings that were emerging in spring. Other palatable species such as *Lupinus perennis, Cirsium pitcheri, Liatris spicata, Parthenocissus quinquefolia, Cakile edentula, Corispermum hyssopifolium, Taxus canadensis, Tsuga canadensis, Celtis tenuifolia* and *Thuja occidentalis* also suffered. In forest communities at Rondeau Provincial Park on Lake Erie many tree species such as *Liriodendron tulipifera, Nyssa sylvatica, Gymnocladus dioica, Ptelea trifoliata, Sassafras albidum, Asimina triloba* and *Magnolia acuminata* became abundant within the exclosures, but outside the exclosures no seedlings were present and a browse line was evident on trees.

When preferred species of shrubs and trees declined in the dune forest, deer started to browse less palatable species such as *Juniperus virginiana*. Other types of damage consisted of trampling, creation of bare paths in the sand dunes, dislocation, fragmentation and destruc-

tion of mats of lichens and mosses. Destruction of plants on animal paths in the first and second dune ridges led to the creation of blowouts on the windward sides of dune ridges. Some annual and biennial species of sand dunes such as *Cakile edentula* and *Cirsium pitcheri* were also browsed. Browsing of *Cirsium pitcheri* plants resulted in significant reductions in root dry weight that delayed onset of flowering and over-winter survival of the species (Phillips and Maun 1996).

In greenhouse and field studies Gedge and Maun (1992, 1994) simulated the effects of grazing by artificially removing fruits and clipping parts of plants. Experiments to mimic grazing do not precisely duplicate natural herbivory. For example, hormonal response may differ in manipulated plants. However, artificial manipulation was easy to perform, could be precisely controlled and replicated and it was assumed that it adequately simulated natural herbivory. The data showed that when all fruits of *C. edentula* were removed as they were formed, the period of flowering was significantly prolonged from 27 to 51 days and each plant produced a significantly greater number of fruits (more than 4 times) per plant than control (Fig. 10.2a, b). Fruit removal at 1 week or 4 weeks after anthesis did not prolong the duration of flowering (Fig. 10.3a) while at 2 and 3 weeks after anthesis, the period of flowering was significantly extended by approximately 12 days. Defruiting 1 week after anthesis stimulated the production of new

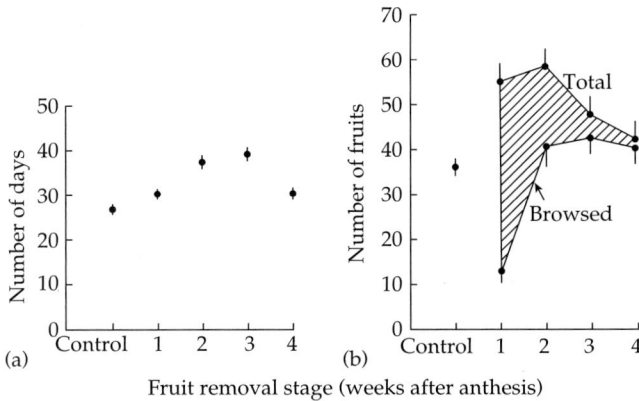

Figure 10.3 Effects of removal of fruits at different stages of growth after anthesis on (a) duration of flowering and (b) number of fruits per plant of *Cakile edentula* under greenhouse conditions (after Gedge and Maun 1992).

Table 10.2 Mean (± SE) length of time (days) from planting to anthesis of first flower and number of fruits per plant of *Cakile edentula* following clipping at different heights in a greenhouse. Plants were 4.5 weeks old at clipping (adapted from Gedge and Maun 1994)

Clipping treatment	Days to anthesis	Number of fruits
Control	41.6 ± 0.84a	31.5 ± 4.9a
Top 25%	49.4 ± 1.50b	24.0 ± 3.9ab
Top 50%	52.5 ± 0.87bc	20.5 ± 3.5ab
Top 75%	56.0 ± 0.80c	16.1 ± 1.9b

Means in each column followed by a different letter are significantly different at P < 0.05 according to Tukey's HSD test.

Table 10.3 Mean (± SE) number of fruits and biomass per plant of *Cakile edentula* under different clipping treatments given on 4–5 July. Plants were growing naturally on a beach at Port Franks, Ontario (adapted from Gedge and Maun 1994)

Clipping treatment	Survival (n)*	Number of fruits	Biomass (g)
Control	13	179.0 ± 25.9a	5.9 ± 1.41a
Top 25%	15	123.4 ± 23.8ab	3.8 ± 0.65ab
Top 50%	13	73.9 ± 17.5bc	3.2 ± 0.37ab
Top 75%	9	39.3 ± 13.0c	2.7 ± 0.37b

Means in each column followed by a different letter are significantly different at P < 0.05 according to Tukey's HSD test.

* The number of plants (out of 16) that survived until the end of experiment.

flowers and, even though maturity was delayed by only 3 days, the plants produced as many fruits as the controls. When defruiting was delayed to 2 weeks after anthesis (Fig. 10.3b), the plants continued to flower and produced 18 more fruits per plant. However, the removal of fruits at 3 weeks after anthesis lowered their capacity to produce new fruits and at 4 weeks after anthesis, no new fruits were produced.

Clipping of *C. edentula* plants (simulated browsing), by removing the top 25, 50 or 75% of plant at pre-anthesis, significantly pro-

longed the growth period by about 7 to 14 days (Table 10.2). In the 25 and 50% clipping treatments all plants survived but in the 75% clipping, only 9 out of 16 plants survived (Table 10.3). All plants started to grow vegetatively after the treatment until they had accumulated sufficient vegetative mass and then produced new flowers. In the 25 and 50% clipping treatments, plants produced as many fruits and as much biomass as control. However, when 75% of the plant was clipped there was a significant reduction in fruit production and total biomass per plant.

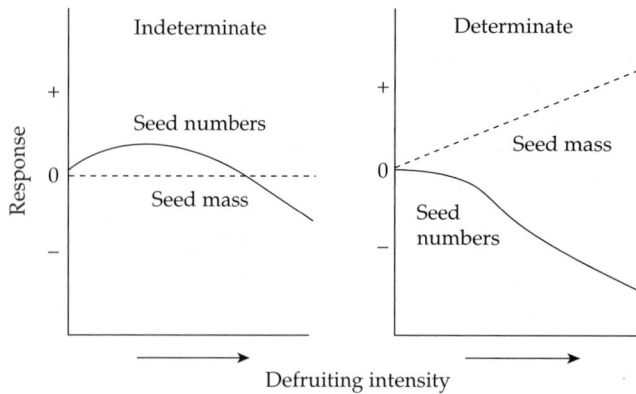

Figure 10.4 A general model of relationship between de-fruiting intensity and seed mass (----) and seed numbers (——) of determinate and indeterminate plants. The response may be positive (+), negative (−), or no effect (0) (after Gedge and Maun 1994).

10.8 The compensatory ability of plants

The data clearly showed that *C. edentula* plants were capable of partial or complete compensation (Gedge and Maun 1992, 1994) similar to those of other indeterminate plants (Stephenson 1992) such as *Gossypium* spp. (Dale 1959), *Phaseolus vulgaris* (Lovett-Doust and Eaton 1982), and *Pastinaca sativa* (Hendrix 1979). In all cases clipping of plants induced tillering and stimulated vegetative growth. Fruit removal also induced the plant to produce more flowers and fruits. The plant diverted its energy to recover from the episode by prolonging the period of flowering and fruiting. This is a natural response of an organism by which it regulates maternal investment after a predatory episode (Lloyd 1980). This characteristic would give an indeterminate parent better opportunity to compensate for the effects of herbivory than a determinate one, because it would have the opportunity to add more units. In contrast fruit predation of a determinate plant would result in fewer maturing fruits and the energy would be diverted to increasing the seed mass of remaining fruits (Maun and Cavers 1971).

A model of response curves of determinate and indeterminate plants clearly shows the relationship between seed numbers and seed mass (Fig. 10.4). In determinate plants seed mass increases because of a decrease in seed numbers, while in indeterminate plants seed mass is not altered but seed numbers increase. In either case there is a trade-off because increased seed mass at the expense of seed numbers may improve seedling quality, change germination time, depth of emergence, and competitive ability (Marshall *et al.* 1985) but would not increase the total seed population and dispersal distance of the species. In the foredunes greater seed mass and range of seed sizes will be adaptive because seedlings from larger seeds will be able to emerge from greater depths of burial than those from small seeds (Maun and Lapierre 1986; Krajnyk and Maun 1982).

The extension of the reproductive period is also a common phenomenon in the animal kingdom, especially the avian fauna (Krebs 1985). For example, the mallard duck (*Anas platyrhynchos*) lays up to 100 eggs compared to 12 eggs in the control group if newly laid eggs are removed from the nest (Klomp 1970), with limit to number imposed by the stored energy reserves (Drent and Daan 1980). In plants the energy for recovery is drawn initially from stored energy reserves and then from photosynthesis. However, restitution depends not only on the energy reserves and hormonal balance but also on the stage at which damage occurred. The earlier the grazing occurred, the greater was the capacity of plants to replace lost foliage and fruits.

In spite of the compensation, grazing produces little or no benefit to the overall fitness of the plant because the browsed material represents an irreversible loss of biomass and new material produced in response is usually not more than the control. Moreover, the increase in the longevity of the plant makes it more vulnerable to herbivory and other environmental assaults.

10.9 Summary

The intertidal zone is directly under the influence of seawater and the meiofauna and macrofauna of this zone is dependent on food from seawater surges as they inundate the seashore. The organisms in this habitat not only have intricate relationships among themselves but also with fauna that probe the sand for food. Intertidal organisms have little impact on terrestrial vegetation. However, storm surges on sea coasts deliver many dead organisms to the beach that provide ready food for scavengers such as gulls, eagles, vultures, and other carnivores. If dead animals are buried in sand they contribute nutrients to plant species on the strand.

Among the reptiles, turtles probably have the greatest influence on dune systems because of nest excavation activities on foredunes that may disrupt plant roots and contribute to burial of plants. Many species of birds use the dune complex for nesting and raising their young, which has both beneficial and deleterious effects on the habitat and dune vegetation. Benefits include deposition of guano, seed dispersal, and incorporation of feathers, eggshells and carcasses of food brought from elsewhere. The deleterious activities are trampling of vegetation, digging and burrowing for nests, uprooting of plants and defoliation that accentuates soil erosion, reduces water retention in soil, decreases the establishment of woody perennials and disrupts successional processes.

Many species of insects inhabit the dune complex and forested areas of sand dunes. Their impact ranges from minor to severe. If the impact is severe on even a single plant species the plant community may be significantly altered over time. However, usually the injury is minor and only a part of the population of plant species is affected. Insect outbreaks in the dune complex tend to be rare, probably because of protection by long fetches of sea on one side and forested areas on the other.

The impact of herbivorous animals depends on the animal species and its numbers. Grazing by domestic cattle devastates grasses and young trees and their hooves disrupt lichen and moss mats. The soil surface is destabilized and wind erosion is initiated. Rabbits and white-tailed deer are equally damaging when their populations exceed the carrying capacity of the plant community. High populations of animals decimate palatable species of plants, increase the coverage of unpalatable species and create bare areas in the sand surface where the invasives are able to gain a foothold. Conversely, the removal of a herbivore from a plant community can have unexpected consequences such as increasing, decreasing or adding species. Individual plant species show remarkable resilience to recover from herbivory. Plants compensate for damage by prolonging vegetative growth, branching, growing from root suckers and producing new meristems. Nevertheless, the overall fitness is compromised.

CHAPTER 11

Plant communities

11.1 Introduction

Plant communities of the dune complex are a result of interaction between tolerance of plant species and sandy substrate, high wind velocities, salt spray, sand accretion and environmental heterogeneity. Propagules of many plant species are dispersed by water currents and deposited on the driftline. Most of these species find ideal conditions for germination but seedling establishment, growth and reproduction is denied to all but a few species with ecological amplitude sufficient to withstand the physical stresses associated with sand accretion, erosion and sandblasting in the highly disturbed environment. The distinct differences between habitats from the water's edge to the inland grass-forest ecotone leads eventually to the establishment of ecologically distinct communities consisting of both plants and animals. The distinction is caused by sharp differences in the physical environment that may create sharp zones with abrupt or gradual blending of the two community types. In some locations these zones are relatively stable for long periods before any visible change occurs in the community depending on the recession of the shoreline, availability of new bare areas and the advance of communities towards the sea coast.

The occurrence of plant communities in zones has been documented along sea coasts worldwide. This chapter examines the plant communities of the sand dune complex along seashores of the world. The following information has been assembled from Doing (1985), *Dry coastal ecosystems* Vol. 2 A, B, C, edited by Eddy van der Maarel (1993), Doody (1991) and Thannheiser (1984). It presents data on plant communities and ecology of each zone from various parts of the world. The species complement in the 'foredune complex' in tropical, temperate and other regions around the world may be different, but their response to the prevailing environmental stresses of foredunes is convergent.

11.2 Composition of plant communities

In different world regions the boundaries between vegetation zones of the sand dune complex may not be defined sharply because of climatic variability, geographic location, physiography of the dune system and other factors peculiar to each location. Usually three to six different plant assemblages have been identified on the dune complex along sea coasts and lakeshores. A brief description of vegetation and ecological traits of species in each zone are presented below.

11.2.1 Pioneer zone

This zone stretches from the water's edge to the driftline and is consequently subject to frequent erosion and renewal of the substratum by wind and wave action. Cowles (1899) divided this area into low beach and mid-beach. The low beach area or foreshore is within the normal tidal range and is always wet because it is continually being washed by waves. Here the soil contains the highest amount of salt. The regular submergence by tides, constant wave action, violence of storms and their frequency makes the surface unstable and usually precludes the establishment of vegetation. Algae are the only plant species that grow on sand particles, but because of their requirement for light they can photosynthesize only within the top few millimetres of sand. Surface sand may also contain bacteria, rotifers, copepods and tardigrades.

The mid-beach—also called the strand or pioneer zone—is the area between low beach and driftline. It represents the interface between land and water that receives sand and feeds dune development (Maun 1993). The habitat is inundated by waves only periodically and the area is highly variable from year to year depending on the intensity of seasonal storms and lake or sea levels. The sand or shingle left by waves lacks nutrients and is highly porous with low moisture retention. The only sources of nutrients in this habitat are sea spray and dead remains of organisms that are washed onto the shore. This is essentially a dynamic, diverse and ephemeral environment that is inhospitable to the survival of most perennial species. This habitat is shrinking because of rising sea levels and anthropological encroachment.

Under such conditions the annual growth habit is favoured (Watkinson and Davy 1985); however, in spite of the unfavourable conditions some annual and occasional perennial (P) species are able to survive in this zone. Usually, three to five species on any one beach are able to establish and complete their life cycles (Doing 1985). For example, along the temperate Northwest European sea coasts the typical species include *Cakile maritima, Atriplex laciniata, A. glabriuscula, Honkenya peploides* (P) and *Salsola kali* (Doody 1991). On the shingle beaches of England, *Crambe maritima* (P), *Eryngium maritimum* (P) and *Raphanus raphanistrum* are the major occupants (Willis *et al.* 1959a). In France, Portugal, Spain and the eastern Mediterranean coasts two additional species, *Xanthium strumarium* and *Euphorbia peplis* (P), establish in this zone (Doody 1991). There are many similarities between the strand flora of European dunes and Atlantic sea coast species of eastern Canada and the United States. The major species—*Cakile maritima*—on European coasts is replaced in North America

by *C. edentula* var. *edentula* but *Atriplex laciniata* and *Salsola kali* are common to both regions (Thannheiser 1984). Two more species, *Strophostyles helvola* and *Xanthium echinatum*, are added as one goes southward towards North Carolina and Florida. On the Gulf of Mexico from Tampa, Florida to the coast of eastern Texas, *Cakile, Atriplex* and *Salsola* species persist but *Chamaesyce mesembryanthemifolia* (P) and *Sesuvium portulacastrum* (P) are added to the habitat (Stalter 1993). The pantropical fleshy perennials of the sea coasts of the world, *Ipomoea pes-caprae* and *I. stolonifera*, with occasional roots along stolons expand into the strand by producing very long trailing stems (5–30 m in length).

The life of perennials establishing in the strand or expanding into this zone is precarious because plants may be fragmented by waves at any time during the year. However, these fragments regenerate and establish as new plants. Along the Gulf coast of Mexico, Honduras, Costa Rica, Panama (Moreno-Casasola 1993), Cuba (Borhidi 1993), Brazil (De Lacerda *et al.* 1993), Uruguay, and Argentina (Pfadenhauer 1993) the species complement remains pretty much the same except that three additional species, *Euphorbia polygonifolia, Canavalia rosea* (P) and *Cakile maritima* (introduced) appear on some beaches. Nevertheless, only a few of the listed species may be found at any one place or time dependent on the recent history of wave amplitudes, storm frequencies, disturbance-free period and weather conditions during germination, establishment and growth.

Along the Pacific coast of North America from Queen Charlotte Islands (54°N) to about 43°N the strand vegetation is very poorly developed (Wiedemann 1993). However, along the coast of California *Cakile maritima* (introduced from Europe) and *Atriplex leucophylla* are common (Johnson 1993a). Moving south to approximately 25°N, one finds *Abronia maritima* (P) on the beaches along with *Ipomoea pes-caprae, I. stolonifera* and *Scaevola plumieri* (Johnson 1993b). On the strand of Japanese islands, *Salsola komarovii, Atriplex cordata, Calystegia soldanella* (P) and *Honkenya peploides* (P) are common (Miyawaki and Suzuki 1993). In Africa the strand flora is not well developed mainly because of violent storms during the major growing period (Lee 1993). Under high temperatures and low rainfall conditions that prevail along the Atlantic coasts of Namibia (Penrith 1993) and the Red Sea coast in Saudi Arabia (Zahran 1993), the main source of moisture is sea fog that does not provide sufficient moisture for annual plants to complete their life cycles. Near the equator in the tropical areas where there are no dry periods such as in Java, Malaysia and some parts of Africa, annuals are generally absent from the strand because of competition with perennials for space (Lee 1993). In Australia, the strand flora includes species of *Cakile maritima, C. edentula* and *Salsola kali* that were introduced from Europe and North America (Doing 1985). Along the freshwater Great Lakes *Cakile edentula* var. *lacustris* is the dominant species. Other associated species with low coverage are *Salsola kali, Corispermum hyssopifolium, Euphorbia polygonifolia* and *Xanthium strumarium*. These species and several other endemics (Table 11.1) represent the maritime element because they are typically associated with sand dunes along the Atlantic and arrived at the shores of glacial meltwater lakes about 11 000 years ago, when the ocean extended up the Ottawa River Valley (Guire and Voss 1963; Morton and Venn 1984). They have survived on the sandy shores, dunes and marshes of the Great Lakes ever since.

11.2.1.1 Survival mechanisms

Although the constituent species differ in different regions of the world plants have evolved several common adaptive traits that include a short life cycle, ability to survive under harsh conditions, high fecundity, good dispersal ability, high phenotypic plasticity, seed dormancy and large seed mass. They complete their life cycles within the short relatively disturbance-free period when conditions are

Table 11.1 List of endemic plant species and their preferred habitat (indicated by x) in the vicinity of the Great Lakes (Guire and Voss 1963; Morton and Venn 1984)

Name of species	Preferred habitat	
	Beach	Dune
Agropyron psammophilum	x	x
Cakile edentula var. *lacustris*	x	—
Calamovilfa longifolia var. *magna*	x	x
Carex garberi	—	—
Cirsium pitcheri	—	x
C. hillii	—	—
Gentianopsis virgata	—	—
Hypericum kalmianum	x	x
Hymenoxys acaulis	—	—
Solidago houghtonii	—	Interdunal areas
*S. ohioensis**	x	—

* Usually common on shores, wet rock pavements and in marshes.

most favourable for germination, growth and reproduction. Under high arctic conditions *Cakile arctica* germinates under low soil temperatures and completes its life cycle in the very short growing period. In southern locations *Cakile maritima* depends on spring rainfall to germinate and complete its life cycle before the onset of summer drought. Even though the population size of a species may vary from year to year because of fluctuating mortality, plants are adapted to withstand the variable hazards and a certain proportion of individuals in the population always survives and reproduces successfully. For example, the most abundant and dominant strand species in the world are in the genus *Cakile* (Rodman 1974). The distribution and abundance of species such as *Cakile edentula* var. *lacustris*, var. *edentula* and *C. maritima* are strongly influenced by differential survival, high reproduction, efficient dispersal mechanism and good germination capacity (Payne and Maun 1981, 1984; Keddy 1981; Barbour 1978). Payne and Maun (1981) showed that most plants were clustered in a narrow strip on the driftline

Table 11.2 Mean (± 2 SE) plant density m^{-2}, number of seeds m^{-2}, seed production m^{-2} and per plant and dry biomass m^{-2} of *Strophostyles helvola* plants on the mid-beach and driftline of Lake Erie shoreline at Port Burwell Provincial Park (adapted from Yanful and Maun 1996a)

Zone	Plant density (m^{-2})	No. of seeds		Seed production (g)		Biomass (g)	
		m^{-2}	plant^{-1}	m^{-2}	plant^{-1}	m^{-2}	plant^{-1}
Mid-beach	4.7 ± 3.2b	128.8 ± 42.1a	27.4	5.2 ± 1.8a	1.10	13.3 ± 4.8a	2.83
Driftline	23.3 ± 6.5a	79.9 ± 16.8b	3.4	2.7 ± 0.7b	0.12	12.6 ± 2.1a	0.54

Means in each column followed by the same letter are not significantly different at $P < 0.05$ according to Tukey's test.

and only a small number were scattered on the mid-beach. However, although population density was significantly higher on the driftline, the survivorship of plants was low because of desiccation and intraspecific competition. The species have extremely efficient dispersal mechanisms owing to dimorphic fruits (Chapter 3). An almost identical mechanism of establishment, fecundity, dispersal and distribution was found in the indeterminate legume, *Strophostyles helvola*, on a shoreline along Lake Erie (Yanful and Maun 1996a). The total fruit output per plant and m^{-2} of this species were several times higher on the mid-beach than the driftline (Table 11.2), but following dispersal the propagules were carried laterally and inland by wind and wave action during autumn and early spring months and deposited on the driftline.

Corispermum hyssopifolium has winged fruits and can occur in any bare area on the beach, in blowouts, on lee slopes and stable surfaces (van Asdall and Olmsted 1963). All of these annuals exhibit high phenotypic plasticity which allows them to reproduce under the most difficult circumstances. For example, a plant of *Cakile*, *Corispermum*, *Salsola* or *Atriplex* growing in an unfavourable microhabitat may produce only a few seeds/fruits while another plant of the same species growing nearby under more favourable conditions may produce several hundred seeds. Other adaptive traits are seed dormancy that confers longevity and larger seed mass that allows seedlings

to emerge from greater depths of burial by sand.

11.2.2 High beach

This habitat is more favourable to plant growth than the strand because of less frequent disturbance by high waves. However, sand cast on the mid-beach by high waves is blown inland by wind and deposited on plants, either burying them or forming embryo dunes. A large number of plant species belonging to the Poaceae, Convolvulaceae, Asteraceae, Cyperaceae, Caryophyllaceae, Fabaceae and Aizoaceae occupy this zone in temperate, subtropical and tropical regions (Tables 11.3 and 11.4).

The major dune-forming species of the high beach in Europe and the Mediterranean are the grasses, *Ammophila arenaria*, *A. baltica*, *A. arundinacea*, *Elytrigia junceiformis*, *Elymus farctus* and *Leymus (Elymus) arenarius*. *Sporobolus pungens* becomes a component of the plant community in the eastern Mediterranean. By far the most common species, however, is *A. arenaria*. In Atlantic North America it is replaced by *A. breviligulata*. The two species are considered ecological equivalents (Huiskes 1979) in spite of a number of distinct morphological and ecological differences between them (Baye 1990). *Ammophila arenaria* has a rhizomatous tussocking pattern of growth while *A. breviligulata* has a very gregarious creeping rhizomatous growth with dispersed clusters.

Table 11.3 Plant species on the high beach, foredunes, embryo dunes and dune ridges, on coastal sand dunes along sea coasts of Europe, North and South America

---

Europe and the Mediterranean
Ammophila arenaria, A. baltica, A. arundinacea, Elytrigia junceiformis, Elymus farctus and *Leymus (Elymus) arenarius.*
  Associated species: *Calystegia soldanella, Lathyrus maritimus, Eryngium maritimum, Euphorbia paralias* and *Medicago marina.*
  Eastern Mediterranean: *Sporobolus pungens* (Doody 1991).

Atlantic Canada
Above 53°N latitude, *Elymus mollis* (Houle 1998). Below 53°N latitude, *Ammophila breviligulata* (Thannheiser 1984).

Atlantic coast of United States
A. breviligulata down to 36°N lat. and then *Uniola paniculata, Panicum amarum, P. amarulum, Croton punctatus* (Maun and
  Baye 1989).

Gulf of Mexico
U. paniculata. Associated species: *Ipomoea pes-caprae, I. stolonifera, Croton punctatus, Heterotheca subaxillaris, Chamaesyce
  buxifolia, Solidago sempervirens, Suriana maritima, Tournefortia gnaphalodes, Waltheria indica* (Stalter 1993).

Mexico, Central America, Cuba, Venezuela, Brazil
Sporobolus virginicus, Panicum racemosum, Mariscus pedunculatus. Associated species: *I. pes-caprae, I. stolonifera, Croton
  punctatus, Sesuvium portulacastrum, Palafoxia lindenii, Scaevola plumieri, Chamaecrista chamaecristoides, Trachypogon
  gouinii, Canavalia rosea, C. maritima* (Moreno-Casasola 1993; Borhidi 1993; De Lacerda *et al.* 1993).

Uruguay and Argentina
Panicum racemosum, Paspalum vaginatum and *Philoxerus portulacoides* (Pfadenhauer 1993).

Pacific North America
From Queen Charlotte Islands, 54°N lat. to about 40°N lat. *Carex eriocephala.* Southward: *Elymus (Leymus) mollis, Ammophila
  arenaria, Abronia latifolia, Ambrosia chamissonis, Calystegia soldanella, Fragaria chiloensis, Lathyrus littoralis, Artemisia
  pycnocephala, Tanacetum douglasii* (Wiedemann 1993; Johnson 1993a). At 30°N lat., *Abronia maritima, Atriplex leucophylla,
  Mesembryanthemum chilense, Salicornia subterminalis, Senecio californicus*; at 24°N lat., *Sporobolus virginicus* and *Jouvea
  pilosa* (Johnson 1993b).

Japanese coasts
Elymus mollis and *Carex kobomugi.* Associated species: *Artemisia stelleriana, Wedelia prostrata, Iris repens, Glehnia littoralis*
  (Miyawaki and Suzuki 1993).

Ammophila breviligulata is replaced to the south of North Carolina (36°N lat.) by the warm season grasses, *Uniola paniculata, Panicum amarum* and *P. amarulum* (Maun and Baye 1989). Along South American seashores, *Uniola paniculata* is replaced by *Ipomoea pes-caprae, I. stolonifera, Croton punctatus*, along with the dune-forming grasses, *Sporobolus virginicus, Panicum racemosum* and *Paspalum vaginatum*. Along the Pacific coast of North America *Elymus (Leymus) mollis* was the main dune-forming species until the introduction at San Francisco of *A. arenaria* in the 1860s (Wiedemann 1993). Since then, it has spread both northward into Oregon and Washington as well as southward towards

Mexico as far as 30°N lat. Farther south at approximately 24°N the foredunes are continuous ridges built by two grasses, *Sporobolus virginicus* and *Jouvea pilosa* (Johnson 1993b). Because of more hot and dry desert-like conditions at this latitude, the proportion of C_4 species in the community increases sharply. Along Japanese coasts the growth pattern of *Carex kobomugi* is similar to that of rhizomatous grasses like *A. arenaria*. Other prominent species in this community are *Artemisia stelleriana, Wedelia prostrata, Iris repens* and *Glehnia littoralis* (Miyawaki and Suzuki 1993).

On the coastal sand dunes of Sierra Leone, *Schizachyrium pulchellum* is a very prolific

Table 11.4 Plant species on the high beach, foredunes, embryo dunes and dune ridges, on coastal sand dunes along sea coasts of Africa, Asia and Australia

---

Sierra Leone, Ivory Coast, Guinea, Zaire, Ghana
Sporobolus virginicus, Schizachyrium pulchellum, Alternanthera maritima, Remirea maritima, Ipomoea pes-caprae, I. stolonifera, Canavalia rosea, Scaevola plumieri, Cyperus maritimus (Lee 1993).

Namibian coast
Sporobolus virginicus, close to the mouths of ephemeral rivers flowing into the sea (Penrith 1993).

South Africa, the west and south coasts
Cladoraphis cyperoides, Salsola nollothensis, Arctotheca populifolia, Didelta carnosa, Senecio elegans, Scaevola thunbergii (Boucher and Le Roux 1993; Taylor and Boucher 1993).

South Africa, East coast
Ammophila arenaria (introduced), *Agropyron distichum, Sporobolus virginicus, Ehrharta villosa, Scaevola plumieri.* Associated species: *I. pes-caprae* ssp. *brasiliensis, Carpobrotus dimidiatus, Arctotheca populifolia, Gazania rigens* and *Hydrophylax carnosa* (Weisser and Cooper 1993).

Somalia, Kenya and Tanzania
Atriplex farinosa, I. pes-caprae ssp. *brasiliensis, Aerva lanata, Scaevola plumieri, Panicum pinifolium, Sporobolus virginicus, S. pungens, Halopyrum mucronatum, Cyperus maritimus* (Pignatti et al. 1993; Frazier 1993).

Indian subcontinent from Pakistan, coastal states of India and Bangladesh (6000 km)
I. pes-caprae, Cyperus arenarius, Launaea sarmentosa, Aerva persica, A. lanata, Borreria articularis, Convolvulus scindicus, Calotropis procera. Subhumid, humid and wet coasts: *I. pes-caprae, Canavalia maritima, Spinifex littoreus, Cyperus arenarius, C. pedunculatus, Launaea sarmentosa.* Bay of Bengal: *Spinifex littoreus, I. pes-caprae, Canavalia maritima, Hydrophylax maritima, Fimbristylis junciformis, Euphorbia rosea, Geniosporum tenuiflorum, Polycarpaea tuberosa* (Rao and Meher-Homji 1993).

Java coast
Spinifex littoreus, I. pes-caprae, Canavalia rosea, Ischaemum muticum, Euphorbia atoto, Pandanus tectorius, Calotropis gigantea (Hardjosuwarno and Hadisumarno 1993).

Australian coastal dune systems
Tropical, subtropical and warm temperate shores from 17°S to 35°S, *Spinifex hirsutus, S. longifolius, I. pes-caprae, Ammophila arenaria* (introduced), *Festuca littoralis, Canavalia maritima, Fimbristylis littoralis, Arctotheca populifolia, Scirpus nodosus, Calystegia soldanella, Remirea maritima* and several other species (Doing 1985).

---

species that produces stolons up to 3 m long with short shoots (Lee 1993). In South Africa, the west, south and east coasts differ considerably in species complements, primarily because of different rainfall patterns and rates of sand movement, but there are also many similarities (Lubke 2004). Lubke showed that artificially planted populations of *Ammophila arenaria* become established within about 6–10 years on the foredunes and do not competitively exclude native species such as *Elymus distichus, Arctotheca populifolia, Didelta carnosa* and *Senecio elegans,* which also establish in these planted stands. These dunes apparently stabilize within about 20–30 years and subsequently many species, e.g. *Psoralea repens, Chironia baccifera, Helichrysum patulum, Ehrharta villosa* and *Passerina rigida,* establish in the dune system as new ridges are created (Boucher and Le Roux 1993; Taylor and Boucher 1993). On the East Cape *Scaevola plumieri* is an important dune-forming species that produces hummocks that eventually coalesce into linear dune ridges (Lubke 2004; Weisser and Cooper 1993). Dominant members of the foredune community in Somalia, Kenya and Tanzania include *Scaevola plumieri, I. pes-caprae* ssp. *brasiliensis, I. stolonifera* and *Aerva lanata*

(Pignatti *et al.* 1993; Frazier 1993). Coastal sand dunes of the Indian subcontinent in Pakistan, as well as coastal states of India and Bangladesh (6000 km) support many different plant communities because of a wide range of climates. In the Indus delta (25°N) 11 months of each year are dry while further south (10°N) conditions are rather humid and tropical with wet conditions for 6–9 months (Rao and Meher-Homji 1993). Major foredune dominants here and Java coast are *I. pes-caprae, Spinifex littoreus, Cyperus arenarius* and *C. pedunculatus.* Along the Australian coasts species of *Spinifex hirsutus, S. longifolius, I. pes-caprae, Ammophila arenaria* (introduced) and *Festuca littoralis* are important in foredune habitats (Doing 1985). On the high beach/foredunes of the Great Lakes, *Ammophila breviligulata* is virtually the only species but two other species, *Calamovilfa longifolia* and *Equisetum hyemale,* are present with low coverage.

11.2.2.1 Survival mechanisms
The species of foredunes exhibit properties that are fundamental to their survival:

1. In temperate regions all fordune species are geophytes (e.g. the ubiquitous *Ammophila*

and other grasses) with buds well protected by soil during the cold winter season. Foredune species in warm tropical, subtropical and dry regions are stoloniferous hemicryptophytes (e.g. the pantropical *Ipomoea* species and warm season grasses) with buds barely embedded in the soil surface.

2. A life history trait commonly exhibited by species (Table 11.5) on the high beach and embryo dunes on all coastal shorelines of the world is that plants advance towards disturbance-prone driftline and mid-beach zones either as rhizomes or as stolons (Fig. 11.1). This is a convergent trait of high adaptive significance that promotes shoreline stabilization. The evolution of this growth form may have been stimulated by erosion of substrate, sand movement and accretion, and fragmentation of populations through wave action. In fact, regular disturbance may also be essential for the proliferation and continued occupancy of these species. Horizontal growth may develop due to increased water availability towards the shoreline, proximity to the freshwater beneath the beach surface, lack of competition in open primary bare areas and access to nutrient-rich detritus washed up on the driftline. They also

Table 11.5 Partial list of dune-forming species that produce rhizomes, stolons or suckers along different shorelines of the world

Family	Species and occurrences
Gramineae	*Ammophila breviligulata, A. arenaria* (NA and Eu); *Phragmitis communis* (NA and Eu); *Calamophila baltica* (Eu); *Leymus arenarius, L. mollis* (NA and Eu); *Distichlis stricta* (NA and Eu); *Elymus farctus* (Eu); *Panicum racemosum* (Brazil); *Ischaemum anthephoroides* (Japan); *Spinifex hirsutus, S. sericeus* (Australia); *Spinifex littoreus, Sporobolus virginicus, S. pungens, Zoysia matrella* (India, Malay Peninsula); *Ehrharta villosa, Agropyron distichum* (South Africa); *Schizachyrium pulchellum* (Sierra Leone, West Africa).
Cyperaceae	*Carex arenaria* (Eu); *C. kobomugi* (Japan); *C. eriocephala* (NA).
Convolvulaceae	*Ipomoea pes-caprae, Ipomoea stolonifera* (pantropical); *Calystegia soldanella* (Eu).
Portulacaceae	*Sesuvium portulacastrum* (Australia, tropics and subtropics).
Caryophyllaceae	*Honkenya peploides* (NA and Eu).
Goodeniaceae	*Scaevola plumieri* (Tropical America, Africa, South India, Galapagos islands).
Fabaceae, Leguminoseae	*Canavalia rosea, C. cathartica* (pantropical, tropics and subtropics).
Salicaceae	*Populus balsamifera, Salix exigua* (NA).

NA, North America; Eu, Europe.

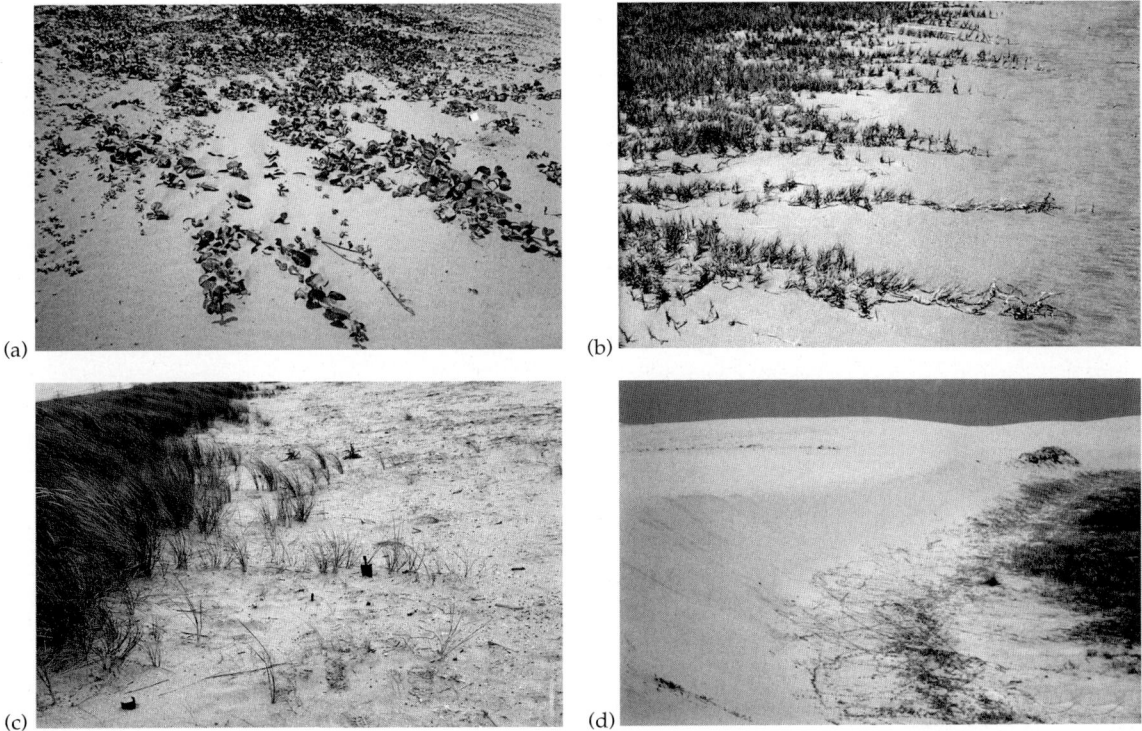

Figure 11.1 Rhizomatous and stoloniferous growth of dominant perennial plant species along the beaches of sea coasts: (a) *Ipomoea pes-caprae*, (b) *Distichlis spicata*, (c) *Ammophila breviligulata* and (d) *I. stolonifera* (photographs by A Maun).

grow upward rapidly in response to burial by blowing sand.

3. Fragmentation of rhizomes and stolons allows dispersal to new locations and establishment of new populations. There is evidence that the viability of buds is not lost during dispersal in salt water (Baye 1990). For more details refer to Chapter 8.

4. Foredune species are characterized by the ability to occupy a bare area rapidly (Maun 1985). Within one year a clump of *Ammophila breviligulata* expanded its area to approximately 19 times its original size (Fig. 11.2).

5. Seedling establishment in most cases is stochastic and related more to the amount and regular distribution of rainfall.

11.2.3 Dune ridges

The habitat here is slightly different from the high beach because there is no stalling effect of the first dune ridge on wind velocity. The high wind velocity causes erosion on the crest and deposits sand on the lee slope. Dune ridges are thus characterized by sand burial and low soil moisture. At some places deposition may exceed one metre per year in which case no vegetation survives and bare areas develop. Because of this deposition, the rhizomes of dune-building grasses that ordinarily grow laterally are orientated vertically and if burial depth is below their threshold of survival, they emerge from the new sand surface (Fig. 11.3).

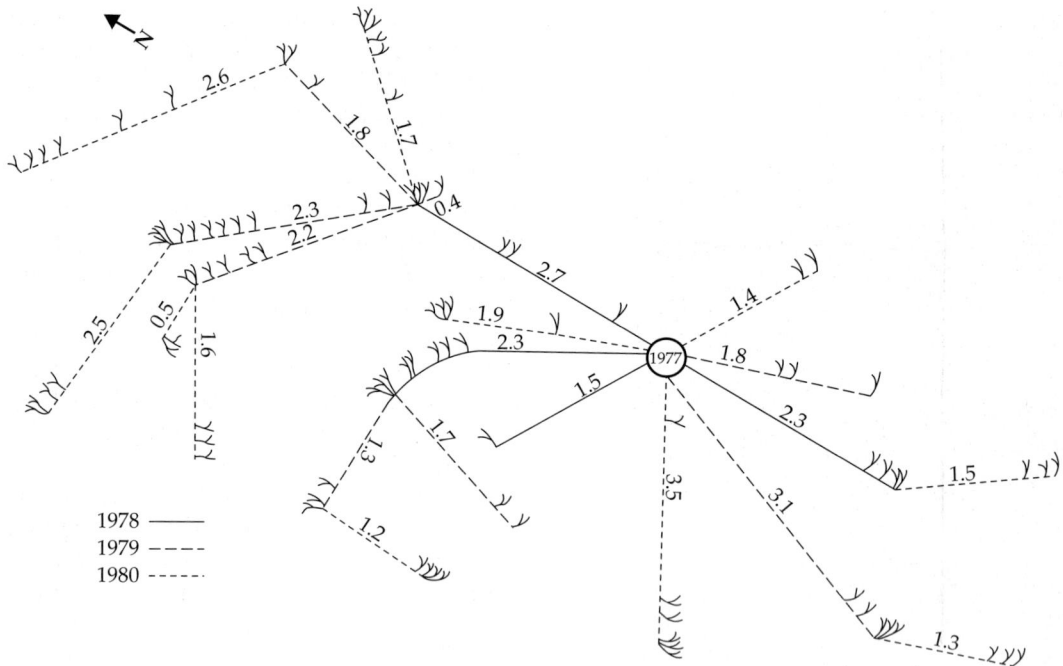

Figure 11.2 The expansion of rhizomes (in metres) of a single clump of *Ammophila breviligulata* for three years after its planting (after Maun 1985).

In Europe, the vegetation on the ridge is occupied by the same dune-building grasses and associated species that were present on the high beach along with a few others such as *Artemisia campestris* ssp. *maritima*, *Echinophora spinosa*, *Carex arenaria* and *Ononis repens*. At Newborough Warren, *Carex arenaria* is very wide spread on the dunes while *O. repens* is confined to lower parts of *Ammophila arenaria* (Willis *et al.* 1959a, b). Along the Atlantic coast in Canada, dune ridges are occupied by *Elymus mollis*, *A. breviligulata*, *Lathyrus japonicus*, *Solidago sempervirens*, *Artemisia stelleriana*, *Festuca rubra* and *Hudsonia tomentosa* (Thannheiser 1984). Atlantic coastal dunes of the United States support *A. breviligulata* as the dominant species except in the south of North Carolina where *Uniola paniculata* replaces it. Associated species in Florida and the Gulf of Mexico are *Scaevola plumieri*, *Canavalia rosea*, *Andropogon virginicus*, *Croton punctatus*, *Cassia*

fasciculata, *Myrica cerifera*, *Opuntia stricta*, *Hydrocotyle bonariensis* and *Coccoloba uvifera* (Doing 1985; Stalter 1993).

On the coastal sand dunes of South and Central America, dune ridges are stabilized by *Sporobolus virginicus*, *Mariscus pedunculatus* and *Panicum racemosum* along with *Schizachyrium scoparium*, *Palafoxia lindenii*, *Scaevola plumieri*, *Chamaecrista chamaecristoides*, *Trachypogon gouinii*, *Vigna luteola*, *Senecio crassiflorus*, *Digitaria connivens* and *Hydrocotyle bonariensis* (Moreno-Casasola 1993). Along the Pacific coast of North America *Carex eriocephala*, *Elymus mollis*, *Ammophila arenaria*, *Sporobolus virginicus* and *Jouvea pilosa* are major dune species (Wiedemann 1993; Johnson 1993a, b). In Japan, coastal dune ridges support *Elymus mollis*, *Carex kobomugi*, *Ischaemum anthephoroides*, *Glehnia littoralis*, *Vitex rotundifolia*, *Lathyrus japonicus* and *Rumex japonicus* (Miyawaki and Suzuki 1993). On the

Figure 11.3 Vertical growth of *Calamovilfa longifolia* following natural burial in sand on the first dune ridge along Lake Huron at the Pinery (photograph by A Maun).

coastal sand dunes of Sierra Leone, Ivory Coast, Guinea, Zaire, and Ghana, the first dune ridge resembles a coastal thicket constituting *Stylosanthes erecta, Cyperus maritimus, Euphorbia glaucophylla, Pennisetum purpureum* and *Andropogon gayanus* (Lee 1993). The major dune-forming species along the coasts of South Africa is *Scaevola plumieri* (Lubke 2004). Along the coastline of India *Spinifex littoreus* continues to be a dominant species on the dunes along with two subdominants, *Borreria articularis* and *Hydrophylax maritima* (Rao and Meher-Homji 1993). Along the Java coast, dune ridges are occupied by *Cocos nucifera, Erythrina variegata, Hibiscus tiliaceus* and *Pandanus tectorius* (Hardjosuwarno and Hadisumarno 1993). In coastal dunes of Australia, the species

complement on the dune ridge is very similar to that of the embryo dunes with *Spinifex hirsutus, S. longifolia, Ammophila arenaria* (introduced) and *Festuca littoralis* as the dominant members (Doing 1985).

At the Pinery along Lake Huron the dominant species on the first dune ridge (3 to 5 m tall) is *Calamovilfa longifolia*, a C_4 species (Elfman *et al.* 1986), with a cover value of about 30% (Baldwin and Maun 1983) along with low coverage of *Schizachyrium scoparium, Ammophila breviligulata, Populus balsamifera, Artemisia campestris* and several other species (Table 11.6). Some woody species such as *Prunus pumila, Salix exigua* and *Juniperus virginiana* also occur as occasional trees or shrubs on the dune ridge. The relative coverage of *A. breviligulata* is only about 0.5% in most locations, except in microhabitats with greater sand deposition where its coverage increases to >80% because of its ability to withstand greater amounts of burial than other species. The dominant grasses on the top of second dune ridge are *C. longifolia* and *S. scoparium* along with occasional shoots of *A. breviligulata*. This ridge is taller than the first dune ridge and the windward side of the ridge is subject to eroding winds that remove sand from there and deposit it on the crest and leeward side of the dune ridge. However, in some areas where the dune has been stabilized woody species consisting of *Juniperus virginiana, Prunus virginiana, Vitis riparia, Arctostaphylos uva-ursi* and *Quercus prinoides* have established.

11.2.4 Lee slopes

The fourth zone with a distinctive plant community occurs on the rather steep lee slopes of dune ridges. Here the wind velocity is low, salt spray is negligible and sand accretion is intermittent. Depending on orientation to the sun at different latitudes, the soil temperature is usually higher and soil moisture lower than on the windward sides of dune ridges. At Braunton Burrows, *A. arenaria*, the dominant species of foredune and dune ridges, provides

Table 11.6 Species list and mean % cover of plants on first dune ridge (100 years), slack (400 years) and heath–grassland zone (800 years) during 1979 on Lake Huron dunes at Pinery (Baldwin and Maun 1983). P indicates presence of species

Name of species	First dune ridge (100 years)	Slack (400 years)	Grassland/ transition zone (800 years)
Artemisia campestris	P	1.1	0.9
Corispermum hyssopifolium	P	0	0
Arenaria stricta	0	0	0.2
Euphorbia polygonifolia	P	0	0
Schizachyrium scoparium	1.1	1.2	12.4
Andropogon gerardii	0	0	7.6
Ammophila breviligulata	0.5	0	0
Calamovilfa longifolia	30.6	18.3	7.2
Stipa spartea	0	0	0.4
Sorghastrum nutans	0	0	2.0
Smilacina stellata	0	0	0.6
Asclepias tuberosa	0	0	P
Lithospermum caroliniense	0	P	P
Carex spp.	0	P	0.2
Liatris cylindracea	0	0	4.3
Lichen spp.	0	0	3.1
Moss spp.	0	0	26.7
Prunus pumila	P	0	0
Arctostaphylos uva-ursi	P	P	24.6
Juniperus communis	0	0	0.3
Cornus stolonifera	P	P	P
Juniperus virginiana	0	P	0.2
Populus balsamifera	P	P	0
Litter	44.0	36.4	82.5
Bare sand	44.0	58.1	4.9

Figure 11.4 Establishment of *Schizachyrium scoparium* (= *Andropogon scoparius*) on the lee slopes of first and second dune ridges at the Pinery (photograph by A Maun).

a sparse cover on the lee slopes and stabilization of the sand surface was performed by two perennials, *Carex arenaria* and *Ononis repens* (Willis *et al.* 1959a). Many other annuals and biennials are also components of this community. Along the Great Lakes the south-facing lee slopes of both the first and second dune ridges are dominated by the clump-forming C_4 species, *Schizachyrium scoparium* (Fig. 11.4) along with several other species: *Lithospermum caroliniense, Asclepias viridiflora, Arctostaphylos uva-ursi* and *Juniperus communis*. Since this habitat receives direct insolation from the sun the soil temperature is usually about 2–4°C higher than the north slopes and the surface soil dries up faster than other aspects (Baldwin and Maun 1983).

11.2.5 Slacks

Slacks are very different from the sand dunes because of proximity to the water table, higher soil moisture levels and coarser texture of the sand. According to Grootjans *et al.* (2004) the slacks are maintained by precipitation and calcareous groundwater discharge: the acidic precipitation augments the calcareous surface water and groundwater. The water table in the slack may fluctuate up to 2 m each year (Grootjans *et al.* 1998) depending on the amount of rainfall and its distribution, changes in sea level, soil texture and hydrology of the watershed. In Europe, the slacks usually become flooded by rainfall during winter and early spring and dry out during summer. Along the northern Atlantic coast of North America slacks freeze during winter, flood due to snow melt and spring rains and dry out during summer. In tropical regions, e.g. along the Gulf of Mexico, at La Mancha, Veracruz the water table rises in summer because of heavy rains and the slack is flooded if the amount of rainfall exceeds 50 cm (Moreno-Casasola and Vázquez 1999; Vázquez 2004). Over a 3-year period, regular flooding occurred from July to October each year, while winters were relatively dry and water levels fell below the sand surface.

Slacks support a characteristic type of vegetation very different from the sand dune ridges mainly because of high soil moisture, anoxic conditions, high or low pH levels, very low nutrient levels—especially nitrogen and phosphorous—and wet and drying cycles (Grootjans *et al.* 1998). However, there is variability among slacks because of the peculiar hydrological conditions of the dune watershed, evaporative demands, amount of precipitation, seepage from the dunes, and groundwater discharge. Generally three types of slacks, (i) dry, (ii) wet, and (iii) flooded, have been identified on the basis of hydrological systems.

In *dry slacks* the water table normally remains more than one metre below the sand surface. Along the Polish Baltic sea coast, dry slacks lack halophytic flora because freshwater lens overlying dense seawater prevents capillary rise of salt-laden water from the water table (Zoladeski 1991). The slack vegetation consisted of *Ammophila arenaria* in dry sand bars, *Agrostis stolonifera* in moist sand in depressions, *Calluna vulgaris* under acidic conditions, as well as shrub and forest communities dominated by *Pinus palustris*. A dry slack in Australia was occupied by *Spinifex hirsutus* (Maze and Whalley 1992a).

In *wet or moist slacks* where the water table is shallow, pioneer species are primarily microbial and include cyanobacteria, algae, diatoms and eubacteria (Grootjans *et al.* 1998). These organisms produce extracellular polysaccharides that create a thin crust on the sand surface which not only protects against desiccation but also glues sand particles together, thus acting as mulch. Microbes also facilitate invasion by new organisms by increasing soil nutrients and organic matter. Similar slacks at Braunton Burrows have an almost continuous plant cover with two rosette-forming species, *Plantago coronopus* and *Leontodon leysseri*, as dominants and many other grasses and sedges such as *Agrostis stolonifera, Festuca rubra, Carex flacca* and *C. nigra* (Willis *et al.* 1959a).

In *flooded slacks* the water table is high and there is standing water for at least part of the year. The soil reaction is basic and many calcicole species such as *Schoenus nigricans, Epipactis palustris, Carex flacca, Pedicularis palustris, Juncus alpinoarticulatus, Scirpus palustris* and *Littorella uniflora* take a foothold (Grootjans *et al.* 1998). These species have well-developed aerenchyma and can survive waterlogged conditions; however, they are also adapted to dry conditions when the water table falls below the sand surface. The persistence of these species depends on the maintenance of basic conditions and high Ca^{++} and HCO_3^- content. However, extended rains following extreme dry conditions acidify the soil and succession may proceed in a different direction. Such a change in soil reaction allows an increase in acid-loving species such as *Viola canina, Potentilla palustris, Carex arenaria, Cirsium dissectum, Hydrocotyle vulgaris* and *Eriophorum angustifolium* (Grootjans *et al.* 1998).

Figure 11.5 At the Pinery, the low-lying area (slack) between the first and second dune ridges is occupied by *Calamovilfa longifolia*, *Schizachyrium scoparium* and *Artemisia campestris* (photograph by A Maun).

These plants promote build up of organic matter and nutrients that pave the way for establishment of strong competitors such as *Calamagrostis epigeios*, *Salix repens*, *Alnus glutinosa* and *Betula pubescens*.

At the Pinery along Lake Huron, the *slack* between the first and second dune ridges is dry (Fig. 11.5) and groundwater flows from the foredunes and three adjacent dune ridges towards the Lake (Schincariol *et al.* 2004). Only once every 15–20 years when the lake level rises by about 1.5 m (as in 1986) the soil surface in the slack becomes moist. This presumably provides an opportunity for seedlings of *Salix*, *Betula* and *Populus* species to establish if seeds are available during these periods, but none of these species established in 1986–1987. *Calamovilfa longifolia*, *Schizachyrium scoparium* and *Artemisia campestris* are by far the most dominant species here (Table 11.6).

11.2.6 Transition to forest

This zone occurs behind the open and migrating dune ridges and acts as a transition between grassy vegetation and dune scrub or woodland. The habitat is xerophytic and stable with low wind velocities, little or no sand movement, burial by sand or salt spray. In Europe, this area

is very diverse with large number of lichens, mosses, grasses and shrubs. Dominants include *Tortula ruraliformis*, *Festuca rubra*, *F. ovina*, *Corynephorus canescens*, *Deschampsia flexuosa*, *Carex arenaria*, *C. maritima*, *Lotus corniculatus*, *Corema album* and *C. conradii* along with shrubs such as *Hippophae rhamnoides*, *Juniperus communis* and *Salix repens* (Doody 1991). In Canada and the eastern United States, the major species of this zone are *Schizachyrium scoparium*, *Deschampsia flexuosa*, *Arctostaphylos uva-ursi* and *Hudsonia tomentosa*. More species such as *Muhlenbergia capillaris*, *Oenothera humifusa*, *Cenchrus tribuloides* and *Hydrocotyle bonariensis* are found southward from North Carolina. Along Mexican, South American coasts, this zone is not well developed. Herbaceous species consist of *Panicum racemosum*, *P. sabulorum*, *Andropogon arenarius*, *A. selloanus*, *Conyza blakei*, *Polygala cyparissias*, *Ononis vaginalis* and *Crucianella maritima* (Moreno-Casasola 1993). In the Indian subcontinent, including Sri Lanka and Indonesia, this zone is seldom observed because it has either been brought under cultivation or converted into resorts (Rao and Meher-Homji 1993). In Australia this zone is occupied by *Acacia sophorae*, *A. ligulata*, *A. rostellifera*, *Olearia axillaris*, *Rhagodia baccata*, *Calocephalus brownii*, *Hibbertia scandens*, *Stephania japonica* and *Scirpus nodosus* (Doing 1985).

At the Pinery the transition zone occurs behind the second dune ridge in the wind shadow of the second dune ridge and acts as a transition between open grassland and forest (Fig. 11.6). The diversity of species and evenness increases steadily in this habitat for approximately 1000 years. The habitat is xerophytic and the dominant species are the low evergreen shrubs *Arctostaphylos uva-ursi* and *Juniperus communis* with other elements of the flora characteristic of tall grass prairie of the mid-west US and Canada. Prominent herbaceous species at the Pinery along Lake Huron and Miller Dunes along Lake Michigan (Poulson 1999) include *Schizachyrium scoparium*, *Andropogon gerardii*, *Sorghastrum nutans*, *Calamovilfa longifolia*, *Panicum virgatum*, *Stipa*

Figure 11.6 Grassland vegetation in the transition zone between the dune complex and forested area on Lake Huron sand dunes at the Pinery (photograph by A Maun).

Figure 11.7 Patches of developing vegetation consisting of *Juniperus virginiana, Rhus aromatica, Symphoricarpos albus, Pinus strobus* and *Carex eburnea* at the Pinery along Lake Huron (photograph by A Maun).

spartea, Koeleria cristata, Liatris cylindracea, L. spicata, Arenaria stricta, Artemisia campestris, Lithospermum caroliniense, Smilacina stellata, Arabis lyrata and a significant coverage of mosses (27%) and lichens (3%) (Table 11.6). *Arctostaphylos uva-ursi* is by far the most common species that expands over the surface horizontally, outcompeting adjacent herbaceous vegetation. This community forms a vegetative cover over the sand surface, prevents soil erosion, allows retention of soil moisture and creates ideal conditions for the germination, survival and establishment of woody vegetation. The interface between grassland and transition into forest communities begins in the stabilized low-lying areas with higher soil moisture content. Yarranton and Morrison (1974) proposed a 'nucleation model' of succession whereby woody plant species grow as small patches in the grassland zone. Usually seedling establishment starts at the base *of Juniperus virginiana* trees and over time develops into groves of woody and herbaceous vegetation consisting of *Juniperus virginiana, Vitis riparia, Quercus velutina, Q. rubra, Pinus strobus, P. resinosa, Rhus aromatica, Prunus virginiana, Symphoricarpos racemosus, Carex eburnea* and *C. lanuginosa* (Fig. 11.7). Gradually, these groves expand and coalesce thus forming closed forest

of the persistent stage. Similar establishment of woody vegetation within the grassland flora also occurs along Lake Michigan (Cowles 1899). At Miller Dunes, nucleation of bird-dispersed species occurs around *Populus deltoides* trees that provide perches to birds (Poulson 1999).

The foregoing community of grasses and shrubs eventually leads to the development of a vast and rich vascular flora of trees, shrubs and herbaceous plants (Table 11.7). The species complement in different regions of the world is diverse and varies according to its location (latitude), amount and distribution of rainfall, length of the dry season, biotic conditions, plant community, age of the dune system and soil development. In most regions of the world these areas have been logged and planted with trees of greater economic importance or logged and converted to farmland and pastures for livestock. Based on Raunkiaer's growth forms, plants in the dune complex of the Great Lakes could be described as therophytes on the mid-beach/driftline, geophytes on the strand, first and second dune ridges, sporadic chamaephytes such as *Arctostaphylos uva-ursi* and *Juniperus communis* on the lee slopes and mixtures of geophytes, chamaephytes, hemicryptophytes and phanerophytes in the grassland/transition zone.

Table 11.7 Dune scrub and woodland species on coastal sand dunes along sea coasts of the world

Europe and the Mediterranean
Quercus robur, Q. ilex, Pinus sylvestris, P. contorta, P. pinea, P. pinaster, P. halepensis, Picea sitchensis, Crataegus monogyna, Hippophae rhamnoides, Juniperus phoenicea, J. oxycedrus, Salix repens, Sambucus nigra, Rosa canina, Myrica gale and many others (Doody 1991).

Atlantic coast of Canada and United States
Picea mariana, Pinus nigra, Myrica gale, Gaylussacia baccata, Juniperus horizontalis, Hudsonia tomentosa, Corema conradii (Thannheiser 1984); Florida coast: *Coccoloba uvifera, Opuntia stricta, Sabal palmetto, Suriana maritima, Tournefortia gnaphalodes* (Stalter 1993).

Gulf of Mexico
Gulf of Mexico from Tampa Florida to the Mexican border: *Pinus taeda, Pinus elliottii, Quercus virginiana, Q. myrtifolia, Sabal palmetto, Prunus* spp., *Ilex vomitoria, Myrica cerifera* (Stalter 1993).

Mexico, Central America, Cuba, Venezuela, Brazil
Acacia farnesiana, A. macracantha, Dalbergia ecastaphyllum, Indigofera hartwegii (Moreno-Casasola 1993); Cuba: *Thrinax radiata, Bursera simaruba* and *Ficus* spp. are common (Borhidi 1993); Brazilian coast: *Clusia hilariana, C. fluminensis, C. lanceolata, Cassia australis, Myrcia lundiana, Allagoptera arenaria, Axonopus barbigerus, Byrsonima sericea, Vriesea neoglutinosa* (De Lacerda *et al.* 1993).

Japanese coasts
Quercus dentata, Q. phillyraeoides, Q. mongolica, Pinus thunbergii, Juniperus conferta, J. chinensis, Vitex rotundifolia, Rosa rugosa (Miyawaki and Suzuki 1993).

Sierra Leone, Ivory Coast, Guinea, Zaire, Ghana
Dalbergia ecastaphyllum, Hibiscus tiliaceous, Phoenix reclinata, Chrysobalanus orbiculari, C. ellipticus (Lee 1993).

Namibian coast
Zygophyllum clavatum, Psilocaulon salicornioides (Penrith 1993).

South Africa
Passerina rigida, Helichrysum asperum, Stipagrostis zeyheri which gradually develops into a dune thicket of *Brachylaena discolor, Colpoon compressum, Eugenia capensis, Rhus nebulosa* (Weisser and Cooper 1993).

Somalia, Kenya and Tanzania
Cocos nucifera, Acacietum tortili-bussei, Rhynchosia velutina, Jatropha crinita, Mariscus somalensis, Pandanus kirkii, P. rabaiensis, Dodonaea viscosa (Pignatti *et al.* 1993; Frazier 1993).

Indian subcontinent from Pakistan, coastal states of India and Bangladesh (6000 km)
Subhumid, humid and wet coasts: Bay of Bengal: *Borreria articularis, Hydrophylax maritima* (Rao and Meher-Homji 1993).

Java coast
Barringtonia asiatica, Cycas rumphii, Hernandia peltata, Terminalia catappa, Woodfordia fruticosa (Hardjosuwarno and Hadisumarno 1993).

Australian coastal dune systems
Eucalyptus papuana, Melaleuca dealbata, Terminalia petiolaris, Mimusops elengi, Pouteria sericea, Acacia holosericea, A. tumida, Gyrocarpus americanus, Lysiphyllum cunninghamii (Beard and Kenneally 1993).

11.3 Quantitative analysis of communities

The Pinery sand dune system of Lake Huron ranges in age from 100 to 4800 years. Morrison and Yarranton (1973, 1974) determined presence or absence of species at 200 points along a transect from the beach to the oldest dunes using a modified point sampling apparatus on each of 15 dune ridges of different ages. They then calculated richness, evenness, species diversity and applied Goodall's (1953) test of heterogeneity to the data. Diversity of

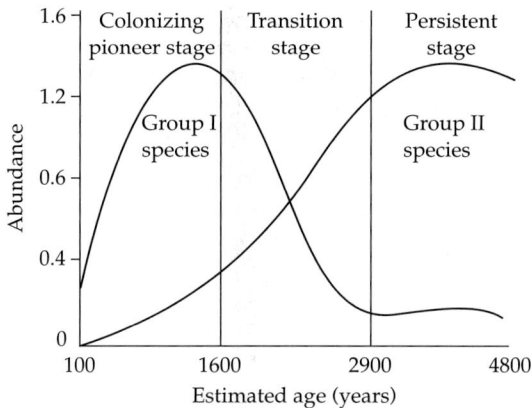

Figure 11.8 Smoothed curves of frequencies of colonizing (Group I) and persistent (Group II) species of plants on the Pinery sand dunes (Morrison and Yarranton 1974).

vegetation increased with distance from the shore, mainly because of an increase in species richness and evenness. The succession on the dunes was divided into three periods: the colonizing pioneer stage up to 1600 years, the period of transition (grassland to forest) from 1600 to 2900 years and the persistent stage of forest communities from 2900 to 4800 years. Inverse classification, based on species frequency, clustered all plant species in the colonizing pioneer stage of the dune complex into one group and a plot of the means, variances and variance/mean ratios in relation to estimated age suggested that this group reached its maximum abundance after 1600 years and then declined rapidly until 2900 years (Fig. 11.8). What are the reasons for this change in vegetation? Four main factors—micro-environmental changes, allogenesis, recurrent disturbance and stability of the habitat—may be the most probable causal agents.

11.4 Summary

Diverse plant communities exist on the sand dune complex along sea coasts because of major differences in climate, rainfall patterns,

length of growing season and dry periods. Along a sea coast, wave action exerts a strong influence on the strand. High waves deposit sandy material, propagules of plants, animal carcasses and other types of detritus on the shoreline. Even though the strand is a harsh environment for the establishment of plants, three to five species of annuals, biennials and perennials such as *Cakile, Salsola, Euphorbia, Atriplex*, and *Honkenya* spp. are able to establish and complete their life cycles in temperate regions. The survival of annual species depends on several adaptive traits such as high phenotypic plasticity, good dispersal ability, high fecundity, short life cycle from germination to maturity, seed dormancy and lager seed mass. In southern locations, the complement of species changes to species of genera *Ipomoea, Sesuvium* and *Canavalia*. The life of perennials is, however, precarious because occasional violent wave storms can wipe out whole populations in some years. The detritus washed up on the shoreline serves as an important source of nutrients for these species.

The area above the driftline is more favourable for plant growth because of lower disturbance by wave action, but high wind velocities are a major environmental hazard in this zone because they transport sand from the beach and deposit it on plants. This zone is occupied by dune-forming species that grow through sand deposits and build embryo dunes. Prominent species in this zone in various parts of the world are *Ammophila arenaria, A. breviligulata, A. baltica, A. arundinacea, Elytrigia junceiformis, Elymus farctus, Leymus (Elymus) arenarius, Carex kobomugi, Schizachyrium pulchellum, Scaevola plumieri, Ipomoea pes-caprae, I. stolonifera, Spinifex* species and many others. These species also grow horizontally beneath the soil surface or at the soil surface by producing rhizomes or stolons. High wave action frequently fragment these rhizomes or stolons and disperse them to new locations.

The vertical growth in response to sand deposition eventually leads to the formation of dune ridges parallel to the sea coast. The lee slopes of dune ridges usually support a different type of vegetation than dune summits because of decreased wind velocity, lower moisture content and higher temperatures. Behind the first dune ridge there is often a slack or low-lying area between dune ridges. In the slack, high winds have been responsible for eroding sand down to the capillary fringe. Generally three types of slacks—dry, wet and flooded—are found in dune systems. They support different vegetation types depending on the soil moisture, oxygen levels, pH, nutrient content and wet/dry cycles. High wind velocities are still important environmental stresses in this zone. However, farther inland the habitat becomes more stable, sand movement is no longer a limiting factor and soil organic matter and soil moisture have increased. This area develops a diverse community consisting of lichens, mosses, grasses and shrubs and creates ideal conditions for the establishment of woody vegetation. Once a forest community establishes, the shade is cast on the forest floor and species capable of growing under shade are able to flourish. The species complement in different regions of the world is diverse because of differences in latitude, rainfall, length of dry or wet periods, biotic conditions, age of dune systems and disturbance history.

CHAPTER 12

Zonation and succession

12.1 Introduction

There is ample evidence that a progressive change in the intensity of an important environmental factor leads to the formation of zones or belt-like communities in which the plant species reflect a fairly distinct range of tolerance for that factor (Daubenmire 1968). Zonation has been defined as a sequence of vegetation *in space* and succession as a sequence of vegetation *in time* (McIntosh 1980). A zone is an area occupied by a plant community that is distinctly different from other zones and can be readily recognized by a change in dominant vegetation. Striking examples of zonation are found in salt marshes, mountain slopes and ponds because of soil salinity in salt marshes, decrease in temperature on mountain slopes and increase in water depth in ponds (Daubenmire 1968; Chapman 1976; Partridge and Wilson 1988). Similarly, it has long been known (Beck 1819) that sand dunes along sea coasts exhibit a zonation pattern extending from the beach to inland dunes. The zones are discrete and occur in parallel series with distinctly different species composition that is related to the ability of plant species to withstand the environmental factors prevailing in that zone (Doing 1985). Many later studies using transects from the shoreline to the inland dunes have confirmed that the taxa are not randomly distributed; they peak at definite distances from the beach (Oosting and Billings 1942; Boyce 1954; Martin 1959; Barbour 1978; Barbour *et al.* 1985).

Succession in coastal dunes is an example of primary succession because the sandy material deposited on the shoreline by waves is inert. The term is generally used to denote a directional change in species composition and physiognomy of vegetation at the same site over time (Drury and Nisbet 1973). However, only the very early stages of dune succession can be observed during the life time of a plant ecologist and the later stages are usually inferred from plant communities represented on older sand dunes. It is hypothesized that the autogenic influence of early colonizers alters environmental conditions in the habitat and facilitates the establishment of new species better adapted to live in the altered habitat. The change in environmental factors may include a decrease in light intensity, change in light quality, increase in soil organic matter and soil moisture, enhancement of nutrients (see Chapter 2) and stabilization of the substrate. The classic study of plant distribution on Lake Michigan sand dunes by Cowles (1899, 1901) initiated the study of sand dune succession. He writes:

The condition of equilibrium is never reached and when we say that there is an approach to a mesophytic forest, we speak only roughly and approximately. As a matter of fact we have a variable approaching a variable rather than a constant. These conditions do not destroy the validity of physiographic classification, but rather they require an enlargement of the conception.

Many varied ecological studies on transformation of one community to another have been conducted in sand dunes around the world during the last 100 years. Detailed accounts of the latest developments and critiques of the theory of sand dune succession are found in Lichter (1998, 2000), Poulson (1999), Poulson and McClung (1999), Tilman (1985, 1988, 1993), Walker and del Moral (2003) and Miyanishi and Johnson (2007). In this chapter attempts are made to discuss various aspects of zonation and succession in coastal sand dune complexes.

Succession as a term has been defined in different ways. If the species change in a plant community is influenced by the component organisms it is referred to as autogenic succession. In contrast, allogenic succession is driven by abiotic forces such as salt spray, burial by sand, wind velocity and burrowing activity of animals. However, any community in a dune complex is a consequence of both biotic and abiotic forces and it would be naive to label an entire sere as autogenic or allogenic. If we define succession as a unidirectional change in composition of vegetation it may not apply to communities of the dune complex. However, if succession is viewed as a universal process of species change (Cooper 1926) with no connotations of directionality or equilibrium, then zonation is really a pattern manifested by both abiotic and biotic factors and may be regarded as reflecting stages in the overall process of succession.

12.2 Zonation: possible causes

What is the cause of this zonation? Ecologists have pondered over this question for more than 100 years. Two major environmental factors, burial in sand and salt spray, have been implicated as stresses because both salt spray and burial have negative impacts on plant growth above a certain threshold concentration of salt and level of burial in sand. Arguments for and against the hypotheses advanced by ecologists are presented in the following pages.

12.3 Is zonation determined by soil salinity and salt spray?

To answer this question researchers sought evidence for two aspects of this problem: (i) was there a gradient in salt spray and soil salinity from the beach to inland dunes and (ii) does the tolerance of component species

in successive zones decrease with a decrease in the intensity of salt spray and/or seawater inundation? In answer to the first question, it was shown by many authors (Chapter 8) along both the Atlantic and Pacific coasts that there was indeed an ecologically and statistically significant salt spray gradient which was correlated with distance from the tide line to the inland dunes. Elevation of dune ridges, topography of habitat and plant height all modified the amount of salt intercepted in the microhabitat but the basic relationship of decreased salt content with distance remained.

To answer question two, phytometer studies were conducted to test the hypothesis that plant species occurring close to the beach should be more tolerant of salinity than those farther inland. Since the distribution of three beach taxa, *Spartina patens, Uniola paniculata* and *Schizachyrium scoparium*, of the Carolina coast correlated well with their tolerance level, Oosting and Billings (1942) concluded that the zonal distribution of major species was largely controlled by the extent of exposure of the habitat to wind-borne salt spray. I will call this the salt hypothesis. The hypothesis was accepted and is reported in major ecology textbooks (Daubenmire 1968; Krebs 1994). They assumed that since salt spray concentrations decrease with distance from the sea coast, the salt tolerance of plant species should also decrease with distance. Barbour and De Jong (1977) questioned the assumption and validity of salt hypothesis and said that 'single factor hypotheses were too naive to describe the zonation of all species on a sand dune'. They artificially exposed 12 dominant taxa of Pacific coast dunes to different levels of salt spray and found that three or four taxa showed either much less or much more salt tolerance than expected from their position on the salt spray gradient. Some species were very sensitive to salt spray yet they occurred very close to the beach where they were receiving much higher concentrations of salt. They concluded that tolerance to salt spray does not explain species zonation on coastal dunes for all beach taxa. Similarly, at the Cape Hatteras seashore along the Atlantic coast, the distribution of species did not correspond with their salt tolerance index and the salt spray gradient (van der Valk 1974). Thus, the degree of susceptibility was not the controlling factor. Here, I present further evidence that salt spray may not be the primary cause of zonation.

12.3.1 Soil salinity

As explained in Chapter 8, the salinity of soil on the beach and strand is very low, ranging from only 0.003–0.13% (Table 8.4), is well below the critical inhibitory level of about 1–1.5% for susceptible species and many orders of magnitude lower than highly resistant species (Table 8.8). It is only very close to the sea coast that the beach habitat may receive rather high amounts of salt in the soil following inundation of the habitat by high waves, especially during storms in the autumn and spring months, thus raising the soil salinity levels. Plants, especially the seaward portions of populations of rhizomatous perennials growing on the upper beach, are frequently inundated by storm waves that may be critical stressful events for their growth and survival. However, according to Olsson-Seffer (1909) plants are not affected because the water runs off before it has time to sink through the layer of freshwater lens above the seawater along sea coasts. They may be temporarily exposed to brackish water and since their roots do not penetrate deeper than the freshwater layer in the soil, they are not affected. However, since functional plant roots are shallow they will be exposed to injurious salt water concentrations.

Baye (1990) conducted seawater immersion experiments to test the regeneration capacity of apices of shoots, dormant buds of rhizomes and roots of two major species, *Ammophila breviligulata* and *A. arenaria*, of sea coasts of North America and Europe, respectively. Dormant tillers and rhizomes of both species were subjected to six treatments of 0 (control), 13, 26, 52, 104 and 1700 hrs of immersion in

cold artificially produced seawater (3.4% salinity) using instant ocean salt. Control plants were immersed in cold tap water. The data showed that both species were capable of vegetative regeneration after immersion of tillers and rhizomes in cold seawater. In *A. breviligulata*, the percentage regeneration of apical meristems of shoots and buds on immersed rhizomes was not significantly different from control and ranged between 80 and 100% in all immersion treatments. In *A. arenaria*, the immersion of buds actually showed a marked stimulation in the regeneration of buds. For example, in the control treatment there was only 10% regeneration compared to 50–66% in the salt immersion treatments ranging from 13–104 hrs of immersion. However, buds on rhizomes subjected to 1700 hrs of immersion failed to regenerate. The regeneration of tillers was also not affected. The apical meristem of tillers resumed growth within a few days of planting and the rate of leaf emergence per tiller did not differ significantly between treatments. The percentage regeneration following seawater inundation was significantly higher in *A. breviligulata* than *A. arenaria* in all immersion treatments. Three other species of coastal strands, *Leymus mollis* (Pavlik 1983), *L. arenarius* (Bond 1952), and *Elymus farctus* (Harris and Davy 1988), also regenerate prolifically following seawater transport of their fragments.

The second factor that protects and prolongs the life of *A. breviligulata*, and probably other species with similar growth habits from injury is the connection of a seawater-inundated ramet with other ramets in the habitat with no inundation. Baye (1990) tested the effect of inundation by a 2% salt solution on ramets originating from rhizomes linked to other ramets or unlinked independent ramets of *A. breviligulata*. He found that inundation was lethal to rhizome-severed independent plants but in treatments where the rhizome was still linked to the parent plant all salinized plants survived. The leaf extension rate (cm day^{-1}) of seawater-inundated over-wintered shoots of *Ammophila breviligulata* and *A. arenaria* was not significantly different from control. This trait would be of high adaptive value to expand and colonize saline sites close to the shoreline. In effect clonal integration allows the species to withstand high soil salinity in part of the population without the corresponding evolutionary costs of developing resistance.

12.3.2 Seasonality of salt spray storms

The major salt storms and inundation of beaches by high waves along sea coasts are seasonal. For the evolution of a trait for resistance in an organism, the environmental stress must be persistent and recurrent. For that reason adaptation in response to salt spray has been limited. In temperate regions with cold winters most storms with high wind velocities, strong wave action and salt spray occur in the autumn and winter months when plants are dormant (van der Valk 1974). The dormant buds especially of geophytes are well protected below the soil surface and are not influenced by seawater inundation during the dormancy period. Then in spring rainfall is abundant: it washes the salt off the plants and leaches it out of the rhizosphere of well-drained sandy soils of coastal dune systems so that plant roots and dormant buds do not experience the high salinity levels. Similarly, the reduced levels of salt in the soil are not a limiting factor for seed germination, seedling establishment and growth of seedlings and adult plants of the sandy beaches and foredunes. During the summer season rain leaching may be less effective, but since major salt spray storms with gale force winds are episodic and occasional during the summer growing period, plant species are not damaged. In subtropical and tropical regions with no dormancy, hurricanes with high wave action and persistent salt spray will exert a strong stress on plant species.

12.3.3 Escape from salt exposure

Salt spray loads in microhabitats within the same habitat are not uniform and may differ considerably. The plants may be killed when growing in an exposed position but some plants may survive because of their location in a protected microhabitat or by escaping in the shadow of nurse plants.

12.3.4 Relative humidity

Relative humidity is an important factor in the entry of salt into the leaf following a salt spray event. The salt enters the leaf only when it is in liquid form. If the relative humidity is <60%, the salt forms a crust over the leaf but does not enter the leaf blade: salt in this form is not very effective in causing injury. Any factor such as high humidity, dew formation, light drizzle and fog would convert the crust into a liquid, thus enhancing the probability of foliar uptake of salt. Martin (1959) observed that on clear days, after the spray treatment on plants the leaves rarely remained wet for more than one hour and a thin crust of salt was formed on the leaves. However, on cloudy days with high relative humidity the salt absorbed moisture from the air due to its deliquescent properties and became a liquid. He reported that plants in the resistant group grew in areas where the salt spray dried quickly and the bathing salt solution was renewed only for a short period each day, suggesting that plants were not exposed to the salt spray for a long enough period to be injuriously affected by salt.

12.3.5 Plant adaptations

As shown in Chapter 8, the amounts of salt spray may be high enough to kill sensitive species, but most species occurring there are well adapted to withstand it. To confirm this the concentration (%) of Na and Cl ions in the foliage and twigs of plant species growing on the foredunes along the sea coast was determined by collecting plants along three transects drawn perpendicular to the sea coast at Basin Head, Prince Edward Island on two dates in 2003 and one date in 2004. The

Table 12.1 The concentration (%) ± SE of Na and Cl ions in dried samples of foliage and twigs of plant species growing on the foredunes along the sea coast at Basin Head, Prince Edward Island on three dates in 2003 and 2004 (A Maun and I Krajnyk unpublished data)

Species	August 2003		October 2003		July 2004	
	Na	Cl	Na	Cl	Na	Cl
Beach						
Cakile edentula	2.43 ± 0.12a	3.73 ± 0.06a	0.73 ± 0.09b	1.08 ± 0.22b	0.87 ± 0.23b	3.02 ± 0.36b
Ammophila breviligulata	0.27 ± 0.03b	0.88 ± 0.17c	0.67 ± 0.09b	1.33 ± 0.33b	0.13 ± 0.04cd	0.83 ± 0.06c
Solidago sempervirens	2.03 ± 0.07a	1.85 ± 0.23b	2.17 ± 0.22a	2.95 ± 0.27a	2.57 ± 0.12a	3.58 ± 0.13a
Dune ridge						
Picea mariana	0.09 ± 0.00c	0.18 ± 0.01d	0.09 ± 0.00d	0.21 ± 0.01c	0.30 ± 0.06c	0.41 ± 0.04d
Gaylussacia baccata	0.20 ± 0.00bc	0.42 ± 0.02cd	0.30 ± 0.06c	0.55 ± 0.04bc	0.09 ± 0.00d	0.31 ± 0.05de
Myrica gale	0.17 ± 0.03bc	0.40 ± 0.01cd	0.30 ± 0.00c	0.52 ± 0.03bc	0.23 ± 0.03cd	0.33 ± 0.04de
Juniperus horizontalis	0.09 ± 0.00c	0.18 ± 0.01d	0.09 ± 0.00d	0.18 ± 0.01c	0.09 ± 0.00d	0.20 ± 0.01f
Arctostaphylos uva-ursi	0.09 ± 0.00c	0.38 ± 0.02cd	0.09 ± 0.00d	0.33 ± 0.02c	0.09 ± 0.00d	0.25 ± 0.01def
Hudsonia tomentosa	—	—	—	—	0.09 ± 0.00d	0.26 ± 0.01def
Slack						
Corema conradii	0.09 ± 0.00c	0.14 ± 0.01d	0.09 ± 0.00d	0.12 ± 0.01c	0.09 ± 0.00d	0.13 ± 0.01f
Pinus nigra (planted)	0.10 ± 0.00bc	0.20 ± 0.01d	0.13 ± 0.03cd	0.20 ± 0.04c	—	—

Means in each column followed by the same letter are not significantly ($P < 0.05$) different according to Anova followed by Tukey's tests.

data are presented in Table 12.1. The dominant plant on the beach strand was *Ammophila breviligulata*. Its Na content ranged between 0.13–0.67% and Cl content between 0.83–1.33%. These values are within the range of tolerance of salt-resistant glycophytes. Two strand species, *Cakile edentula* and *Solidago sempervirens*, contained rather high amounts of both Na and Cl ions but the plants were healthy and showed no symptoms of injury. Importantly, both species are succulents and sequester salts in their vacuoles. The deciduous shrubs *Myrica gale* and *Gaylussacia baccata* were growing primarily on the lee of *Picea mariana*. Since they lose their leaves in autumn and new growth in spring takes place from dormant buds, they are able to avoid the salt spray injury. Even *Picea mariana* contained low amounts of Na and Cl ions. The trees were short in stature (~2–4 m in height) and did receive rather high amounts of salt in the spring and autumn months that was evident from the absence of needles on the sides of trees facing the sea coast. However, during July and August trees were healthy and growing well. Nevertheless they are pruned back by high winds and salt during winter. The pine trees (*Pinus nigra*) planted in the slack were taller than *P. mariana* but the needles on seaward branches of many trees were missing. The rest of the species on the first dune ridge and slack had rather low to very low concentrations of Na and Cl ions (Table 12.1). All these species are decumbent, are covered by snow in winter and probably receive low amounts of salt spray deposits.

12.3.6 Zonation on lacustrine shores

As shown earlier, zonation also occurs on sand dunes along shorelines of freshwater Great

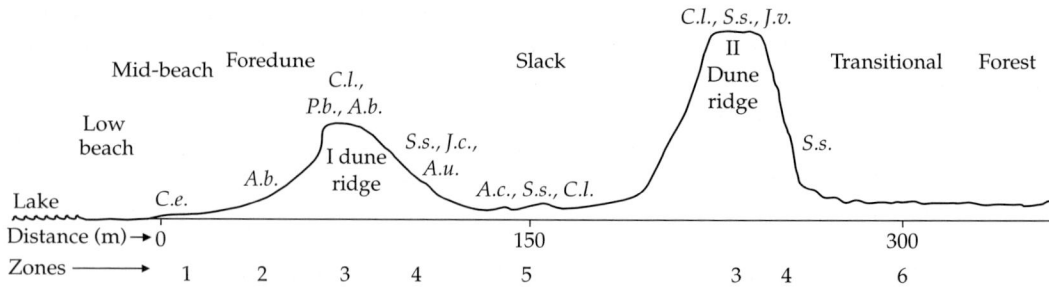

Figure 12.1 Zonation of vegetation from the beach to the rear of dunes along the shoreline of Lake Huron at the Pinery (modified from Maun 1998). For a complete list of species in each zone refer to Chapter 11. More examples of zonation patterns along sea coasts of the world are presented by Doing (1985) and *Dry coastal ecosystems*, 1993, Vol. 2 A, B, C, edited by Eddy van der Maarel.

Lakes Huron and Michigan with many of the same species (Cowles 1899). Six zones are distinct on the dune complex at the Pinery within about 300 m of the shoreline (Fig. 12.1).

12.4 Is zonation caused by burial in sand?

We critically examined the 'burial hypothesis' that species zonation is caused by differential sand accretion and differential susceptibility of plant species of a community to burial in sand. We used the same types of approaches as those used by Oosting and Billings (1942), (i) relate species distribution with natural sand accretion on a sand dune system and (ii) determine survival, distribution and relative abundance of plant species in a community following artificial exposure to abiotic stresses.

12.4.1 Natural burial and species distribution

As shown earlier (Fig. 7.1) the total deposition of sand decreased as one moved inland from the shoreline to inland dunes (Martin 1959). Davidson-Arnott and Law (1990) measured sand supply and seasonal variation in sand deposition on a lacustrine shoreline along Lake Erie. They showed that there was a moderate deposition of about 0.25 and 0.35 m^3 m^{-1} of beach width per month from May to the end of October, except during July when there was a slight deflation. The deposition increased substantially in November to 0.95 m^3 m^{-1} of beach width. At the second site deposition exceeded 2.0 m^3 m^{-1} of beach width in October and November. The significant increase in sand deposition during the autumn months was caused by an increase in both the frequency and magnitude of onshore winds. Vegetation produced a significant surface roughness during the autumn, thus reducing wind velocity and turbulence and trapping sand being transported by winds. However, by December the effectiveness of vegetation to trap sand had declined because of the burial of leaves and culms of grasses and the sand was moving farther inland onto and over the main foredune ridge. There was considerable variation in the amount of sand deposition or deflation on different parts of the strand, and there was a distinct decrease in the amount of sand deposition with distance from the shoreline.

Perumal (1994) installed steel stakes (5 cm apart) on six transects running perpendicular to the shoreline and extending inland to about 50 m from the foredune of the first dune ridge. The amount of sand accumulated (+) or eroded (–) at each stake was recorded once a month for two years, except during winter when the dunes were covered with snow. The percentage cover and frequency of each plant species was recorded in a 1 m × 1 m quadrat at each steel stake using the modified Braun Blanquet technique (Westhoff and van der Maarel 1978). The data revealed a close correlation among plant species, vegetation cover, sand movement and dune shape. The principal components analysis (PCA) ordination showed a species gradient which was related to the topography of dunes and the amount of sand deposition in each region of the dunes. The spatial distribution of plants was correlated with the amount of sand accretion. Four clusters were clearly evident: (i) all species of the community, (ii) *Ammophila breviligulata* and *Calamovilfa longifolia*, (iii) *A. breviligulata* alone, and (iv) bare areas. Almost all species survived limited sand accretion of a few centimetres. However, as sand accretion increased the susceptible species were eliminated and resistant ones proliferated until burial exceeded the survivability levels of all species and a bare area was created.

A similar study by Moreno-Casasola (1986) on a semi-mobile sand dune system along the Gulf of Mexico showed that the floristic gradient resulting from a relevé ordination was correlated with the amount of sand movement at each relevé and the topography of the sand dune. She found *Croton punctatus* and *Palafoxia lindenii* on the crests and leeward slopes where sand movement was the highest, *P. lindenii* and *Chamaecrista chamaecristoides* occurred in more mobile dunes with less sand accretion than the first group and *C. chamaecristoides* occurred alone in areas with both sand accretion and erosion. Bach (2001) found that upon burial 2 of the 13 species—*Thuja occidentalis* and *Prunus pumila*—showed an increase in relative abundance while the remaining 11, including species well adapted to sand dunes

such as *Populus balsamifera*, *P. tremuloides*, *Potentilla fruticosa* and *Hypericum kalmianum*, showed a decrease in the number of stems per quadrat. Burial imposes a strong stress on production by altering normal growth conditions and exposing plants to extreme physiological limits of tolerance. Another recent study (Dech and Maun 2005) on a burial gradient on the leeward slopes of the Pinery sand dunes along Lake Huron showed that burial altered woody and herbaceous species richness, diversity, and community dominance on the basis of their tolerance on the burial gradient.

12.4.2 Artificial burial and species coverage

Perumal (1994) selected one homogeneous area approximately 15 × 25 m at each of two locations, the Pinery along Lake Huron and Port Burwell along Lake Erie, and divided it into 1 m × 1 m or 0.5 × 0.5 m squares. An estimate of percentage cover (pre-burial coverage) of each species was then recorded for each square using the modified Braun Blanquet technique (Westhoff and van der Maarel 1978). Burial treatments consisting of 0 (control), 5, 10, 20, 40 and 80 cm of sand deposition were then assigned to each square. Following a recovery period of one month the presence or absence and coverage of species for all burial treatments in both locations were recorded once a month for 15 months except during winter.

The calculated chi-square test showed that the plants were randomly distributed at both the Pinery (calculated $\chi^2 = 38.4$, table value $= 43.8$) and Port Burwell (calculated $\chi^2 = 28.7$, table value $= 66.3$) before the imposition of burial treatment. However, when the χ^2 test was repeated using the final post-burial coverage values, the null hypothesis was rejected at both sites (Pinery: calculated $\chi^2 = 63.5$, table value $= 43.8$); (Port Burwell: calculated $\chi^2 = 137.3$, table value $= 66.3$). Thus, there was an association between the species abundance and burial depth. The species clearly showed differential tolerance to burial

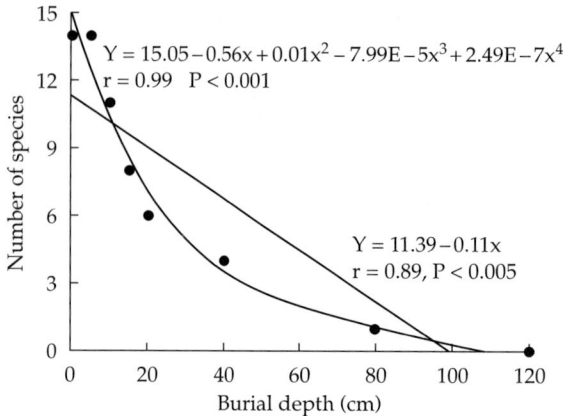

Figure 12.2 The number of plant species remaining in the sand dune plant community after 15 months of burial at different depths in sand. Linear and polynomial regressions were fitted to the data and both curves are represented (Maun and Perumal 1999).

and the number of species declined with an increase in burial depth (Fig. 12.2). A quartic relationship gave a better fit, indicating that a majority of species were killed by small amounts of burial in sand.

The percentage coverage of each plant species in response to different depths of burial in sand is presented in Fig. 12.3. *Euphorbia polygonifolia* plants were decumbent and completely overwhelmed by the 5 cm burial treatment and their cover was reduced to almost 0%. However, by the end of summer the plants had recovered to the level of control. *Corispermum hyssopifolium* showed a slight decline in coverage in the 5 cm burial treatment but were killed at higher depths of burial. *Cakile edentula* plants were not affected by the 5 cm burial treatment but at the 10 cm burial depth the percentage cover was reduced to very low levels. *Strophostyles helvola* showed stimulation in growth in the 5 and 10 cm burial treatments. Buried plants of *Melilotus alba*, *Xanthium strumarium* and *Elymus canadensis* exhibited an increase in percentage cover in the 5, 10 and 15 cm burial plots as compared to the control. *Poa compressa* and *Schizachyrium scoparium* plants survived in the 5, 10 and 20 cm

burial depths. Although the live shoots were more luxuriant than control the plants did not show any significant change in their percentage coverage compared to control. *Calamovilfa longifolia* survived burial to depths of 40 cm but the percentage cover only increased in the 10 cm burial plot. In the 20 cm burial treatment there was no increase in coverage. *Agropyron psammophilum* and *Panicum virgatum* had about 28% cover in the plots before burial. Following burial to 20 cm depths the coverage increased above the level of control. However, at 40 cm burial even though the plants survived, the coverage was significantly lower than control. *Ammophila breviligulata* exhibited an increase in its percentage coverage with the imposition of burial treatments. Partially buried plants recovered quickly and started to show enhanced growth within the first month and soon had a cover of almost twice that of control plots. Only in the 80 cm burial treatment was the recovery slow.

All species showed that below a certain threshold level of burial specific to each species there was either no change or stimulation in plant growth (Fig. 12.3). However, as the level of burial increased the species coverage declined and plants failed to emerge from the burial deposit. Superimposition of data on species tolerance created six clusters. The four annuals did not emerge from 15 cm of burial. The three biennial species were eliminated at 20 cm of sand deposit. Among the perennials, *Poa compressa* and *A. scoparius* dropped out at 40 cm depth, *C. longifolia, Agropyron psammophilum* and *P. virgatum* were killed at 80 cm and only one species, *Ammophila breviligulata*, was left at 80 cm. In an earlier report (Maun and Lapierre 1984) it was shown that even *A. breviligulata* failed to emerge from 120 cm of burial and a bare area was created. These studies clearly showed that burial in sand altered the species composition of the original plant community in relation to the amount of burial. As burial increased the tolerant species proliferated and expanded into the space vacated by susceptible species.

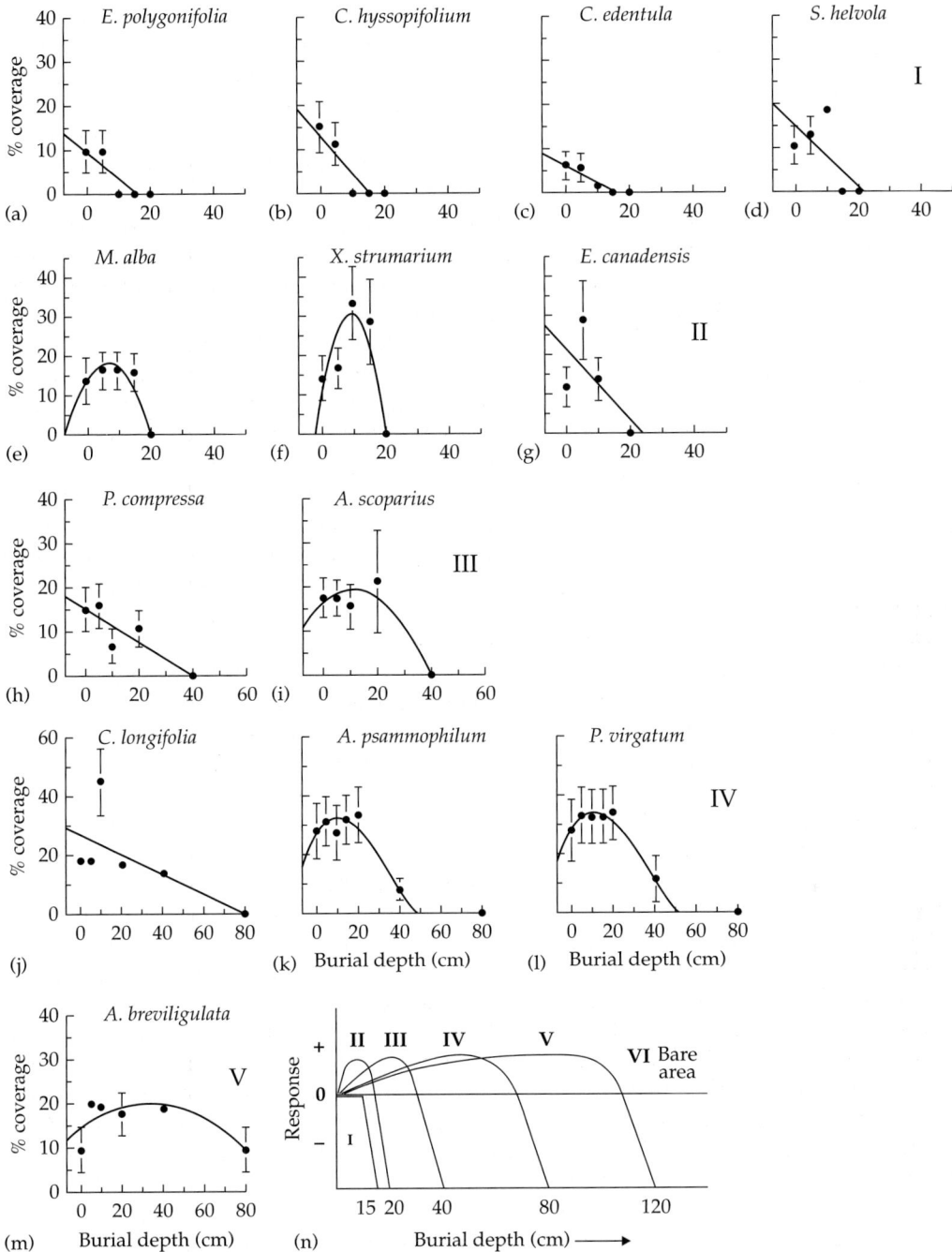

Figure 12.3 Effects of different depths of burial on percentage cover of each species of naturally growing plant communities on sand dunes of Lakes Huron and Erie. A model of species response curves in relation to the depth of burial in sand obtained by the superimposition of data on coverage of species from 12.3a–m is shown in 12.3n. (+) and (−) represent increase or decrease in species coverage, respectively. Roman numerals I to VI are zones created by the elimination of species in response to increasing depths of burial (Maun and Perumal 1999).

12.5 Causes of zonation: a critical evaluation

In studies on salt spray and burial by sand the assumption was that since salt spray concentrations and burial depths decrease with distance from the sea coast, the salt and burial tolerances of plant species growing in subsequent zones should also decrease. This is perhaps a wrong assumption because the environmental factors such as wind velocity, sandblasting and shifting of substrate also decrease with an increase in distance from the coast. At very low concentrations, salt spray is a weak nutrient solution that stimulates the growth of plants. As shown earlier, strand plants possess many traits by which they avoid entry of salt into the plant. Nevertheless, salt does enter the leaves at high relative humidities and given the levels of salt spray on sea coasts the majority of the species are capable of surviving the salt spray episodes of each zone. Mortality that may have been caused by salt spray alone is rarely observed. According to Barbour (1970b) strand plants can only be classified as intolerant halophytes or salt-resistant glycophytes.

From the foregoing discussion it is clear that strand plants may resist stress from salt spray and burial in three ways: escape, avoid or tolerate. Plant species that *avoid* salt stress exclude the factor from their tissues by various morphological features that reduce or inhibit its penetration into the plant, thus maintaining low salt concentrations within cells. In contrast, avoidance will not be an option for a plant that gets buried in sand. It must either emerge from the sand or suffer mortality. Plant species that *tolerate* salt stress allow the penetration of salt into their tissues and maintain dynamic equilibrium with their external environment. Tolerance is a more primitive form of adaptation because lower plants such as mosses and lichens resist frost, heat and drought solely by tolerance. As explained earlier, several mechanisms of salt tolerance are prevalent in plants of sandy sea coasts. Strand plants tolerate burial by an etiolation response,

elongation of the apical meristem and leaf petioles, activation and growth of dormant buds on rhizomes, stolons, roots and stems, and formation of roots on buried parts. For details refer to section 6.8.

The statement by Barbour and De Jong (1977) that single factor hypotheses can not describe species zonation on coastal dunes should also apply to burial by sand as a factor. Burial events do not deposit sand evenly over plants on foredunes. Strand plants are not only adapted to survive partial burial but some also are stimulated by it. However, burial above a certain threshold level kills them. When a plant is buried in sand it has no choice but to tolerate it. Thus, strand plants cope with salt spray stress by using all three strategies: escape, avoid and tolerate. In contrast, plants exposed to burial must tolerate the stress because they can not escape or avoid it.

Both salt spray and sand burial are important in species zonation on sand dunes along sea coasts. However, which one of these factors is more important? In an earlier report, Maun and Perumal (1999) suggested that burial by sand may be the most important factor in zonation. Wilson and Sykes (1999) argued against it on the basis of ordination and regression analysis of data on tolerance of 30 species to experimental salt spray, burial in sand, salinity of rooting medium and continuous darkness. They concluded that since mean variation in simple regression for the two salt treatments (salt spray and root medium salinity) was 27% compared to 13% for the two burial traits (burial and darkness), salt was more important than burial. I will argue against their interpretation of data. First, it is abundantly clear from Chapter 8 that natural salinity in the rooting medium is too low (Kearney 1904; Boyce 1954; Barbour and De Jong 1977) to be of any consequence to strand species. It should not even be considered as a variable affecting strand plants. Second, salt spray had an R^2 value of 72% only at one site (Kaitorete) with values ranging from 1 to 4 at the other three sites. Combining it with salt in the rooting medium increased the mean variation to 27%. Third, even though the authors used 66% burial in sand to evaluate the impact, it consistently had high R^2 values (26, 10, 19, 15) at the four sites. Fourth, continuous darkness should not be considered as a factor because following burial a plant experiences more than only darkness in the soil. Its leaves lose chlorophyll, respiration increases, it is attacked by bacteria and fungi, moisture increases, temperature decreases and the plant diverts all its energy to the growing apices and dormant buds of the plant. Combining burial treatment with continuous darkness treatment was not justified because it decreased the contribution by burial. Finally, the utmost caution should be exercised in the extrapolation of data from controlled experiments to natural field conditions. Plants growing under field conditions have different tolerances than those grown in the greenhouse and under hydroponic conditions.

All evidence supports the hypothesis that burial by sand is the major causative factor in the zonation of vegetation along sea coasts. On the dune complex many confounding variables and their interactions are affecting the plant communities of various zones. However, on the leeward sides of blowouts burial is the dominant factor affecting the plant community (Dech 2004). On a freshwater dune complex at the Pinery with no salt spray, Dech and Maun (2005) showed clear evidence that on the lee slopes of blowouts burial alone was responsible for altering tree and herbaceous species distributions on the basis of their tolerance to burial. Even though woody species took a long time to show signs of stress and eventual death, they did show differential susceptibility and change in plant community composition similar to that of herbaceous under-storey species.

12.6 Succession: basic assumptions

It is assumed that differences in plant communities on dunes of different ages from the foredunes to inland dunes are age-dependent and are chronosequences. That is, given enough time a foredune community, exhibiting only zonation due to present successional processes, will change by proceeding through stages in a chronosequence into communities present in older dunes. The soil development, macroclimate and source of propagules are also assumed to have been the same over the development period of succession. In other words, the spatial sequence of vegetation of a site across a chronosequence reflects only its present stage of development, with younger dunes representing a developmental stage through which the older dunes have passed. It is also assumed that pioneer plant species facilitate colonization and establishment of later successional species at the community scale because they add humus and improve soil development so that the conditions become more favourable for the establishment and survival of later successional species (the principle of facilitation). Tilman (1985) termed this the 'holistic theory' in which the successional change is orderly, predictable and deterministic. He further refined the idea by proposing the 'resource ratio hypothesis' which states that the pattern of species replacement in infertile sandy soils during succession can be explained on the basis of the relative availability of two main elements, nitrogen and light. In the early stages of primary succession, nitrogen is the most limiting element and later successional species that require high amounts of nitrogen would not be able to compete with early successional species requiring rather low levels of nitrogen. However, as soil resources improve the later successional species will establish and shade out the early successional species because of their superior ability to capture the available light. The resource ratio hypothesis thus makes two assumptions: (i) there is a directional change in the ratio of limiting resources and (ii) each plant species exhibits a superior competitive ability at a certain proportion of limiting resources.

More recently, a 'reductionist neo-Darwinian approach' based on Darwinian doctrine of dynamism was proposed, whereby succession is considered as merely sequential dominance of species based on differences in colonization potential, competitive ability, life history traits and rates of growth. The hypothesis is based on Gleason's (1926) interpretation of the plant community as a fortuitous assemblage of populations of species (the individualistic concept) with each assemblage independent of others both in time and space. Two models of succession are based on autogenic influence of individual species and environmental constraints. In the *inhibition model*, the autogenic influences of early successional species are considered inhibitory to the colonization of late successional species with current occupants of a site inhibiting the colonization of invading species by pre-empting space or chemical inhibition. After all, according to Crawley (1993) 'possession was nine-tenths of the law'. Later successional species will succeed only if they can overcome the inhibition of the current occupants through a chance release of space and resources through death or damage. Inhibition at the community scale can also occur by abiotic forces. For example, promotion of fire by graminoids in black oak savannah scorches tree trunks, top kills juveniles with vigorous sprouting and stimulates growth of graminoids and forbs (Poulson 1999). The *tolerance model* states that late successional species establish because they are more tolerant of lower levels of resources to reach maturity. The earlier successional species neither facilitate nor inhibit later ones.

12.7 Factors other than time that affect sand dune succession

There is increasing evidence that most assumptions regarding sand dune chronosequences are violated (Miyanishi and Johnson 2007). Changes in dune vegetation and soil properties across a chronosequence are a function of many other factors beside time: climatic history, composition of sandy material deposited by waves, past history of disturbance by

animals, fire, washovers, sand storms, fluctuation in sea levels and colonization constraints of plant species.

12.7.1 Climate

The successional hypothesis assumes that the climate has not varied significantly during the developmental history of plant communities in the chronosequence, however, this is not a valid assumption. Climatic conditions may have been quite different at various times during the successional history of the sand dune system. Ayalon and Longstaffe (2001) used isotopes to construct the climatic history during the chronosequence of sand dune succession of Pinery sand dunes and found drastic changes in rainfall and weather patterns during the last 4900 years. Global warming and sea level rise will probably have unpredictable consequences on sand dune succession in the future.

12.7.2 Texture of sand

The texture of material deposited by waves at different time periods may have been different. For example, on the Pinery sand dunes along Lake Huron mass abundance of silt and clay particles in soils was higher in older dune ridges, probably because of weathering of original deposits, influx from wind-blown material or the original deposits may have had greater amounts of silt and clay fractions (VandenBygaart and Protz 1995). Presently, the mean amount of sand deposited by waves is about 8 cm per year (Maun 1985), but judging from the height of old dunes (5 to 40 m) deposition must have been much higher in previous time periods. Evidence of massive sand deposition with dropping lake levels from 1880–1892, 1910–1930 and severe droughts of 1908–1910, 1917–1919 and 1930–1937 must have significantly altered the successional progression on Lake Michigan dune systems at Miller, Indiana (Poulson 1999) and Dunes Park (Cowles 1899). In the early 1900s, severe drought and low lake levels deposited large amounts of sand on the dune systems at Miller, Indiana and Dunes Park (Poulson 1999; Cowles 1899). Thus, the sand budget hypothesis recently proposed by Miyanishi and Johnson (2007) must be examined critically.

12.7.3 History of disturbance

Development of plant communities on dunes may have been arrested or accelerated by natural disturbances to the next stage. In some cases a recurrent disturbance may be maintaining an existing community. For example, promotion of fire by graminoids in *Quercus velutina* savannah stimulates growth of graminoids and forbs but scorches tree trunks. The leader of young saplings may be killed but the plant usually survives by re-sprouting from buds located deep on the rootstock (Poulson 1999). Poulson showed that the *Quercus velutina* stage in the successional diagram of Olson (1958a) is actually a fire climax rather than a temporal stage in the chronosequence. The species is relatively shade tolerant and deep rooted and adult trees can survive low-intensity fires. Similarly, selective grazing or browsing of palatable plants alters plant species composition and competitive balance within the plant community. According to Crawley (1983) mammals grazing herbaceous vegetation or browsing woody plants have a stronger impact on plant population dynamics than insects. For example, white-tailed deer do considerable damage by trampling vegetation, dislodging lichens and mosses, loosening the sand and creating paths that lead to the initiation of blowouts (Maun and Crabe 1995). However, the impact really depends on the severity of damage and the length of time it continues. Even removal of a herbivore may have far-reaching and sometimes unpredictable consequences. For instance, the decimation of rabbit populations by myxomatosis allowed *Crataegus monogyna* populations to expand into *Ammophila* dominated stands in a dune complex in Wales (Hodgkin 1984).

Phytophagous insects similarly impact dune vegetation and can selectively decrease growth and kill plants. Damage to *Salix cordata* on a foredune by flea beetle (*Altica subplicata*) had a significant impact on the successional pattern of the plant community within one year by permitting an increase in the relative abundance of neighbouring non-host plants, *Juncus, Carex, Aster* and *Artemisia* (Bach 1994, 2001). Plots treated with pesticides had 2.5 times higher density m^{-2} of monocot plants and 1.5 times higher density of dicot plants than controls. Periodic ice storms and hurricanes that break limbs and uproot trees also alter the relative composition of plant communities on a regular basis.

12.7.4 Allogenic influences

In the early stages of dune succession allogenic mechanisms (abiotic forces external to the influence of the plant community) mediated by wind, salt spray and burial by sand are predominant and mask the effects of facilitative or inhibitive autogenic changes brought about by plants and other biotic organisms. Washovers and violent sand storms may dump large amounts of sand on plant communities on sea coasts thus reversing or deflecting sand dune succession. With time, however, the impact of abiotic forces declines and that of biotic factors increases, thus advancing the plant community to the next stage in succession (Fig. 12.4).

12.7.5 Colonization constraints

Lichter (2000) showed experimentally that sand dune succession is mediated by availability constraints of plant propagules. He concluded that the replacement of early successional species (primarily grasses) by the late successional species (pines, oaks) is limited by seed dispersal, seed and seedling desiccation and seed predation by rodents rather than intrinsic changes associated with dune stabilization and soil development. He showed that soil nutrients were not a limiting resource

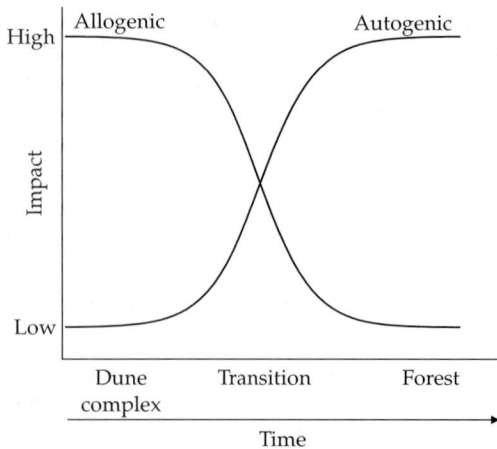

Figure 12.4 The impact of allogenic and autogenic forces from the early stages of succession in foredunes to the later stages of succession in backdunes.

Figure 12.5 Established seedlings of *Populus deltoides* in a moist depression on a beach along Lake Huron (photograph by I Krajnyk).

because chance seed dispersal, favourable weather conditions and low rodent densities allowed late successional species to establish in younger dune ridges. Similarly, Chapin *et al.* (1994) in a study on primary succession following deglaciation at Glacier Bay, Alaska showed that the pattern of succession was determined by the availability of propagules and life history traits of plant species.

Poulson (1999) included the colonization constraints of plant species as one aspect of the individualistic concept of Gleason (1926) which states that 'every species of plant is a law unto itself, the distribution of which in space depends upon the peculiarities of migration and environmental requirements'. He showed that on the Miller Dunes along Lake Michigan, *Ammophila breviligulata* and *Populus deltoides* established independently of each other in different microhabitats, rather than the former acting as a facilitator of the latter as suggested by Olson (1958a). We have also observed something similar on the Pinery sand dunes of Lake Huron. Receding lake levels enlarged the beach and created damp depressions in the sand parallel to the first dune ridge, where

seeds of *P. deltoides* originating from the trees on the first dune ridge lodged, germinated in spring and established as seedlings (Fig. 12.5). *Populus deltoides* does not normally establish on dune ridges because it can only germinate and establish in moist places such as moist slacks and beach margins. The lateral spreading habit of *A. breviligulata* allows it to expand into the newly exposed bare area closer to the lakeshore, giving the mistaken impression that *A. breviligulata* established first on the beach and facilitated the establishment of *P. deltoides*. Actually, as shown by Poulson and McClung (1999) on Miller Dunes, *A. breviligulata* retards succession and provides an example of autogenic inhibition. At the species replacement scale there was no support for direct facilitation or for tolerance but clear support for indirect facilitation and indirect inhibition (Poulson 1999).

12.7.6 Stability of the dune habitat

Tilman's (1985) resource ratio hypothesis assumes that 'the habitat is relatively stable' and the only changes are facilitative effects of species. Stability is an essential prerequisite for an increase in species richness and coverage. This is not true, however, in open foredunes along sea coasts and lakeshores where plant communities are subject to recurrent disturbance. Progression to the forest stage is more related to factors such as rate of retreat of the shoreline, amount of sediment deposition on the shoreline, wind velocities, life history traits of plant species and above all the length of time taken for the stability of substrate. Mobile substrate must become stable and achieve protection from wind and salt spray (Wells and Shunk 1938) before the autogenic factors become the principal force in community dynamics.

12.8 Relationship of zonation to succession

Should the zonation of vegetation in a dune complex be considered part of a successional sequence? Oosting (1954) showed that zones along the North Carolina coast were not indicative of the existence of succession because they were primarily controlled by abiotic factors caused by exposure to wind, salt spray, sand accretion and depth to the water table, all of which decline farther inland. The future of these zones depends on geological changes, retreat or advance of sea coast, and impact of biotic and abiotic factors whose direction and intensity can not be anticipated. These zones were relatively stable unless violent storms disturbed the vegetative cover. Similarly, Doing (1985) stated that 'in most cases the zonation does not coincide with a succession' and hence a successional interpretation was not valid. On a subarctic dune along Hudson Bay, colonized by *Leymus mollis* and *Lathyrus japonicus* populations, no discernible changes have been recorded for many years (Imbert and Houle 2000).

Nevertheless, successional autogenic effects of plants must have also partly contributed to the creation of a zone. Moreover, within a zone seed germination and seedling establishment are regular events that indicate the existence of succession. Succession is clearly evident on

some sea coasts, especially those prograding at a faster rate as in South Africa (Lubke 2004), where a new dune ridge was formed every five years and a forest community was well established in about 120 years. Similarly, at Wilderness State Park along Lake Michigan, open dune-building species were replaced by shrubs and grassland in about 100 years and they were subsequently replaced by pine forest within approximately 345 years, clearly showing evidence of successional replacement of species (Lichter 1998). Thus, it is possible that the stable zones seen by Oosting (1954) and Doing (1985) may actually be influenced by autogenic forces and progressing very slowly, especially below the soil, towards the next successional stage. The plant communities are dynamic and changes in vegetation, whether externally or internally induced, are therefore successional (Cooper 1926). As soon as the influence of abiotic factors declines or is modified by plants and other biotic organisms the effects of facilitative autogenic changes become evident and the successional progression becomes quite visible. Such zones might have existed in the past and, following a decrease in allogenic influences, have now advanced to more mature stages in succession. Thus different zones in a sand dune complex are actually stages in succession.

12.9 Summary

It has long been known that plant communities on sand dune complexes along sea coasts exhibit a zonation pattern from the beach to grassland-forest transition. Zones exhibit a spatial pattern on the sandscape that may be unique to each coastal dune system in terms of plant and animal life. Succession is a process of change in composition of species over time that is mediated by a variety of mechanisms. Two major causes of zonation, salt spray and burial by sand, have been implicated by dune ecologists. Both of these possible causes have been debated over a long time. The evidence of salt spray as an agent of zonation is based on correlation between tolerance level and distribution of species on the dune complex. Usually the more tolerant taxa are located in areas receiving the maximum amounts of salt spray; however, it is not always the case because several susceptible species are also found in areas receiving high salt spray levels. Other evidence suggests that soil salinity is very low, salt spray storms are seasonal, plants are capable of withstanding salinity, environmental conditions favour survival and plant species are capable of withstanding salt spray loads and soil salinity.

Burial by sand as a factor in zonation has been examined using the same two approaches. First, relate species distribution with natural sand deposition that decreases with distance from the sea coast and second, examine survival, distribution and relative abundance of species following artificial burial by sand. The first approach clearly shows a close correlation between species richness, vegetation cover and sand deposition. The species distribution is also correlated with their levels of tolerance. However, the actual levels of burial in accreting habitats are not uniform. Some microhabitats receive much more deposition than others and some plants growing close to obstructions receive lower sand loads than others because of deflection of wind. Artificial burial experiments confirm the results of natural burial and a clear distinction is evident in the tolerance levels of different species in the plant community. Thus all evidence supports the suggestion that burial by sand may be the major causal agent of zonation in dune systems. Nevertheless, as argued by leading dune ecologists, single factors may not be responsible for a phenomenon as complex as zonation because of confounding of many other

variables. Given enough time, a bare sand surface will develop into progressively more diverse communities of plants and animals until a climax community is established. Many hypotheses have been advanced over the years and the jury is still out in deciphering a large number of variables affecting successional progression not only in different ecosystems but also within ecosystems. The discrepancy is primarily related to basic assumptions and many confounding factors. For example, primary succession in coastal sand dunes is influenced by many factors other than time such as past climatic history, sand texture at different time periods, disturbance history, sea level fluctuations, colonization constraints of plant species, stability of the dune habitat and many others.

Finally, is there a correlation between zonation and succession, or are they separate entities? In some coastal locations some dune ecologists conclude that zones do not coincide with succession because of abiotic factors such as wind velocity, salt spray and burial by sand. They exert a greater influence than biotic factors originating from the plant community itself. However, the autogenic influence of plants growing in a zone and other biotic agents can not be denied. It probably takes a longer period of time for a zone to proceed to the next stage in succession in some coastal locations. On some sea coasts where the coast is receding at a faster rate, successional stages can be seen quite clearly and a forest community is well established in relatively short periods of time. The dynamic changes in vegetation caused by abiotic and/or biotic factors are part of the overall successional sequence in the dune complex. Thus, zonation and succession often can not be separated.

CHAPTER 13

Dune systems in relation to rising seas

Anwar Maun and Dianne Fahselt

13.1 Introduction

Beaches and associated dunes are constituted of unconsolidated materials, such as sand, and thus are low-strength land forms less robust than rocky cliffs (van der Meulen *et al.* 1991). It is estimated that 70% of sand-based coastlines in the world are presently subject to erosion (Bird 1985; Wind and Peerbolte 1993). However, natural dune systems are inclined to adjust after stress without permanent damage (Brown and McLachlin 2002), and when stabilized by plant cover they offer a first line of coastal defence against assault from wave action (Wind and Peerbolte 1993; Broadus 1993; De Ronde 1993). Natural self-sustaining dune systems interact with the sea and closely reflect changes in sea levels.

13.2 Sea levels over time

At any given time no single sea level characterizes all oceans, that is, the resting position of the ocean surface, or geoid, is not uniformly elevated over the earth. Eustatic sea levels, free of influence from tides, waves and storms, thus vary from place to place as well as over time. Satellite altimetry, which permits more accurate as well as more numerous observations than older tide-gage methods of measuring sea levels, shows that the ocean is actually a spheroid modified by depressions and elevations. For example, in parts of the Indian Ocean sea levels are as much as 70 m lower than the global mean and in the North Atlantic 80 m higher (Carter 1988).

Climate is governed by long-term periodic variations in the earth's orbit that effect changes in solar radiation and, consequently, also in sea levels (Bartlein and Prentice 1989; Woodroffe 2002). As a result, ice ages repeatedly alternate with periods of interglacial warming in which ice masses contract and sea levels increase. Most of the time that has passed since the Cambrian period—approximately 500 million years—sea levels, although fluctuating on several timescales, have been higher than they are today. Because of the difficulties in documenting conditions so far in the distant past estimates of these sea levels vary considerably, but those shown in Fig. 13.1, based on different kinds of evidence, are representative of attempts at reconstruction (personal communication RA Rohde 2008).

In the last two million years massive extensions of ice sheets have taken place approximately every 100 000–150 000 years, and in the interglacial period immediately preceding the present one sea levels were higher than they are currently (Warrick 1993). Consequently more land was submerged as underwater shelves (Milliman and Emery 1968; Ray and McCormick-Ray 2004). During the last major glacial maximum (LGM), which ended approximately 21 000 years BP, mean temperatures were lower than presently, possibly by 4–5°C, and the ocean surfaces 120–135 m below current levels (Mitrovica 2003; IPCC

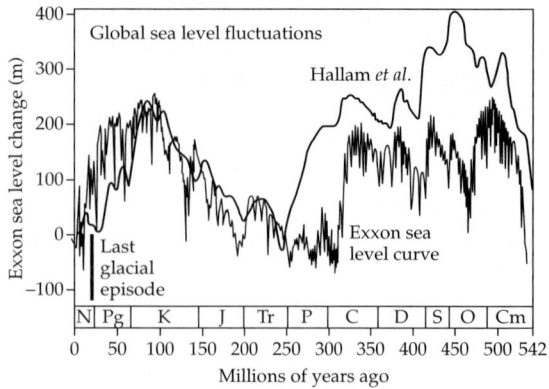

Figure 13.1 Estimates of sea-level fluctuations since the Cambrian period derived from exposed geological sections plus extent of flooding of continental interiors (Hallam 1989) and seismic profiles (Exxon data from Haq *et al.* 1987; Ross and Ross 1987; Ross and Ross 1988); http://en.wikipedia.org/wiki/ Image: Phanerozoic_Sea_Level.png.

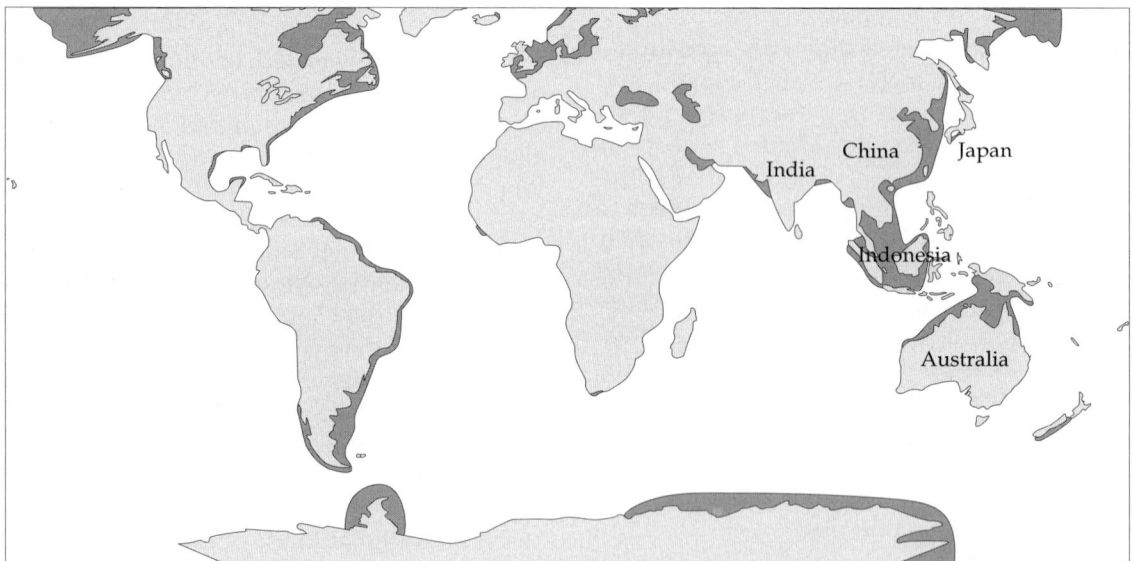

Figure 13.2 Estimated extent of land masses in the world at the last glacial maximum (LGM) with areas since covered by seas indicated as dark shading (modified from Hancock 2002).

and Solomon *et al.* 2007b). Melt water from these last major ice sheets has already covered portions of the continental margins that were emergent at LGM.

A detailed reconstruction showing land exposed at the height of the last ice age (Isayeva and Velicho 1998) reveals many differences from the present: the United Kingdom was not an island, Japan was peninsular, considerably more land was emergent in unglaciated parts of the Arctic than presently and a wide land

bridge connected Alaska and Siberia. One large land mass emergent at LGM, Doggerland, is now totally submerged beneath the North Sea (Coles 2000). A simple outline map (Fig. 13.2) similarly indicates land exposed at LGM, as well as the furthest advance of ice, and also illustrates the scattered nature of areas thus far flooded by increasing quantities of water in the ocean (Hancock 2002). For example, more area has been submerged by rising seas along shorelines on the east side of both North and

South America than on the west, and inundation is not uniform along a single coastline. Unevenness is partly due to topography, with low-lying coastal plains more vulnerable to rising seas than ridges or mountain ranges.

Submerged shelves slope down to 200 m or more below sea level (Ray and McCormick-Ray 2004) and the remains of flooded canyons, cliffs, valleys, river systems and hills are in some cases discernable underwater. Ancient coasts are still recognizable off the shores of England and Australia (Pethick 1984), but sand-based shoreline features that are quickly modified by erosion and siltation are poorly maintained after submergence (Carter 1988; Coles 2000). However, submerged fossil dunes, cemented by their carbonate content, remain identifiable on sea shelves in many locations (Pye and Tsoar 1990). In addition, marine archaeology has revealed massive structures under 5–40 m of water, some of which appear to have been man-made, and the depths where these are found suggest civilizations as long as 8–10 thousand years ago (Hancock 2002).

The extent of shoreline alteration due to high seas does not depend solely on the increased mass of sea water and continental topography. Altered drainage from continents, uneven lithospheric loading, tectonic alterations of the shape of ocean basins and subsidence of the earth's crust are some factors contributing to the differential effects of increased volume of ocean water on shorelines worldwide (Fairbridge 1961; Warrick 1993; Woodroffe 2002). Some factors interact. Overlying ice masses strongly influence the elevation of land because continents are depressed during glacial advances and rebound as pressure is released upon melting. Continental recovery from the last ice age is still in progress, and isostatic rebound is particularly evident along arctic and subarctic coasts in North America. In many areas uplift is >5 mm yr^{-1} (Milliman and Emery 1968; Carter 1988). In locations where isostasy allows continents to rise at the same rate as ocean surfaces the relative positions of sea and land remain unchanged (De Ronde 1993), thus offsetting

the shoreline impact of rising seas. However, flooding due to the increased mass of water in oceans is even more extensive when adjacent land masses sink, as is the case along southern coasts of the North Sea in Europe (Allen 2000). Over time the extent of terrestrial environments may thus increase, decrease or remain unchanged in relation to sea levels, with natural variations occurring over a range of time and space scales (Carter 1988).

Projecting the effects of sea level rise is complicated because different rates of relative increases may occur even within limited geographic areas. In the UK, partly because of the differential effects of past ice loading, coasts in the north subjected to a heavy burden of ice during the last major glaciation are now undergoing isostatic uplift, while those in the south-east are submerging in relation to sea level (Murray-Wallace 2007). The sensitivity of Canadian shorelines differs greatly from one location to another, with the most vulnerable areas mainly concentrated in three eastern provinces and along the coast of the Beaufort Sea (Shaw et al. 1998). Susceptibility appears to vary between islands off the east coast of North America because extensive erosion of coastal dunes is expected on both Prince Edward Island and Îles-de-la-Madeleine, while only minimal changes are forecast for Sable Island. Different trends in sea levels over time have already been reported from a number of locations in USA (Fig. 13.3).

Melting of ice has in general been taking place since the LGM, but the rate of global melt generally decreased over approximately the last 6000 years (Shennan 2007), as did the magnitude of overall annual increase in sea level (Rampino 1979; Donnelly 2006). Nevertheless, the rise continued in some locations (Pirazolli 1996) has meant that the coast of Connecticut, for example, was submerged almost 3 m in the last 3000 years (Bloom and Stuiver 1963). Acceleration in the rise of eustatic sea levels worldwide began in the 1890s (Clark 1986; Carter 1988; IPCC and Solomon et al. 2007b). Over the last century

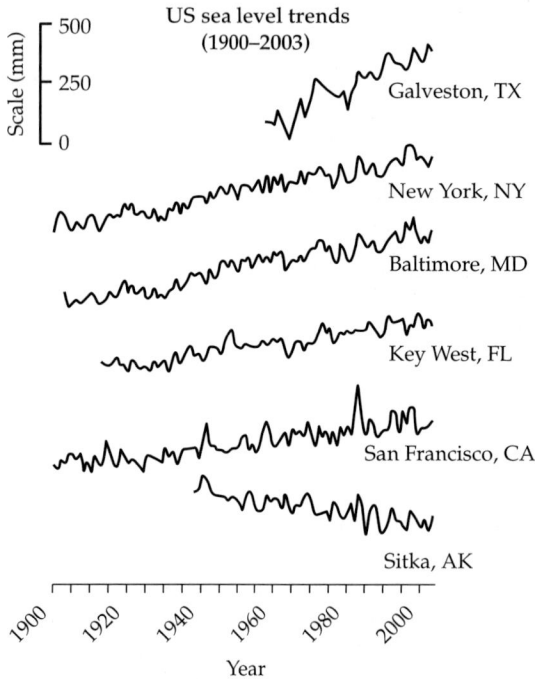

Figure 13.3 US sea level trends. Source: Monthly and annual mean sea level station files from the Permanent Service for Mean Sea Level (PSMSL) at the Proudman Oceanographic Laboratory (United States Environmental Protection Agency); http://www.epa.gov/climatechange/science/recentslc.html.

annual average global temperatures increased markedly (Fig. 13.4) and the annual rates of increase in sea level (Fig. 13.5) have been considerably more than in the immediately preceding centuries (Warrick 1993; Woodworth 1993; Donnelly 2006). The consensus of IPCC was that 1.7 mm y^{-1} represented the global average increase from 1893–1993 (IPCC and Solomon *et al.* 2007b). The rate estimated for the next ten years, 1993–2003 was 3.1 mm y^{-1}, with about half predicted to come from an increase in the mass of sea water and half to thermal expansion of the oceans under higher global temperatures. With expected increases in greenhouse gas emissions and higher temperatures the predicted mean global sea level rise was 4 mm y^{-1} for the century ahead (IPCC and Solomon *et al.* 2007b).

The effect of rising sea levels upon shorelines is influenced not only by the rate of relative increase, but also by the magnitude of the overall increment. This, in turn, depends largely on the degree of deterioration of large terrestrial ice fields (Murray-Wallace 2007). Much less than 1% of the total ice on earth is on smaller islands and in scattered glaciers, and it is estimated that, if the entire global supply of ice on polar islands and in mountain glaciers were to melt, ocean levels would increase on average 30–60 cm (Woodworth 1993). However, complete melting of the largest masses of terrestrial ice, that is, those in polar ice fields in Greenland and Antarctic, would augment the mean height of oceans considerably as they would further increase sea levels by 8 and 60–65 m, respectively (Woodworth 1993; Dowdeswell and Hagen 2004; Bentley 2004).

Although it is considered unlikely that major ice fields will be suddenly lost within the next century, even small changes in temperature could produce considerable melting and impact sea levels greatly. Almost 20 years ago the US Environmental Protection Agency advised Congress in 1989 that global warming could increase sea levels 0.5 to 2 m by 2100 (Titus 1989). More recent projections have suggested that global sea level will increase 15–95 cm (IPCC and McCarthy *et al.* 2001) or 44 cm (IPCC and Solomon *et al.* 2007b) in roughly the same period. Since trends can vary markedly from place to place, however, all forecasts are necessarily speculative (Brown and McLachlin 2002). Important parameters, such as those relating to dynamics of ice fields, are still imperfectly understood (Walsh *et al.* 2005; Shepherd and Wingham 2007). Even IPCC projections may be unreliable as they attempt to harmonize a number of climatic scenarios, some of which diverge widely from one another (Walsh *et al.* 2005). The range of future projections is shown as 2–5 m in the next century in Fig. 13.6. However, based on models that take into account the volume contributions of detached glacier fragments as they slide into the oceans, a more plausible estimate of mean sea level increase in the present

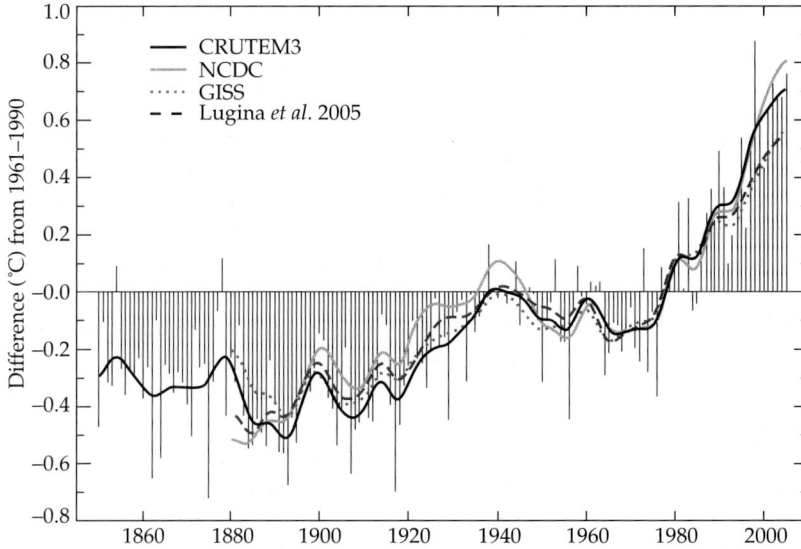

Figure 13.4 Annual anomalies of global land-surface air temperature (°C), 1850–2005, relative to the 1961 to 1990 mean for Climatic Research Unit land-surface air temperature, version 3 (CRUTEM3) updated from Brohan *et al.* (2006). The smooth curves show decadal variations. The black curve from CRUTEM3 is compared with those from NCDC (Smith and Reynolds 2005; blue), Goddard Institute for Space Studies (GISS) (Hansen *et al.* 2001; red) and Lugina *et al.* (2005; green) (Climate Change 2007, Figure 3.1).

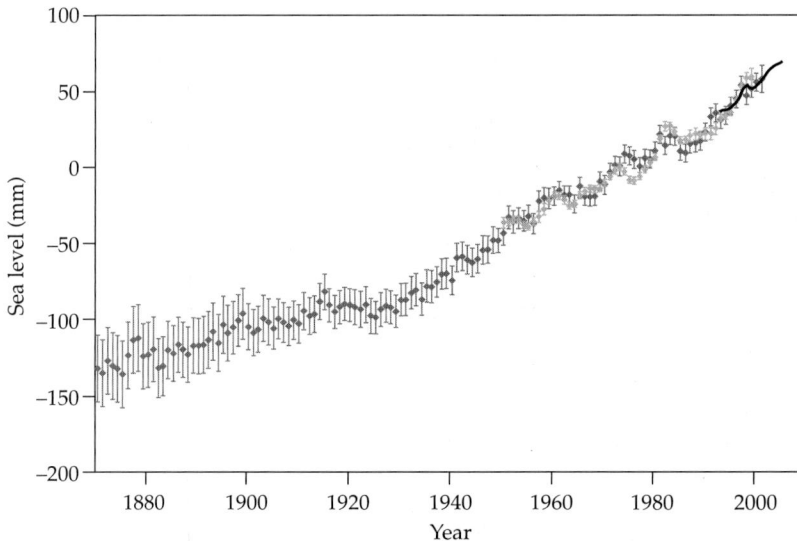

Figure 13.5 Annual averages of the global mean sea level (mm). The red curve shows reconstructed sea level fields since 1870 (adapted from Church and White 2006); the blue curve shows coastal tide gauge measurements since 1950 (from Holgate and Woodworth 2004) and the blue curve is based on satellite altimetry (Leuliette *et al.* 2004). The red and blue curves area deviations from their averages for 1961–1990, and the black curve is the deviation from the average of the red curve for the period of 1993–2001. Error bars show 90% confidence limits (Climate Change 2007, Figure 5.13).

century is now thought to be 0.8 m, or twice the IPCC projection (Pfeffer *et al.* 2008).

Sand-based coastal dune systems are formed under and altered by both marine regression and transgression (Packham and Willis 1997), but this chapter focuses mainly on possible dune responses under transgression associated with rising sea levels. Imminent changes in sea level induced by greenhouse gas emissions are of immediate concern, as these could potentially dominate normal variation (Warrick 1993) and increase stress on coastal environments.

13.3 The special significance of high water surges

Global mean water level is not as important in reshaping shorelines as temporary high water fluctuations such as large waves or storm surges (Woodworth 1993). When sea levels rise in relation to adjacent land masses, coastline erosion is usually increased (Carter 1991; Raper 1993; Brown and McLachlin 2002), partly because greater depth of water is translated into increased wave height and energy (Clayton 1993). Waves damage foredunes and even older dune ridges, and with time such disturbances may be enlarged by onshore winds to produce bare areas oriented along the path of prevailing winds. On high-energy coasts an entire ridge or even a whole dune system can be obliterated in the course of a single storm. Although waves off the coast of California are higher than 4 m less than 2% of the time, these large waves move more than 20% of all sediments (Larson and Kraus 1994; Larson *et al.* 2002).

Along the Welsh coastline the configuration of dune systems is closely correlated with mean high water (Saye and Pye 2007), and in Northern France tide gage data also suggests a strong relationship between dune front erosion caused by episodic high water events and the frequency of storm surges

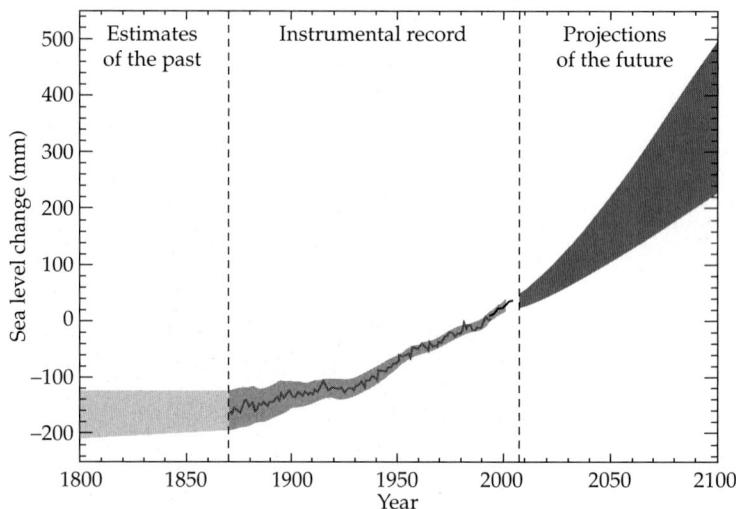

Figure 13.6 Time series of global mean sea level (deviation from the 1980–1999 mean) in the past and as projected for the future. For the period before 1870, global measurements of sea level are not available. The grey shading shows the uncertainty in the estimated long-term rate of sea level change. The red line is a reconstruction of global mean sea level from tide gauges, and the red shading denotes the range of variations from a smooth curve. The green line shows global mean sea level observed from satellite altimetry. The blue shading represents the range of model projections for the SRES (Special Report on Emission Scenarios) A1B scenario for the twenty-first century, relative to the 1980 to 1999 mean, and has been calculated independently from the observations. Beyond 2100, the projections are increasingly dependent on the emissions scenario. Over many centuries or millennia, sea level could rise by several metres (Climate Change 2007, FAQ 5.1, Figure 1).

(Vasseur and Hequette 2000). Storms and tsunamis are a major cause of coastal reconfiguration in many parts of the world, e.g. Sri Lanka (Liu *et al.* 2005) and Hong Kong (Yim 1993). The North Sea and Atlantic coasts of North America are also subject to devastating storm surges (Flather and Khandker 1993; Pugh 2004). Beaches and natural dunes are routinely breached on the Norfolk coast of England (Clayton 1993), and generally along the eastern seaboard of the USA (Dean 1999) as well as maritime coasts of Canada (Shaw *et al.* 1998). Extensive incorporation of marine shell fragments in coastal dunes attests to the high energy of waves. Along the northern coast of Australia, for example, infrequent storms force bottom depositions inshore and inland where they are deposited as ridge sequences well above the zone of normal wave processes (Nott 2006). A highly multiplicative association has been confirmed between long-term erosion of sandy beaches and sea level rise (Zhang *et al.* 2004); a finding which suggests that expected increases in sea level will continue to stress coastal environments.

13.4 Natural erosion and building under high sea levels

Although shoreline features may be damaged by higher oceans, they may also equilibrate, sometimes in a relatively short period of time. Entrapped sand released by shoreline erosion under high seas may increase the height of existing ridges (Carter 1991). Maintenance of sand dunes in spite of rising seas has been well documented in some locations. For example, dunes on Sable Island, habitually in a state of flux with sand constantly moving between beaches, sand bars and coastal dunes, have changed little in 7000 years. It is expected that the total area occupied by dune systems on the island will slightly decrease under even higher seas, but otherwise dunes will continue essentially unchanged (Shaw *et al.* 1998). Fourteen of the fifteen dune systems studied along a Welsh coast that experienced increased sea levels over the last 100–120 years exhibited

a net increase in area while only one experienced net loss (Saye and Pye 2007). It is projected that dune habitat in this area, barring unforeseen changes in sediment supply and coastal processes, will be maintained under rising seas in the next century.

On the other hand beaches and dunes often retreat landward as a result of erosion (Martínez and Psuty 2004; Greaver and Da Sternberg 2006) and then equilibrate (Clayton 1993), with released sediment from destruction of old shorelines used to build new sand-based land forms onshore. Under conditions of rising seas, aeolian sand deposited over vegetated or semi-vegetated terrain (Figs. 13.7 and 13.8) may develop into transgressive dune fields (Hesp 1999). In eastern Canada sandy

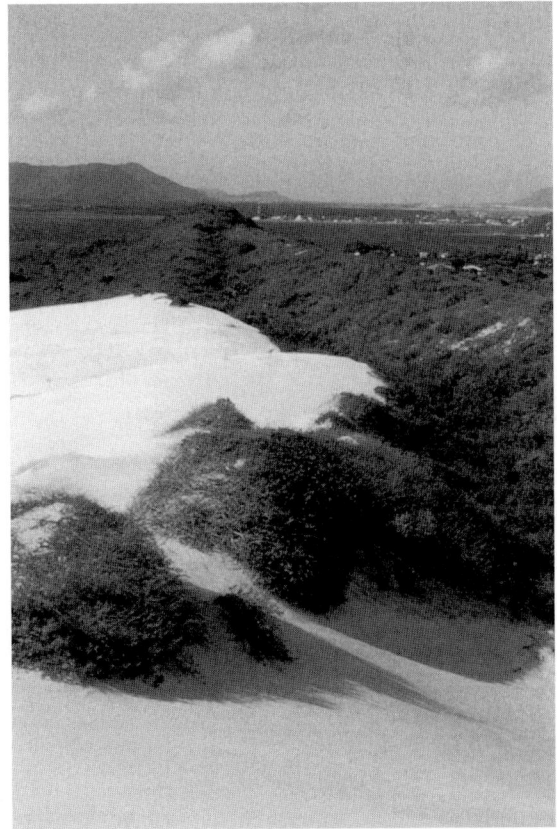

Figure 13.7 Active sand precipitation ridges engulfing vegetated relict ridges in a dune field along the Brazilian coast at Florianopolis (Hesp 2004).

beaches have migrated landward, in one case at 0.5–1.8 m y^{-1} across woodland underlain by peat (Shaw *et al.* 1998). A parallel example is in Doñana National Park in Spain, where stands of pines in slacks were engulfed by 30 m dune ridges migrating landwards (Garcia Novo 1979). On coasts in south-western Australia active sand ramps facilitate movement, not only inland but upward to cliff tops where older vegetation is smothered under newly forming dunes (Hesp and Short 1999).

If a large volume of sediment is suddenly mobilized open dune fields may be created, e.g. the Sands of Forvie in north-east Scotland. In strong tropical storms particulates are relocated hundreds of metres further inland (Nott 2006) through either alluvial (Brammer 1993) or aeolian sedimentation (Carter 1988). Mobilized sand may be deposited in the form of flat splays or washover fans (Hesp 1999; Nott 2006). In British Columbia, Canada, sediment released from retreating beaches under rising seas was transferred high upshore where it accreted in fields of driftwood (Walker and Barrie 2003).

Some constructions based on released sediments are located offshore as well. Sand, for example, may be transported laterally in long-shore currents which are initiated when waves strike shorelines at an angle. Although longshore transport is not well understood, it is evident that sediments from one area may be

carried from the site of origin and incorporated into new sand-based structures downstream (Clayton 1993; Saye and Pye 2007). Mobilized sediments tend to move within semi-closed systems, or circulation cells, along the coastline (Packham and Willis 1997; Inman 2005), such as those described off the coasts of the British Isles and California (Larson *et al.* 2002). Within each sediment cell factors such as rainfall, evaporation, currents, nature of suspended particles and geology all influence the formation or enhancement of dunes and other sedimentary features. However, if the offshore sea shelf is particularly steep, sediments may be removed from further participation in coastal processes by downwelling (Inman and Masters 1991) and finally lost by deposition in submarine canyons (Shepard 1973; Larson *et al.* 2002).

Materials from drowned coastal features may also be reworked into linear offshore strands parallel to coastlines. Such barriers, for example, those along the north Norfolk coast in Great Britain (Clayton 1993) and the coast of eastern North America (Carter 1988), are often based on coarser sediments such as gravel and shingle and contribute significantly to the security of mainland coasts. These features seem less prevalent off high-energy shores, such as the coast of south-east Australia, than lower-energy coasts like the Atlantic seaboard of the United States (Carter 1988). The offshore position of barriers may stabilize or, if wave action continually moves sediment from the outward face to the landward side, shoreward migration may take place.

Bottom sediments re-suspended by storm waves are another major source of building material in active coastal regions. For example, during transgression in Newfoundland sand from the inner shelf was moved onshore (Shaw and Forbes 1990). In the Netherlands also, configuration of the coast seems to be dependent upon continuing onshore transport of sandy sediments with increased wave activity (Wind and Peerbolte 1993). Massive surges of sea water may push coarse matter

Figure 13.8 Aerial view of transgressive dune field in the Veracruz region on the east coast of Mexico (Hesp 2004).

landward from sea shelves to form shingle beaches (Pethick 1984).

13.5 The future of coastal dune landforms

Higher sea surface temperatures, higher frequency and greater intensity of storms and increased geographical areas subject to storm disturbances will probably accompany climatic warming (Schlesinger 1993; Raper 1993) and intensify damage to dune systems. If the present rate of rise in sea levels increases even modestly there will be greater mobilization of sand, more dune development inland, more frequent loss of foredunes, more frequent scarping and increased incidence of blowouts. Flux in the system will increase. High frontal dunes will be eroded less than low (Pye and Tsoar 1990), but may be destabilized by storm waves nevertheless (Fig. 13.9). Complete removal of the first dune ridge by rising seas will increase the activity of the second, which will then become the first, and at the same time the slack between first and second ridges will be transformed into beach. Dune fields will continue to shift landward, as long as migration is not restricted by insurmountable barriers, either natural or engineered (Feagin *et al.* 2005; Greaver and Da Sternberg 2006). However, the usefulness of global general-

Figure 13.9 Storm wave damage on high frontal dunes along the Sefton coast, Merseyside, UK (Pye and Soar 1990).

izations is limited, and projections must be regionalized to be meaningful (Kattsov *et al.* 2005) because subsidence and emergence of land may be as important to the maintenance of shorelines as rise in sea level (Titus 1993).

Because extrapolation from the past is not necessarily a valid way of predicting either future sea levels or the effects on shorelines under a rising sea level scenario, modelling is an important tool for projecting coastal responses. In order to construct meaningful models, however, intensive *in situ* assessment and monitoring is first required, along with an understanding of glacial hydrology, as well as ice, tide and wave dynamics. Dune-recession models may be evaluated by comparing modelled projections to actual field observations, and in one study, although long-term projections were not borne out in the field, short-term dune recession was almost identical to that predicted by modelling (Leont'yev 2003).

Along with the relative positions of sea and land, several other factors are pivotal to the evolution of coastal systems: the rate of change in sea level, the robustness and orientation of land form features, the intensity and direction of waves and tides, the type and availability of sediments, degree of aeolian transport and nature of offshore–onshore exchange (Clayton 1993; Brown and McLachlan 2002; Saye and Pye 2007). All must be represented in models, along with feedback mechanisms which either maintain existing equilibria or promote reorganization leading to a new stable state. The consequences of rising seas must also be considered over varying timescales and parameters relevant to coastal systems as well as interactions among them. Thus a multidisciplinary approach is needed to forecast the coastal effects of increasing sea levels. The myriad of relevant variables (Trenhaile 2003) demands multivariate analyses and sophisticated modelling tools that can be linked with global information systems (Green and King 2003). To increase accuracy many sample scenarios are considered (Kattsov *et al.* 2005); in making climatic projections IPCC varies uncertain parameters systematically and widely (Forster *et al.* 2007). Owing to the role of plants in determining dune form, function and stability, many projections could probably be improved by inclusion of biotic variables, especially vegetational parameters (Day *et al.* 1993; Dech *et al.* 2005). Models assuming a low rise in sea level at Galveston, Texas (0.09 m) in the next century suggested that plant communities typical of all stages of succession would be able to establish, while models based on a greater increase (0.48 m) indicated that dunes would be restricted in size and offer establishment opportunities for ephemeral annuals but not perennial dune-building species (Feagin *et al.* 2005).

13.6 Responses of dune vegetation

The degree of destabilization is largely dependent upon the energy of the impacting waves, both along freshwater and saltwater coasts, but vegetational cover may be maintained if coastal erosion is gradual. Precipitation dunes of sand transported inland are often revegetated rapidly, as in the case of regressive dune fields in Mexico (Fig. 13.10). Much evidence suggests that shoreline communities tend to be modified or shifted by rising seas, but not totally lost. Species at the seaward edge of

Figure 13.10 Rapid stabilization of a transverse dune in the El Quixote transgressive dune field on the Gulf of Mexico by the sand dune colonizers, *Croton punctatus* and *Chamaecrista chamaecristoides* (Hesp 2004).

dune communities may be eliminated from their present environments and re-established at higher sites inland (Vestergaard 1997), with migration depending upon availability inland of suitable sandy sediments. As sea levels change, plants position themselves in relation to their resistance to wind, wave action, salt and abrasion and other physiological and ecological parameters, such as water uptake requirements (Greaver and Da Sternberg 2006). Communities are little altered when beach-dune complexes are shifted inland as long as similar habitat is available (Brown and McLachlin 2002). Extensive erosion, however, may alter communities (Roman and Nordstrom 1988), as in Maryland where high-energy overwash favours herbaceous species while less wave disturbance permits development of shrub thickets and salt marshes.

In the Netherlands, where most of the shoreline is characterized by dune systems, existing foredunes will undoubtedly be affected by a sea level rise of 60 cm in the next century (Noest 1991; van der Meulen *et al.* 1991). Whole dune communities are expected to migrate inland (Davy *et al.* 2006). It is also expected that the relative rise in sea level projected within the next century could also produce horizontal movement of vegetation along the Danish coast (Vestergaard 1997). Currently dune systems are of the prograding type in which minimal amounts of eroded sediments are simply added to existing stabilized ridges, and even the youngest ridge is approximately 150 years old (Vestergaard 1991). However, higher seas will destabilize dunes, interrupt soil development, reduce organic content of the soil and probably displace dune communities inland (Vestergaard 1991).

When land is elevated to a greater extent than sea levels, however, vegetation zones may shift seaward with dune species establishing on raised beaches or strands (Compton and Franceschini 2005; Martini 1975). If sea levels fluctuate, landward–seaward movements of native dune plants can be expected to reflect the back-and-forth migration of the beach

(Feagin *et al.* 2005). That vegetation can spontaneously re-establish in different locations is indicated by the observation that within 24 years man-made dunes along the coast of Denmark develop communities characteristic of natural dunes (Vestergaard 2006).

South-eastern Long Island, New York, USA, currently exhibits vertical vegetational zonation, with uplands occupied by pine barrens (*Pinus rigida*) and a zone of second growth mixed hardwood forests, including *Quercus velutina, Q. alba, Robinia pseudoacacia* and *Carya glabra,* just downslope. At still lower elevations the heath under-storey diminishes in importance and is replaced by mesic shrubs, such as *Viburnum dentatum* and *Lindera benzoin,* while the lowest edge of forest involves shrubs such as *Vaccinium corymbosum* and *Nyssa sylvatica.* Finally, intertidal areas are occupied by salt marsh species such as *Spartina alterniflora, S. patens, Distichlis spicata* and *Juncus gerardii* (Clark 1986). Pollen analysis of the sediments underlying coastal marshes shows that over time present wetland communities have displaced previous upland and transitional vegetation types. This indicates that rising seas have altered growing conditions to the point where they became generally unsuitable for forest species. Disturbances such as fire and agricultural clearing evidently increased opportunities for seedling establishment at higher elevations and thus facilitated the gradual shifting of vegetation upslope. The idea that marine transgression has forced inland migration in this area is supported by analysis of spacial, as well as age–class, distributions of the most abundant tree species in extant forests (Clark 1986).

An opportunity to monitor vegetational response immediately following high water damage presented itself along the Great Lakes shores in Canada (Maun 2004) where water levels oscillate and reach a high every 15–20 years. In 1986–1987 water levels in Lake Huron rose approximately 1.25 m, an increase of the same order of magnitude of increases that have sometimes been used for modelling effects of

sea level change over a century. In March–April, 1987, wave action produced major modifications in the configuration of the shoreline and the foredune was totally obliterated, along with *Calamovilfa longifolia, Ammophila breviligulata* and any annuals established on it. The first dune ridge was also partly removed leaving a steep scarp along it (see Chapter 1, Fig. 1.12), and a driftline developed 2 m from the base of the scarp leaving a new zone of bare beach. Within the driftline were fruits of all the annuals normally present, fragments of all perennials including *A. breviligulata, C. longifolia,* and propagules of a few species not previously observed in the system, such as *Potentilla anserina, Tussilago farfara, Fraxinus* spp., *Acer* spp., *Juglans* spp., and many others. By the following year when lake levels were receding, re-colonization of damaged dunes was already well underway, and within four years all typical species had re-established. Even tussocks of *C. longifolia,* which the storm had left perched on top of the scarp on the first dune ridge, had subsequently slid down the scarp face after being undercut by exposure to strong winds. They then re-established at the base of the scarp along with many other species. After ten years the vegetation had fully recovered (see Chapter 1, Fig. 1.13).

In each of several breaches in the first dune ridge caused by storm waves vegetation was totally removed, although it was sometimes relocated into a landward slack. Delayed effects of the storm included blowout development in the breaches and over the next 5 years secondary blowouts appeared on the downwind sides of these. The eroded sand was deposited inland and in areas of high deposition *A. breviligulata* increased and became dominant. However, burial killed plants of several other species, even apparently healthy large trees in full foliage, such as *Quercus* spp.

Changes in ten blowouts in the second dune ridge along Lake Huron between 1973 and 1998 were determined by multispectral analysis of digitalized colour air photos to permit classification of land surfaces into cover types, such as bare sand, herbaceous vegetation and trees. The cover type in some blowouts did not change in 25 years, but in others the cover changed from bare sand to vegetation or from vegetation to bare sand. Blowouts in fact showed a continuum of recovery responses (Dech *et al.* 2005), with at least some re-establishment of dune vegetation in each of them and a mean area of 18.5% colonized per blowout after 25 years. Six exhibited net regression and four net colonization. Thus, breaches and blowouts do not necessarily expand indefinitely. The time-scale for recovery of blowouts in the Pinery was variable, with normalization in some not accomplished in 25 years, but in others within 20 years or even less.

While the habitat of some species is destroyed by high waves and storms, corresponding sand depositions provide opportunities for plants with differing physiological responses to burial. The growth of many dune species is enhanced by sand deposition, with burial increasing chlorophyll content significantly (mg g^{-1} fresh weight) in some dune species and maximum quantum efficiency of photosystem II in others (Perumal and Maun 2006). In *Agropyron psammophilum*, *Elymus canadensis* and *Panicum virgatum* both parameters increased. Elevated carbon dioxide exchange rates provide a further indication of improved photosynthetic performance under sand depositions. After experimental burial to one-third or two-thirds of plant height, total biomass increased significantly in all species tested and in most species leaf thickness was also increased (Perumal and Maun 2006). Sand movement thus influences species presence and relative abundance by favouring plants that thrive under burial.

An essential characteristic of plants that survive sand deposition appears to be tolerance to anoxia and, curiously, the trees and shrubs capable of surviving sand burial are those that also tolerate wetter than normal conditions. Plants of these species develop adventitious roots near the soil surface that enhance the supply of oxygen to underground structures (Dech and Maun 2006), and they also tend to re-allocate resources internally, moving carbon from deeply buried root systems to emergent shoots. Woody plants resistant to sand burial include *Picea mariana*, *Salix* spp., *Betula* spp., and *Cornus stolonifera*, all of which are known to establish in newly developing dune communities and increase in importance as sand accumulates. On the other hand, almost all oaks, e.g. *Quercus velutina*, *Q. rubra* and *Q. muehlenbergii*, as well as *Juniperus communis*, are intolerant to burial. Among herbaceous plants, *A. breviligulata* best tolerates sand burial, and survives for 1 year under an 80 cm of sand. No species in Lake Huron dunes can survive burial under more than 120 cm but a number tolerate up to 80 cm, e.g. *C. longifolia*, *Agropyron psammophilum* and *Panicum virgatum*. Some of the least tolerant herbaceous species, however, can withstand burial by no more than 15 cm, e.g. *Corispermum hyssopifolium*, *Euphorbia polygonifolia*, *Cakile edentula* and *Strophostyles helvola*.

Along the North Sea and Atlantic coasts of Europe vegetation differs between erosional and accumulational sites (Pluis and De Winder 1990). Microbial crusts, predominantly green algae or cyanobacteria, in some cases reduce sand mobility in erosional sites and hyphae of mycorrhizal fungi also bind sand particles. The roots of annuals and perennials anchor more deeply than microbes, with *Agropyron junceiforme* and *Elymus farctus* (=*Ammophila arenaria*), the first vascular plant stabilizers, followed by *Leymus arenarius* (=*Elymus arenarius*) on embryo dunes. In accumulation sites where sand is being trapped, the stabilizer species establishing are those that withstand or require burial. In secondary dunes or mobile dunes, *Ammophila arenaria* is the most common of these (Vestergaard 2006). *Ammophila arenaria* and *Carex arenaria* may extend vegetatively into a blowout from its edges, while *Sedum acre* with a low profile establishes in the bottom. Growth rate and height of plants is correlated with sand activity, with the result that slow-growing low plants occur in areas with net erosion and little accretion.

Slack vegetation may be particularly influenced by changes in levels of groundwater levels associated with increases in relative sea levels (Vestergaard 1997). If the dune zone is narrowed by shoreline retreat, the size of the catchment area will be reduced and soil moisture may be decreased. This could lead to the decline of moisture-requiring plant species typical of slacks, because many plant species are sensitive to even small changes in available moisture (Noest 1991). Species composition may thus be altered, with xerophytes displacing more mesophytic species (Vestergaard 1991).

Although unconfined subterranean water levels (phreatic water) may fall as a result of coastline retreat, it can also rise, depending on the width of the dune body, erosion, coastal management (Noest 1991) and especially the degree to which moisture is replenished by rising seas (Vestergaard 1997). In the event that groundwater supplies increase with rising seas, dune species typically found in slacks may be displaced by shrubs or tall grasses (Davy *et al.* 2006). If diminishing fresh water is replaced by rising seas the species composition may change to reflect increasingly brackish conditions (Vestergaard 1991; Davy *et al.*

2006). Noest (1991) modelled the effects of sea level rise in the Netherlands on dune slack vegetation over the next century and found by Canonical Correspondence Analysis that the impact could be considerable if the dune field is narrowed excessively, with brackish species, such as *Centaurium pulchellum*, *Glaux maritima* and *Scirpus maritimus*, potentially favoured by increasing salt spray (Noest 1991).

The extent of effects of sea level rise on dune vegetation can also be gaged generally from the successional stage of vegetation. The degree to which a complete series of vegetation zones is maintained from beach to inner dunes is an indication of impact, as is species composition and diversity of plant species (Vestergaard 1997). In extreme cases habitat may be altered so severely that survival is contraindicated. On the coast of Florida, USA, erosional forces removed protective dunes on a barrier island and exposed inland species such as *Pinus* spp. directly to wind and wave action (Carter 1988, Fig. 13.11). As a result the forest was flooded out. Similarly, a drowned forest off the coast of Ireland (Carter 1988), estimated to be 4000 years old, is now reduced to stumps in the littoral zone exposed at low tide (Fig. 13.12).

Figure 13.11 A sand barrier off the Atlantic coast of Florida altered by overwash, with pines once located on the leeward side now exposed to conditions on the seaward face following removal of materials from protective beach and foredunes by high seas (Carter 1988).

Figure 13.12 The remains of an approximately 4000-year-old forest in western Ireland inundated by rising seas. The site now is located in the lower intertidal zone and subject to regular submersion (Carter 1988).

13.7 Episodic rise in sea levels?

With climatic warming and rapid rise in sea levels, many coastal features existing today could suddenly be totally submerged and added to the sea shelf (Ray and McCormick-Ray 2004), where they have been for most of the previous geological time. However, because not all parameters associated with increases in sea level are sufficiently well understood, controversy persists as to whether the expected increases will be abrupt (Murray-Wallace 2007).

Although sea levels currently appear to be increasing relatively gradually, sudden or episodic change is also within the realm of possibility. In fact, meltwater pulses (Fig. 13.13) are now considered a proven feature of the present postglacial period (Fairbanks 1989; Shaw

1989; Peltier 2002; Hancock 2002). Cataclysmic release of water seems to have taken place about 15 000 years BP, but may have occurred more than once. It could be caused not only by continued melting initiated after the LGM, but also by collapse of major ice sheets in the northern hemisphere or accelerated melting in Antarctica (Zong 2007). It has been suggested that in the past global sea levels may have increased as much as 0.23 m within a week (Shaw 1989). Interestingly, accounts of cataclysmic floods are known in all parts of the world, with at least 500 flood myths reported in widely diverse cultures (Murray-Wallace 2007). Similarities in these stories support the geological evidence that indicates there has been sudden catastrophic release of meltwater

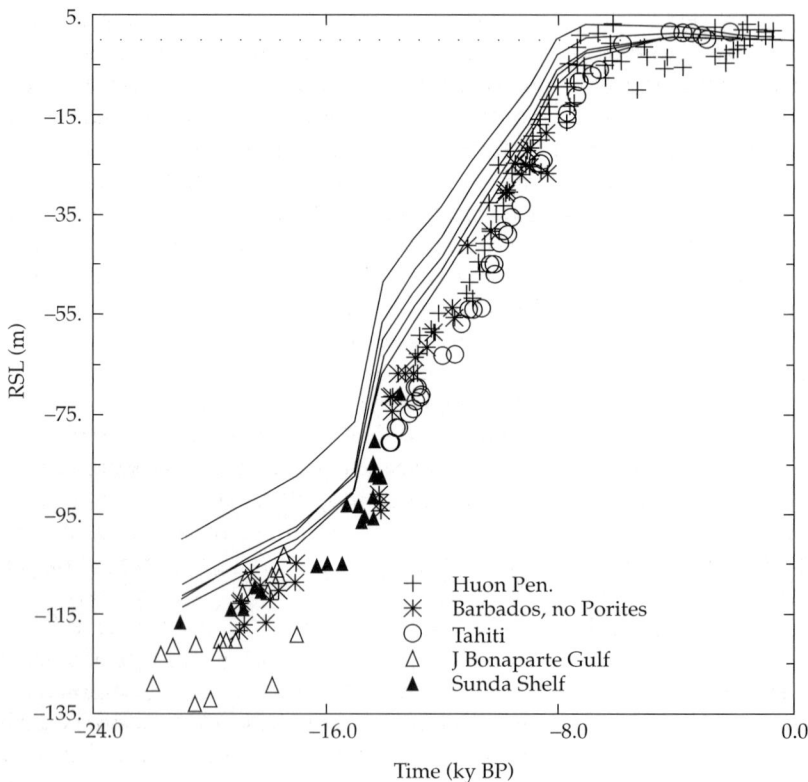

Figure 13.13 Relative Holocene sea levels as inferred from coral data obtained in the equatorial Pacific Ocean; also shown are theoretical predictions for the same locations (Peltier 2002).

from large continental ice sheets during the present interglacial period.

Recent observations in polar ice suggest that rapid change in sea levels is not only possible, but plausible in the near future. Although the snowfall-driven growth of ice masses which accompanies global warming helps to mitigate loss through melting, there has been a net annual loss of ice in both Antarctica and Greenland (Shepherd and Wingham 2007). The periphery of the Greenland ice sheet is thinning, and accelerated discharge has more than doubled the ice sheet mass deficit in the last decade (Rignot and Kanagaratnam 2006). The monthly mean extent of ice has exhibited a steady downward trend from 1979–2006, and decline is apparent in each of the twelve months. The particularly strong decrease in mean September ice cover reflects longer summers, and although estimates of future rates of ice-loss vary widely, areas in the Arctic that now freeze over in September will probably tend to become ice-free in that month some time during the next century (Walsh *et al.* 2005). As ice melts and the albedo of the northern oceans decreases, progressively less ice will be available to protect coastlines against severe winter storm damage. Erosion of more exposed shorelines has already started and it will be accelerated (Serreze *et al.* 2007), with shoreline recession expected on arctic islands (Leontiev 2006).

Satellite observations now indicate that glacier movement has surprisingly accelerated 20–100% in the last decade (Shepherd and Wingham 2007), possibly due to increased internal melt water and reduced basal friction between ice and land surfaces (Bindschadler 2006). Fast-flowing glaciers such as Jakobshavn Isbrae in Greenland strongly affect discharge from ice sheets at subdecanal timescales, suggesting that response to climate change could indeed be rapid (Joughin *et al.* 2004). Increased melting of outlet glaciers will reduce the buffering effect normally offered to continental ice by adjacent floating ice shelves, making ice masses more vulnerable to climatic warming than previously realized (Bindschadler 2006).

Current predictive models that do not include consideration of ice shelf removal, meltwater lubrication of forward movement, ice front retreat and glacier ungrounding only indicate the lower limits to the potential contribution of ice sheets to sea level rise (Rignot and Kanagaratnam 2006). It has been suggested that to better describe the range of future sea levels a large percentage, e.g. 50%, should be added to predicted mean increases and then corrected for expected changes in elevation of land (IPCC and Nicholls *et al.* 2007a). Certainly there is no longer a clear consensus for the next century that increases in sea level due to melting of Greenland and Antarctic ice sheets will be small or slow (Shepherd and Wingham 2007), and in fact multi-metre rises in sea level have been envisioned on a century timescale (Hansen 2007).

Unfortunately scientists are pressured to downplay the potential effects of global warming, as worrisome trends in loss of ice mass in Antarctica are not reflected in projections for the future, even though these may well produce multi-metre increases in sea level on a century timescale (Hansen 2007). As new phenomena are recognized, projections regarding the nature of future increases in sea level must be repeatedly revisited (Shepherd and Wingham 2007). If global warming does produce catastrophic increases in sea level along coastlines, habitat for particular biotic communites, including dune systems, could be reduced severely (Feagin 2005).

13.8 Vulnerability to human manipulation

In addition to those human activities that stimulate global warming, many other societal influences may seriously exacerbate the effects of high water on coastal systems (Vestergaard 1997). For example, residential, recreational, agricultural and commercial use of land and water all magnify the impact of high sea levels on coastal dune systems.

Engineered structures such as canals, dykes, levees, and dams directly interfere with the normal delivery of freshwater and sediments to the ocean and, as a result, alter the dynamic offshore processes critical for shoreline stability (Ray and McCormick-Ray 2004). At the mouth of the Nile a substantial series of dunes is at risk in part because sediments retained behind dams are no longer transported to the sea; coastal lands, lacking normal sedimentary input, are subsiding (Broadus 1993; Milliman et al. 1989). A similar situation occurs in the deltas of many other rivers.

Settlements along coastlines involving massive engineered structures and large human populations also promote land subsidence, as does withdrawal of groundwater or oil/gas (Salinas et al. 1986; Milliman et al. 1989; Day et al. 1993; Ray and McCormick-Ray 2004). Weakening of offshore reefs, e.g. by water pollution and mining of coral, increases the risk to shorelines. In Argentina, offshore barriers and mainland coasts are stressed by sand mining as well as urbanization (Schnack 1993). Partial removal of coastal dunes in one tourist area in Sri Lanka opened up the shore to more severe tsunami damage than experienced along immediately adjacent unmanipulated coastline (Liu et al. 2005).

Jetties are rarely successful in reducing the shoreline erosion that they were intended to prevent and may actually cause more problems than they solve. In fact construction of jetties and piers, and any other activity that disrupts normal sediment transport, often leads to increased erosion (Inman and Masters 1991; Brown and McLachlin 2002). Excessive modification of coastal and offshore ecosystems to facilitate shipping and real estate development also destabilizes land masses, for example the delta of the Mississippi River in Louisiana has been known to be sinking for many years (Salinas et al. 1986) and remains extremely vulnerable to hurricanes (Day et al. 1993; Brown and McLachlan 2002).

The welfare of dune systems in close proximity to, and defending, dense human settlements has been of particular concern. Introduction of natural dune plants, such as Ammophila spp., Croton punctatus, Schizachyrium scoparium and Chamaecrista chamaecristoides var. chamaecristoides, can be used to stabilize and strengthen weakened areas (Clayton 1993; Maun and Krajnyk 1989; Martínez and Vásquez 2001; Lonard and Judd 2008), and restricting human recreation and other activities also helps to maintain vegetation pivotal to the maintenance of healthy dunes. However, in seriously devastated or high-risk areas, more drastic measures such as dune-feeding are sometimes contemplated. This procedure involves enhancement of coasts with massive amounts of supplemental sand in an attempt to increase coastal security and minimize the ecological consequences of higher sea levels. Unfortunately, the costs are astronomical (De Ronde 1993; Schnack 1993), and the exercise may well be futile (Trembanis et al. 1999) because beach supplements are easily removed by high waves. On the other hand, supplementation of shorefaces, which involves placement of large increments of sand offshore, may be more effective because such deposits at least tend to damp wave action on the lee side (van Leeuwen et al. 2007). While dikes and beach nourishment have both been used in Denmark, the prevailing approach largely involves preservation of natural coastline, even if it means some loss of land (Vestergaard 2008).

13.9 Extinction of dune plants?

The biosphere is generally expected to be considerably less predictable after 2100 (Woodruff 2001) due to lack of critical information about the parameters involved, and it is thus no simple matter to forecast the effects of sea level rise on dune vegetation. Little is known about tolerance of individual species to changes in sea level, nor to longer length of growing season and higher mean temperatures. Temperature increases of as much as 3–4°C, an amount which is distinctly possible, could eliminate some genotypes and favour others with higher cardinal points for temperatures (maximum, minimum, and optimum). Another reason that tracking of changes and projection of vegetational trends is extremely difficult is the paucity of detailed baseline field information about current ecology and distribution, as well as a lack of updated inventories and maps. Detailed information that is digitalized would be valuable but such data is particularly scarce (van der Meulen *et al.* 1991).

Land plants and flowering plants have persisted over millions of years since first evolving while oceans have risen and fallen repeatedly, but highly technical societies which overpopulated and contaminated the earth have not been documented in previous glacial and interglacial periods. It is usually assumed that humans were not responsible for accelerating climate change in earlier cycles. However, the impact of *Homo sapiens* has clearly been overpowering in the last few centuries, with intensified population pressures forcing unparalleled exploitation and destruction of atmosphere, lithosphere, biosphere and hydrosphere. Mutation rates will probably increase in response to greater exposure to ultraviolet light, industrial pollution, nuclear wastes and depleted uranium widely used in allied military aggression. If human interference is as extensive as it appears to be, and indeed unprecedented, dune species may respond less well to the present interglacial period than to its predecessors.

Current extinction rates have been estimated at 50–500 times background levels, and it is expected that these will increase even more (Woodruff 2001). A widely held view is that many species will become extinct under current accelerated warming (Woodruff 2001). Another possibility is that species will survive but certain genotypes will be lost. The ability to respond to future climatic changes by range shifts depends on the extent of inhospitable territory interrupting natural ranges. Northern species may be lost from a given system while southern elements immigrate into it, with

those species that approach the limits of their geographical distributions most vulnerable to environmental change (Vestergaard 1991). The strongest impact is expected at high latitudes where the amplitude of climatic variation is greatest (Dynesius and Jansson 2000). Coastal communities moving northward would probably be modified compositionally, structurally and genetically under altered selection pressures. Northward movement of species such as *Cakile maritima* and *Eryngium maritimum*, which are confined to coastal systems and disperse mainly along shorelines, would be contraindicated by habitat fragmentation such as frequently occurs along coasts (Clausing *et al.* 2000). Genealogical methods used by Rauch and Bar-Yam (2005) showed that diversity in a population appears to be strongly related to the area of available habitat, and is particularly depressed in linear habitat fragments such as those along rivers or coasts.

Due to genetic drift and inbreeding, the loss of genetic diversity is especially problematic in habitat fragments. Genetic erosion can be demonstrated by assessing variation over time in DNA sequences, such as those in nuclear micro satellites (Woodruff 2001). In Europe baseline genetic properties of two coastal species have been determined through examination of RAPDs (randomly amplified polymorphic DNAs) along with ISSRs (intersimple sequence repeats) distributed throughout the genome (Clausing *et al.* 2000). The real evolutionary significance of nucleotide sequences varies, however, because genetic regions differ greatly in their influence on functionality and phenotype (Stanley 1985).

Under high seas and prolonged shoreline erosion those species that are least burial-tolerant may be eliminated from some areas, however, adaptive dispersal, dormancy, avoidance and tolerance mechanisms may permit many species to persist regardless. Climatic change may elicit adaptation to different temperature regimes, and coastal plants in the future may also develop superior detoxification mechanisms. Phenotypic plasticity, seed banks, bud banks and a capacity for vegetative fragmentation also promote survival. It is possible that, despite high local mortality, continued encroachment by rising seas will rearrange dune vegetation, altering precise locations of communities and distribution of species.

13.10 Summary

It is difficult to predict what the effect of sea level rise on coastal dune systems in the world will be, partly because predictions on the rate and full extent of rise vary so greatly. Even the most conservative estimates, however, suggest that sea levels will be higher within the next century, and that this will in large part be due to increased mass of sea water and floating ice, as well as increased volume, resulting from higher mean annual temperatures.

Effects on coastal sand dune systems will be similar to those of the past, with particulate materials reworked and displaced to other locations, either inland or offshore, by wave action, wind and currents within offshore circulation cells. Some sand will be reconstituted as different land forms, such as shoals or barriers, or it will become submerged offshore and incorporated into the growing continental shelves. Dune vegetation is extremely resilient and well-adapted to the stresses of dune environments, so most species will probably shift along with the dunes themselves. They may evolve, however, in the face of unusual changes in their environment, some of which are man-made. Extinction is therefore possible but unlikely, except in the case of extremely localized genotypes subjected to rapid or catastrophic increases in sea level.

CHAPTER 14

Glossary of terms

Ammophila **problem** The deterioration in vigour of plant populations growing on sand dunes following cessation of burial by sand. This debilitation of populations is not confined to *Ammophila* alone because it occurs in populations of all perennial plant species adapted to live in the sand accreting habitats of foredunes.

Amphibious lifestyle Living part of life on land and part in water.

Andropogon scoparius This plant name has been used in literature for a long time. Recently the name has been changed to *Schizachyrium scoparium*.

Arbuscule Branched, tree-like structures within the cortex made by endomycorrhizal fungi.

Aseptate hyphae Hyphae without cross walls.

Backwash Backward movement of water under the force of gravity when the onshore velocity of swash is lost.

Blowout A hollow, depression, trough or swale within a dune complex created by the blowing out of sand by wind from a part of a dune ridge. Wind-scoured gaps in an otherwise continuous transverse dune.

BP Before the present.

Bulk density Weight per unit volume of soil. It provides a useful measure to estimate soil structure and the extent to which root penetration may be restricted by the soil.

C_3 plants Cool-season plants. They start to grow early in the growing season under cooler and more moist conditions.

C_4 plants Warm-season plants. Their growth starts later in the growing season under hotter and drier conditions.

Channel Blowout or cut in a dune ridge made by waves or wind. They interrupt dune ridges and cut across back-dune areas.

Clonal integration Refers to sharing of resources such as water, nutrients, oxygen, hormones and photosynthates through inter-ramet clonal connections between rhizomes, stolons and lateral roots of plants.

Collapsing waves Collapsing breakers are similar to plunging types but instead of curling the front face of the crest collapses. This happens on moderately steep slopes.

Dissipative shoreline A shoreline with a wide beach in which the energy level of waves is dissipated gradually.

Driftline Area of the beach marked by washed-up detritus and indicating the farthest inland reach of waves.

Drought The potential of the environment to induce water loss from the plant.

Drought avoidance Ability of a plant to exclude drought stress from its tissues by maintaining high vapour pressure in the tissues through morphological adaptations such as pubescence on leaves, thick cuticle, fleshy leaves, closing of stomata, rolling of leaves, or by lowering their metabolic rate to extremely low levels.

Drought tolerance Ability of a plant to withstand the penetration of water stress. For example, lichens and mosses dry out completely under drought conditions and decrease their vapour pressure to very low values.

Dune A wind-blown pile of sand. A rounded hill or a ridge of sand heaped up by the action of wind.

Dune complex A dune complex refers to the open dune systems colonized primarily by grasses, forbs and shrubs and extending from the beach to the forested dune ridges. The light intensity penetrates to the soil level and the area is under the influence of winds. Doing (1985) defined it as a geographical, geomorphological and ecological functional unit with a complete range of dune ridges from the beach to the forest as a direct effect of active transportation of sand and organic material.

Dune ridge A long narrow elevation of land composed of a row of dunes. A straight row of dunes.

Embryo dune Low unconnected mounds or hummocks of sand about 1 to 2 m in height.

Facultatively mycotrophic Plants that can survive with or without mycorrhizae.

Fetch Distance of ocean/water over which wind has blown in the same direction.

Field capacity Moisture retained by soil after the drainage of excess water by the force of gravity.

Gale force winds Strong winds that range in speed from 32–63 miles (51–101 km) per hr.

Halophyte Plants occurring naturally on soils or in water which is too salty for the average plant in the range of 200–500 mM of NaCl concentrations.

Intertidal zone Area of the beach that is covered at high tide and exposed at low tide.

Invasive species An artificially introduced alien species that expands its range after its introduction and replaces indigenous species, thus disturbing the ecological balance.

Leaf area index (LAI) Total surface area of leaves per unit area of ground surface.

Mid-beach Area below the farthest inland reach of waves usually colonized by ephemeral, salt-tolerant plant species. However, perennial species may expand into this area by stolons or plagiotropic rhizomes.

Mycorrhizal inoculum potential The rate at which roots are colonized by hyphae and spores of AM species.

Obligately mycotrophic Plants that require mycorrhizae for survival.

Oceanography A branch of geography that deals with the oceans.

Orbital velocity The time taken by a water molecule to move from crest to trough and back to crest.

Plunging waves The whole front face of the wave steepens until it becomes vertical, curls over the front face and plunges downwards dissipating its energy within a very short distance on shallow to intermediate beach slopes. Plunging breakers have a concave front and a convex back.

Primary dune ridge First dune ridge next to the shoreline.

Reflective shoreline A shoreline with a narrow beach in which the energy level of waves is dissipated quickly over a very short distance.

Saltation The process of sand movement in which at a certain shear velocity of wind, sand particles are ejected from the sand surface into the air stream and propelled forward. They are acted upon by gravity at the same time and lose height and strike the ground at about 10–16° angles with considerable force and eject more grains into the air stream, thus accentuating the sand-moving process.

Sand flats Broad, low, generally level areas of sand, lying between primary and secondary foredunes. They are generally found in high-energy coasts with strong wind velocities and sparse vegetation.

Sand hummock or nebka Mounds or hillocks of sand created around plant species that can tolerate burial by sand.

Saucer blowout Shallow, ovoid, dish-shaped hollows with a steep marginal rim with sand deposited immediately downwind giving it the shape of a saucer.

Schizachyrium scoparium This plant was known for a long time in the scientific literature as *Andropogon scoparius*.

SD Standard deviation.

SE Standard error.

Secondary dune ridge Second dune ridge inland from the first dune ridge.

Slack, low or deflation plane Area created on the lee of dune ridges where wind erodes sand down to the water table. Because of wet sand surface further erosion stops and a deflation plane is created. A valley between two dune ridges. A trough or hollow between the primary and secondary dune ridges.

Spilling waves Spilling starts at the crest of the wave when a small tongue of water moves forward faster than the wave as a whole and foam, bubbles and turbulent water appear at the crest and front of the waves. Spilling waves usually occur on gently sloping flat beaches and break at considerable distance from shore.

Stress In plant ecology, stress has been defined as 'external constraint which limits the rate of resource acquisition, growth or reproduction of organisms' (Grime 1979). The emphasis is on an adverse force that hinders the ability of an organism, community or an ecosystem to reach its maximum potential. Dune colonizers respond to stress in several ways. Almost all species germinate their seeds early in spring when moisture is plentiful, thus escaping the low substrate moisture later in summer.

Surface creep Sand particles too heavy to be ejected by saltation are propelled forward short distances on the sand surface by saltating grains.

Surging waves These breakers are formed on very steep beaches by flat low waves in which the front faces and crests remain smooth and relatively unbroken and gently slide up and down the beach with only minor production of foam and bubbles.

Suspension Very small particles of grains suspended in air and scattered as dust because their velocity of fall is less than the upward eddies of air currents within the average velocity of wind.

Swale A low-lying area between dune ridges.

Swash Rush of water onshore when a wave breaks.

Trough blowout An elongated depression in a dune ridge shaped by wind flowing through it and eroding sand from the surface and pushing it over the steep backslope.

Upper beach, high beach Area normally above the direct tidal influence. However, it may be inundated and destroyed by high wave action at infrequent intervals. Species growing in this zone must be able to withstand burial in sand and salt spray episodes.

Vesicle Intercalary or terminal hyphal swellings formed within the root cortex or between cells. They are storage organs for lipids but may also function as propagules.

Vigour increase An improvement in physiological processes of a plant and increase in density, cover and biomass per plant and per unit area in a plant community.

Water potential A term used to express the energy status of water in the soil, the plant or atmosphere. Pure water has the highest water potential with a value of 0 and a soil saturated with water has a water potential of 0.

Wave height Vertical distance between crest and trough of a wave.

Wave length Horizontal distance between successive crests.

Xerophyte Plants that are able to survive, grow and reproduce under water deficit stress conditions. They can grow in soils that dry out to a depth of 20 cm during the normal growing season.

CHAPTER 15

Literature cited

Abuzinadah RA and Read DJ (1989a). The role of proteins in the nitrogen nutrition of ectomycorrhizal plants. IV. The utilization of peptides by birch (*Betula pendula* L.) infected with different mycorrhizal fungi. *New Phytologist*, **112**, 55–60.

Abuzinadah RA and Read DJ (1989b). Carbon transfer associated with assimilation of organic nitrogen sources by silver birch (*Betula pendula* Roth.). *Trees: Structure and Function*, **3**, 17–23.

Akhtar P and Shaukat SS (1979). Drought resistance and dew utilization in *Sorghum bicolor* (L.) Moench. and *Ipomoea pes-caprae* (L.) Sweet. *Pakistan Journal of Botany*, **11**, 85–91.

Al-Agely AK and Reeves FB (1995). Inland sand dune mycorrhizae: effects of soil depth, moisture, and pH on colonization of *Oryzopsis hymenoides*. *Mycologia*, **87**, 54–60.

Allen EB and Allen MF (1990). The mediation of competition by mycorrhizae in successional and patchy environments. In JB Grace and GD Tilman, eds, *Perspectives on plant competition*, pp. 367–389. Academic Press, New York.

Allen JRL (2000). Morphodynamics of Holocene salt marshes: a review sketch from the Atlantic and southern North Sea coasts of Europe. *Quaternary Science Reviews*, **19**, 1155–1231.

Allen MF (1991). *The ecology of mycorrhizae*. Cambridge University Press, Cambridge.

Allen MF, Allen EB and Friese CF (1989). Responses of the non-mycotrophic plant *Salsola kali* to invasion by vesicular–arbuscular mycorrhizal fungi. *New Phytologist*, **111**, 45–49.

Allen MF, Smith WK, Moore TS Jr and Christensen M (1981). Comparative water relations and photosynthesis of mycorrhizal and non-mycorrhizal *Bouteloua gracilis* H. B. K. Lag Ex Steud. *New Phytologist*, **88**, 683–693.

Alpert P (1999). Clonal integration in *Fragaria chiloensis* differs between populations: ramets from grassland are selfish. *Oecologia*, **120**, 69–76.

Ames RN, Porter LK, St. John TV and Reid CPP (1984). Nitrogen sources and 'A' values for vesicular–arbuscular and non-mycorrhizal sorghum grown at three rates of ^{15}N-ammonium sulphate. *New Phytologist*, **97**, 269–276.

Amsberry L, Baker MA, Ewanchuk PJ and Bertness MD (2000). Clonal integration and the expansion of *Phragmites australis*. *Ecological Applications*, **10**, 1110–1118.

Anderson C and Taylor K (1979). Some factors affecting the growth of two populations of *Festuca rubra* var. *arenaria* on the dunes of Blakeney Point, Norfolk. In RL Jefferies and AJ Davy, eds, *Ecological processes in coastal environments*, pp. 129–143. Blackwell, London.

Ansell AD (1983). The biology of the genus *Donax*. In A McLachlan and T Erasmus, eds, *Sandy beaches as ecosystems*, pp. 607–635. Junk, The Hague.

Antos JA and Zobel DB (1984). Ecological implications of belowground morphology of nine coniferous forest herbs. *Botanical Gazette*, **145**, 508–517.

Antos JA and Zobel DB (1985). Plant form, developmental plasticity, and survival following burial by volcanic tephra. *Canadian Journal of Botany*, **63**, 2083–2090.

Armstrong W (1979). Aeration in higher plants. *Advances in Botanical Research*, **7**, 225–332.

Art HW, Bormann FH, Voigt GK and Woodwell GM (1974). Barrier Island forest ecosystem: role of meteorologic nutrient inputs. *Science*, **184**, 60–62.

Augé RM, Schekel KA and Wample RL (1986). Greater leaf conductance of well-watered VA mycorrhizal rose plants is not related to phosphorus nutrition. *New Phytologist*, **103**, 107–116.

Ayalon A and Longstaffe FJ (2001). d^{13}C and d^{18}O variations in soil organic carbon and pedogenic calcite as indicators of late Holocene climate change, southern Great Lakes area, Canada. *Israel Geological Society Annual Meeting*, p. 8. Eilat, Israel.

Ayyad MA (1973). Vegetation and environment of the western Mediterranean coastal land of

Egypt. I. The habitat of sand dunes. *Journal of Ecology*, **61**, 509–523.

Bach CE (1994). Effects of a specialist herbivore (*Altica subplicata*) on *Salix cordata* and sand dune succession. *Ecological Monographs*, **64**, 423–445.

Bach CE (2001). Long-term effects of insect herbivory and sand accretion on plant succession on sand dunes. *Ecology*, **82**, 1401–1416.

Bagnold RA (1960). *The physics of blown sand and desert dunes*. Methuen, London.

Baker HG (1955). Self-compatibility and establishment after 'long-distance' dispersal. *Evolution*, **9**, 347–349.

Baker HG (1989). Some aspects of the natural history of seed banks. In MA Leck, VT Parker and RL Simpson, eds, *Ecology of soil seed banks*, pp. 9–21. Academic Press, New York.

Baldwin AH and DeRico EF (1999). The seed bank of a restored tidal freshwater marsh in Washington, DC. *Urban Ecosystems*, **3**, 5–20.

Baldwin KA and Maun MA (1983). Micro-environment of Lake Huron sand dunes. *Canadian Journal of Botany*, **61**, 241–255.

Balestri E and Cinelli F (2004). Germination and early-seedling establishment capacity of *Pancratium maritimum* L. (Amaryllidaceae) on coastal dunes in the north-western Mediterranean. *Journal of Coastal Research*, **20**, 761–770.

Baptista TL and Shumway SW (1998). A comparison of the seed banks of sand dunes with different disturbance histories on Cape Cod National Seashore. *Rhodora*, **100**, 298–313.

Barbour MG (1970a). Seedling ecology of *Cakile maritima* along the California coast. *Bulletin of the Torrey Botanical Club*, **97**, 280–289.

Barbour MG (1970b). Is any angiosperm an obligate halophyte? *American Midland Naturalist*, **84**, 105–120.

Barbour MG (1972). Seedling establishment of *Cakile maritima* at Bodega Head, California. *Bulletin of the Torrey Botanical Club*, **99**, 11–16.

Barbour MG (1978). Salt spray as a microenvironmental factor in the distribution of beach plants at Point Reyes, California. *Oecologia (Berl.)*, **32**, 213–224.

Barbour MG and De Jong TM (1977). Response of west coast beach taxa to salt spray, seawater inundation, and soil salinity. *Bulletin of the Torrey Botanical Club*, **104**, 29–34.

Barbour MG and Rodman JE (1970). Saga of the west coast sea-rockets: *Cakile edentula* ssp. *californica* and *C. maritima*. *Rhodora*, **72**, 370–386.

Barbour MG, De Jong TM and Johnson AF (1976). Synecology of beach vegetation along the Pacific coast of the United States of America: a first approximation. *Journal of Biogeography*, **3**, 55–69.

Barbour MG, De Jong TM and Pavlik BM (1985). Marine beach and dune plant communities. In BF Chabot and HA Mooney, eds, *Physiological ecology of North American plant communities*, pp. 296–322. Chapman and Hall, New York.

Bartlein PJ and Prentice IC (1989). Orbital variations, climate and paleoecology. *Trends in Ecology and Evolution*, **4**, 195–199.

Baskin CC and Baskin JM (1998). *Seeds: ecology, biogeography, and evolution of dormancy and germination*. Academic Press, New York.

Baye PR (1990). Comparative growth responses and population ecology of European and American beachgrasses (*Ammophila* spp.) in relation to sand accretion and salinity. PhD thesis, University of Western Ontario, London, ON.

Beard JS and Kenneally KF (1993). Dry coastal ecosystems of northern Australia. In E van der Maarel, ed., *Ecosystems of the World 2B. Dry coastal ecosystems: Africa, America, Asia and Oceania*, pp. 239–258. Elsevier, Amsterdam.

Beck JB (1819). Observations on salt storms, and the influence of salt and saline air upon animal and vegetable life. *American Journal of Science*, **1**, 388–397.

Beena KR (2000). Studies on the interaction of vesicular–arbuscular mycorrhizae and other endophytic fungi with the sand dune vegetation of west coast of India. PhD thesis, Mangalore University, Karnataka.

Beena KR, Raviraja NS and Sridhar KR (2000a). Seasonal variations of arbuscular mycorrhizal fungal association with *Ipomoea pes-caprae* of coastal sand dunes, Southern India. *Journal of Environmental Biology*, **21**, 341–347.

Beena KR, Raviraja NS, Arun AB and Sridhar KR (2000b). Diversity of arbuscular mycorrhizal fungi on the coastal sand dunes of the west coast of India. *Current Science*, **79**, 1459–1466.

Belcher CR (1977). Effect of sand cover on the survival and vigor of *Rosa rugosa* Thunb. *International Journal of Biometeorology*, **21**, 276–280.

Bentley CR (2004). Mass balance of the Antarctic ice sheet: observational aspects. In JL Bamber and AJ Payne, eds, *Mass balance of the cryosphere: observations and modelling of contemporary future changes*, pp. 459–489. Cambridge University Press, Cambridge.

Bergman HF (1920). The relation of aeration to the growth and activity of roots and its influence on the ecesis of plants in swamps. *Annals of Botany*, **34**, 13–33.

Bertrand AR and Kohnke H (1957). Subsoil conditions and their effects on oxygen supply and the growth of corn roots. *Soil Science Society of America*, **21**, 135–140.

Bindschadler R (2006). Hitting the ice sheets where it hurts. *Science*, **311**, 1720–1721.

Bird ECF (1985). *Coastline changes: a new global review*. Wiley, Chichester.

Birse EM, Landsberg SY and Gimingham CH (1957). The effects of burial by sand on dune mosses. *Transactions British Bryological Society*, **3**, 285–301.

Blake AK (1935). Viability and germination of seeds and early life history of prairie plants. *Ecological Monographs*, **5**, 405–460.

Bloch MR, Kaplan D, Kertes V and Schnerb J (1966). Ion separation in bursting air bubbles: an explanation for the irregular ion ratios in atmospheric precipitations. *Nature*, **209**, 802–803.

Bloom AL and Stuiver M (1963). Submergence of the Connecticut coast. *Science*, **139**, 332–334.

Boerner RE and Forman RTT (1975). Salt spray and coastal dune mosses. *The Bryologist*, **78**, 57–63.

Bond TET (1952). Biological flora of the British Isles: *Elymus arenarius* L. *Journal of Ecology*, **40**, 217–227.

Boorman LA and Fuller RM (1984). The comparative ecology of two sand dune biennials: *Lactuca virosa* L. and *Cynoglossum officinale* L. *New Phytologist*, **69**, 609–629.

Borhidi A (1993). Dry coastal ecosystems of Cuba. In E van der Maarel, ed., *Ecosystems of the world 2B. Dry coastal ecosystems: Africa, America, Asia and Oceania*, pp. 423–452. Elsevier, Amsterdam.

Bossuyt B and Hermy M (2004). Seed bank assembly follows vegetation succession in dune slacks. *Journal of Vegetation Science*, **15**, 449–456.

Boucher C and Le Roux A (1993). Dry coastal ecosystems of the South African west coast. In E van der Maarel, ed., *Ecosystems of the world 2B. Dry coastal ecosystems: Africa, America, Asia and Oceania*, pp. 75–88. Elsevier, Amsterdam.

Boudreau S and Houle G (2001). The *Ammophila* decline: a field experiment on the effects of mineral nutrition. *Ecoscience*, **8**, 392–398.

Bowles JM (1980). Effects of human disturbance on the sand dunes at Pinery Provincial Park. PhD thesis, University of Western Ontario, London, ON.

Boyce SG (1951). Salt hypertrophy in succulent dune plants. *Science*, **114**, 544–545.

Boyce SG (1954). The salt spray community. *Ecological Monographs*, **24**, 29–67.

Boyd RS (1988). Microdistribution of the beach plant *Cakile maritima* (Brassicaceae) as influenced by a rodent herbivore. *American Journal of Botany*, **75**, 1540–1548.

Boyd RS (1991). Population biology of west coast *Cakile maritima*: effects of habitat and predation by *Peromyscus maniculatus*. *Canadian Journal of Botany*, **69**, 2620–2630.

Boyd RS and Barbour MG (1986). Relative salt tolerance of *Cakile edentula* (Brassicaceae) from lacustrine and marine beaches. *American Journal of Botany*, **73**, 236–241.

Boyd RS and Barbour MG (1993). Replacement of *Cakile edentula* by *C. maritima* in the strand habitat of California. *American Midland Naturalist*, **130**, 209–228.

Brammer H (1993). Geographical complexities of detailed impact assessment for the Ganges–Brahmaputra–Meghna delta of Bangladesh. In RA Warrick, EM Barrow and TML Wigley, eds, *Climate and sea level change: observations, projections and implications*, pp. 246–262. Cambridge University Press, Cambridge.

Bressolier C and Thomas Y-F (1977). Studies on wind and plant interactions on French Atlantic coastal dunes. *Journal of Sedimentary Petrology*, **47**, 331–338.

Brewer JS, Levine JM and Bertness MD (1998). Interactive effects of elevation and burial with wrack on plant community structure in some Rhode Island salt marshes. *Journal of Ecology*, **86**, 125–136.

Brewer R (1994). *The science of ecology*, 2nd edn. Saunders, New York.

Broadus JM (1993). Possible impacts of, and adjustments to, sea level rise; the cases of Bangladesh and Egypt. In RA Warrick, EM Barrow and TML Wigley, eds, *Climate and sea level change: observations, projections and implications*, pp. 263–275. Cambridge University Press, Cambridge.

Brohan P, Kennedy JJ, Harris I *et al.* (2006). Uncertainty estimates in regional and global observed temperature changes: a new dataset from 1850. *Journal of Geophysical Research*, **111**, D12106, doi:10.1029/2005JD006548.

Brown AC and McLachlan A (2002). Sandy shore ecosystems and the threats facing them: some predictions for the year 2025. *Environmental Conservation*, **29**, 62–77.

Brown J, Colling A, Park D, Phillips J, Rothery D and Wright J (1989). *Seawater: its composition, properties and behaviour*. Pergamon Press, Oxford.

Brown JF (1997). Effects of experimental burial on survival, growth, and resource allocation of three species of dune plants. *Journal of Ecology*, **85**, 151–158.

Brundrett M (2004). Diversity and classification of mycorrhizal associations. *Biological Reviews*, **79**, 473–495.

Brundrett MC (2002). Coevolution of roots and mycorrhizas of land plants. Tansley Review No. 134. *New Phytologist*, **154**, 275–304.

Bustard HR (1973). *Sea turtles: natural history and conservation*. Taplinger, New York.

Byrne M-L (1997). Seasonal sand transport through a trough blowout at Pinery Provincial Park, Ontario. *Canadian Journal of Earth Sciences*, **34**, 1460–1466.

Cameron SD (1998). *The living beach*. Macmillan, Toronto.

Carey AE and Oliver FW (1918). *Tidal lands: a study of shore problems*. Blackie and Son, London.

Carey PD and Watkinson AR (1993). The dispersal and fates of seeds of the winter annual grass *Vulpia ciliata*. *Journal of Ecology*, **81**, 759–767.

Carter RWG (1988). *Coastal environments: an introduction to the physical, ecological and cultural systems of coastlines*. Academic Press, London.

Carter RWG (1991). Near-future sea level impacts on coastal dune landscapes. *Landscape Ecology*, **6**, 29–39.

Carter RWG and Wilson P (1990). The geomorphological, ecological and pedological development of coastal foredunes at Magilligan Point, Northern Ireland. In K Nordstrom, N Psuty and B Carter, eds, *Coastal dunes: form and process*, pp. 129–157. Wiley, Chichester.

Carter RWG, Hesp PA and Nordstrom KF (1990a). Erosional landforms in coastal dunes. In K Nordstrom, N Psuty and B Carter, eds, *Coastal dunes: form and process*, pp. 217–250. Wiley, Chichester.

Carter RWG, Nordstrom KF and Psuty NP (1990b). The study of coastal dunes. In K Nordstrom, N Psuty and B Carter, eds, *Coastal dunes: form and process*, pp. 1–14. Wiley, Chichester.

Cartica RJ and Quinn JA (1980). Responses of populations of *Solidago sempervirens* (Compositae) to salt spray across a barrier beach. *American Journal of Botany*, **67**, 1236–1242.

Cavers PB (1963). Comparative biology of *Rumex obtusifolius* L. and *R. crispus* L. including the variety *trigranulatus*. PhD thesis, University of Wales, Bangor.

Cavers PB and Harper JL (1967). The comparative biology of closely related species living in the same area. IX. Rumex: the nature of adaptation to a sea-shore habitat. *Journal of Ecology*, **55**, 73–82.

Chapin FS III, Walker LR, Fastie CL and Sharman LC (1994). Mechanisms of primary succession following deglaciation at Glacier Bay, Alaska. *Ecological Monographs*, **64**, 149–175.

Chapman ARO (1992). Vegetation ecology of rocky shores. In U Seeliger, ed., *Coastal plant communities of Latin America*, pp. 3–30. Academic Press, New York.

Chapman VJ (1976). *Coastal vegetation*, 2nd edn. Pergamon Press, Oxford.

Chen H and Maun MA (1998). Population ecology of *Cirsium pitcheri* on Lake Huron sand dunes. III. Mechanisms of seed dormancy. *Canadian Journal of Botany*, **76**, 575–586.

Chen H and Maun MA (1999). Effects of sand burial depth on seed germination and seedling emergence of *Cirsium pitcheri*. *Plant Ecology*, **140**, 53–60.

Cheplick GP (2005). Patterns in the distribution of American beachgrass (*Ammophila breviligulata*) and the density and reproduction of annual plants on a coastal beach. *Plant Ecology*, **180**, 57–67.

Cheplick GP and Demetri H (1999). Impact of saltwater spray and sand deposition on the coastal annual *Triplasis purpurea* (Poaceae). *American Journal of Botany*, **86**, 703–710.

Cheplick GP and Grandstaff K (1997). Effects of sand burial on purple sandgrass (*Triplasis purpurea*): the significance of seed heteromorphism. *Plant Ecology*, **133**, 79–89.

Cheplick GP and White TP (2002). Saltwater spray as an agent of natural selection: no evidence of local adaptation within a coastal population of *Triplasis purpurea* (Poaceae). *American Journal of Botany*, **89**, 623–631.

Church JA and White NJ (2006). A 20th century acceleration in global sea-level rise. *Geophysics Research Letters*, **33**, L01602, doi:10.1029/2005GL024826.

Clark JS (1986). Coastal forest tree populations in a changing environment, southeastern Long Island, New York. *Ecological Monographs*, **56**, 259–277.

Clausing G, Vickers K and Kadereit KW (2000). Historical biogeography in a linear system:

genetic variation of sea rocket (*Cakile maritima*) and sea holly (*Eryngium maritimum*) along European coasts. *Molecular Ecology*, **9**, 1823–1833.

Clayton JL (1972). Salt spray and mineral cycling in two Californian coastal ecosystems. *Ecology*, **53**, 74–81.

Clayton KM (1993). Adjustment to greenhouse gas induced sea level rise on the Norfolk coast—a case study. In RA Warrick, EM Barrow and TML Wigley, eds, *Climate and sea level change: observations, projections and implications*, pp. 310–321. Cambridge University Press, Cambridge.

Climate Change (2007). The Physical Science Basis. Contribution of Working Group I to the Fourth Assessment Report of the Intergovernmental Panel on Climate Change. Cambridge University Press, Cambridge, UK, Fig. 3.1, 5.13 and FAQ 5.1, Figure 1.

Coles BJ (2000). Doggerland: the cultural dynamics of a shifting coastline. In K Pye and JRL Allen, eds, *Coastal and estuarine environments: sedimentology, geomorphology and geoarchaeology*, pp. 393–401. *Special Publication*, **175**. The Geological Society, London.

Compton JS and Franceschini G (2005). Holocene geoarchaeology of the Sixteen Mile beach barrier dunes in the Western Cape, South Africa. *Quaternary Research*, **63**, 99–107.

Cooper AJ (1979). Quaternary geology of the Grand Bend–Parkhill area, southern Ontario. *Ontario Geological Survey Report*, **188**, pp. 1–70. Ontario Ministry of Natural Resources, Toronto.

Cooper WS (1926). The fundamentals of vegetational change. *Ecology*, **7**, 391–413.

Cooper WS (1958). Coastal sand dunes of Oregon and Washington. *Geological Society of America Memoir*, **72**, 1–169.

Cordazzo CV and Davy AJ (1999). Vegetative regeneration of *Panicum racemosum* from rhizome fragments on southern Brazilian coastal dunes. *Journal of Coastal Research*, **15**, 520–525.

Corkidi L and Rincón E (1997a). Arbuscular mycorrhizae in a tropical sand dune ecosystem on the Gulf of Mexico. I. Mycorrhizal status and mycorrhizal inoculum potential along a successional gradient. *Mycorrhiza*, **7**, 9–15.

Corkidi L and Rincón E (1997b). Arbuscular mycorrhizae in a tropical sand dune ecosystem on the Gulf of Mexico. II. Effects of arbuscular mycorrhizal fungi on the growth of species distributed in different early successional stages. *Mycorrhiza*, **7**, 17–23.

Cowles HC (1899). The ecological relations of the vegetation on the sand dunes of Lake Michigan. *Botanical Gazette*, **27**, 95–117, 167–202, 281–308, 361–391.

Cowles HC (1901). The physiographic ecology of Chicago and vicinity; a study of the origin, development, and classification of plant societies. *Botanical Gazette*, **31**, 73–108, 145–182.

Crawley MJ (1983). *Herbivory: the dynamics of animal–plant interactions*. Blackwell, Oxford.

Crawley MJ (1993). Succeeding in the sand dunes. *Nature*, **362**, 17–18.

Dahl BE, Fall BA, Lohse A and Appan SG (1975). Construction and stabilization of coastal foredunes with vegetation: Padre Island, Texas. *US Army Corps of Engineers, Coastal Engineering Research Center*. Fort Belvoir, VA.

Dale JE (1959). Some effects of the continuous removal of floral buds on the growth of the cotton plant. *Annals of Botany*, **23**, 636–649.

Dallmeyer DG, Porter JW, and Smith GJ (1982). Effects of particulate peat on the behavior and physiology of the Jamaican reef-building coral *Montastrea annularis*. *Marine Biology*, **68**, 229–233.

Dalpé Y. (1989). Inventaire et répartition de la flore endomycorhizienne de dunes et de rivages maritimes du Québec, du Nouveau-Brunswick et de la Nouvelle-Écosse. *Le Naturaliste Canadien*, **116**, 219–236.

Danin A (1996). *Plants of desert dunes*. Springer-Verlag, Berlin.

Darwin CR (1857). On the action of sea-water on the germination of seeds, *Journal of the Linnean Society of London*, **1**, 130–140.

Daubenmire R (1968). *Plant communities: a textbook of plant synecology*. Harper and Row, New York.

Daubenmire RF (1974). *Plants and environment: a textbook of plant autecology*, 3rd edn. Wiley, New York.

Davidson-Arnott RGD and Law MN (1990). Seasonal patterns and controls on sediment supply to coastal foredunes, Long Point, Lake Erie. In K Nordstrom, N Psuty and B Carter, eds, *Coastal dunes: form and process*, pp. 177–200. Wiley, Chichester.

Davidson-Arnott RGD and Law MN (1996). Measurement and prediction of long-term sediment supply to coastal foredunes. *Journal of Coastal Research*, **12**, 654–663.

Davies JL (1972). *Geographical variation in coastal development*. Oliver and Boyd, Edinburgh.

Davy AJ, Grootjans AP, Hiscock K and Petersen L (2006). Development of eco-hydrological guidelines for dune habitats-Phase 1. *English Nature Research Reports*, **No. 696**.

Day JW, Conner WH, Costanza R, Kemp GP and Mendelssohn IA (1993). Impacts of sea level rise on coastal systems with special emphasis on the Mississippi River deltaic plain. In RA Warrick, EM Barrow and TML Wigley, eds, *Climate and sea level change: observations, projections and implications*, pp. 276–296. Cambridge University Press, Cambridge.

Deacon EL (1949). Vertical diffusion in the lowest layers of the atmosphere. *Quarterly Journal of Royal Meteorological Society*, **75**, 89–103.

Dean C (1999). *Against the tide: the battle for America's beaches*. Columbia University Press, New York.

Dech JP (2004). The ecological significance of sand burial to woody plants on coastal sand dunes. PhD thesis, University of Western Ontario, London, ON.

Dech JP and Maun MA (2005). Zonation of vegetation along a burial gradient on the leeward slopes of Lake Huron sand dunes. *Canadian Journal of Botany*, **83**, 227–236.

Dech JP and Maun MA (2006). Adventitious root production and plastic resource allocation to biomass determine burial tolerance in woody plants from central Canadian coastal dunes. *Annals of Botany*, **98**, 1095–1105.

Dech JP, Maun MA and Pazner MI (2005). Blowout dynamics on Lake Huron dunes: analysis of digital multispectral data from colour air photos. *Catena*, **60**, 165–180.

De Jong TJ and Klinkhamer PGL (1988). Population ecology of the biennials *Cirsium vulgare* and *Cynoglossum officinale* in a coastal sand-dune area. *Journal of Ecology*, **76**, 366–382.

De Jong TM (1979). Water and salinity relations of Californian beach species. *Journal of Ecology*, **67**, 647–663.

De Lacerda LD, De Araújo DSD and Maciel NC (1993). Dry coastal ecosystems of the tropical Brazilian coast. In E van der Maarel, ed., *Ecosystems of the world 2B. Dry coastal ecosystems: Africa, America, Asia and Oceania*, pp. 477–493. Elsevier, Amsterdam.

De Ronde JG (1993). What will happen to The Netherlands if sea level rise accelerates? In RA Warrick, EM Barrow and TML Wigley, eds, *Climate and sea level change: observations, projections and implications*, pp. 322–335. Cambridge University Press, Cambridge.

De Rooij-van der Goes PCEM (1995a). Role of plant-parasitic nematodes and soil-borne fungi in the decline of *Ammophila arenaria* (L.) Link. *New Phytologist*, **129**, 661–669.

De Rooij-van der Goes PCEM (1996). Soil-borne plant pathogens of *Ammophila arenaria* in coastal foredunes. PhD thesis, Landbouw Universiteit, Wageningen.

De Rooij-van der Goes PCEM, Peters BAM and van der Putten WH (1998). Vertical migration of nematodes and soil-borne fungi to developing roots of *Ammophila arenaria* (L.) Link after sand accretion. *Applied Soil Ecology*, **10**, 1–10.

De Rooij-van der Goes PCEM, van der Putten WH and Peters BAM (1995b). Effects of sand deposition on the interaction between *Ammophila arenaria*, plant-parasitic nematodes and pathogenic fungi. *Canadian Journal of Botany*, **73**, 1141–1150.

De Rooij-van der Goes PCEM, van der Putten WH and van Dijk C (1995c). Analysis of nematodes and soil-borne fungi from *Ammophila arenaria* (marram grass) in Dutch coastal foredunes by multivariate techniques. *European Journal of Plant Pathology*, **101**, 149–162.

D'Hertefeldt T and Falkengren-Grerup U (2002). Extensive physiological integration in *Carex arenaria* and *Carex disticha* in relation to potassium and water availability. *New Phytologist*, **156**, 469–477.

D'Hertefeldt T and van der Putten WH (1998). Physiological integration of the clonal plant *Carex arenaria* and its response to soil-borne pathogens. *Oikos*, **81**, 229–237.

Díaz S, Grime JP, Harris J and McPherson E (1993). Evidence of a feedback mechanism limiting plant response to elevated carbon dioxide. *Nature*, **364**, 616–617.

Dirzo R and Domínguez CA (1986). Seed shadows, seed predation and the advantages of dispersal. In A Estrada and TH Fleming, eds, *Frugivores and seed dispersal*, pp. 237–249. Junk, Dordrecht.

Disraeli DJ (1984). The effect of sand deposits on the growth and morphology of *Ammophila breviligulata*. *Journal of Ecology*, **72**, 145–154.

Dodge RE and Vaisnys JR (1977). Coral populations and growth patterns: responses to sedimentation and turbidity associated with dredging. *Journal of Marine Research*, **35**, 715–730.

Dodge RE, Aller RC and Thomson J (1974). Coral growth related to resuspension of bottom sediments. *Nature*, **247**, 574–577.

Doing H (1985). Coastal fore-dune zonation and succession in various parts of the world. *Vegetatio*, **61**, 65–75.

Dong M (1999). Effects of severing rhizome on clonal growth in rhizomatous grass species *Psammochloa villosa* and *Leymus secalinus*. *Acta Botanica Sinica*, **41**, 194–198.

Dong M and Alaten B (1999). Clonal plasticity in response to rhizome severing and heterogeneous resource supply in the rhizomatous grass *Psammochloa villosa* in an inner Mongolian dune, China. *Plant Ecology*, **141**, 53–58.

Donnelly JP (2006). A revised late Holocene sea-level record for Northern Massachusetts, USA. *Journal of Coastal Research*, **22**, 1051–1061.

Doody JP (1989). Management for nature conservation. In CH Gimingham, W Ritchie, BB Willetts and AJ Willis, eds, *Coastal sand dunes*, pp. 247–265. *Proceedings Royal Society of Edinburgh*, **96B**, Edinburgh.

Doody JP (1991). *Sand dune inventory of Europe*. Joint Nature Conservation Committee, Monkstone House, Peterborough.

Dowdeswell JA and Hagen JO (2004). Arctic ice caps and glaciers. In JL Bamber and AJ Payne, eds, *Mass balance of the cryosphere: observations and modelling of contemporary and future changes*, pp. 527–558. Cambridge University Press, Cambridge.

Drent RH and Daan S (1980). The prudent parent: energetic adjustments in avian breeding. *Ardea*, **68**, 225–252.

Drury WH and Nisbet ICT (1973). Succession. *Journal Arnold Arboretum*, **54**, 331–368.

Dynesius M and Jansson R (2000). Evolutionary consequences of changes in species' geographical distributions driven by Milankovich climate oscillations. *Proceedings of the National Academy of Sciences*, **97**, 9115–9120.

Edwards RS and Claxton SM (1964). The distribution of air-borne salt of marine origin in the Aberystwyth area. *Journal of Applied Ecology*, **1**, 253–263.

Eldred RA and Maun MA (1982). A multivariate approach to the problem of decline in vigour of *Ammophila*. *Canadian Journal of Botany*, **60**, 1371–1380.

Elfman B, Maun MA and Hopkins WG (1986). Population biology of *Ammophila breviligulata* and *Calamovilfa longifolia* on Lake Huron sand dunes. II. Ultrastructure of organelles and photosynthetic properties. *Canadian Journal of Botany*, **64**, 2151–2159.

Ellis JC (2005). Marine birds on land: a review of plant biomass, species richness, and community composition in seabird colonies. *Plant Ecology*, **181**, 227–241.

Emery KO and Milliman JD (1978). Suspended matter in surface waters: influence of river discharge and of upwelling. *Sedimentology*, **25**, 125–140.

Ernst A (1908). *The new flora of the volcanic island of Krakatau*. Cambridge University Press, Cambridge.

Ernst WHO, van Duin WE and Oolbekking GT (1984). Vesicular–arbuscular mycorrhiza in dune vegetation. *Acta Botanica Neerlandica*, **33**, 151–160.

Estrada A and Fleming TH (eds) (1986). *Frugivores and seed dispersal*. Junk, Dordrecht.

Etherington JR (1967). Studies of nutrient cycling and productivity in oligotrophic ecosystems. I. Soil potassium and wind-blown sea-spray in a South Wales dune grassland. *Journal of Ecology*, **55**, 743–752.

Faber BA, Zasoski RJ, Munns DN and Shackel K (1991). A method for measuring hyphal nutrient and water uptake in mycorrhizal plants. *Canadian Journal of Botany*, **69**, 87–94.

Fairbanks RG (1989). A 17,000-year glacio-eustatic sea level record: influence of glacial melting rates on the younger Dryas event and deep-ocean circulation. *Nature*, **342**, 637–642.

Fairbridge RW (1961). Eustatic changes in sea level. In LH Ahrens, K Rankama and SK Runcorn, eds, *Physics and Chemistry of the Earth*, **4**, pp. 99–185. Pergamon Press, London.

Farrow EP (1919). On the ecology of the vegetation of Breckland. VII. General effects of blowing sand upon the vegetation. *Journal of Ecology*, **7**, 55–64.

Feagin RA (2005). Artificial dunes created to protect property on Galveston Island, Texas: the lessons learned. *Ecological Restoration*, **23**, 89–94.

Feagin RA, Sherman DJ and Grant WE (2005). Coastal erosion, global sea-level rise, and the loss of sand dune plant habitats. *Frontiers in Ecology and the Environment*, **3**, 359–364.

Fenner M (1987). Seedlings. *New Phytologist*, **106**, 35–47.

Filion L and Marin P (1988). Modifications morphologiques de l'Epinette blanche soumise à la sédimentation éolienne en milieu dunaire, Québec subarctique. *Canadian Journal of Botany*, **66**, 1862–1869.

Finlay RD and Read DJ (1986). The structure and function of the vegetative mycelium of ectomycorrhizal plants. II. The uptake and distribution of phosphorus by mycelial strands interconnecting host plants. *New Phytologist*, **103**, 157–165.

Fishelson L (1983). Population ecology and biology of *Dotilla sulcata* (Crustacea, Ocypodidae) typical for sandy beaches of the Red Sea. In A McLachlan and T Erasmus, eds, *Sandy beaches as ecosystems*, pp. 643–654. Junk, The Hague.

Fisher JD, Law MN and Davidson-Arnott RGD (1987). *The geomorphology of Pinery Provincial Park. Environmental Planning Series*. University of Guelph, Guelph, ON.

Flather RA and Khandker K (1993). The storm surge problem and possible effects of sea level changes on coastal flooding in the Bay of Bengal. In RA Warrick, EM Barrow and TML Wigley, eds, *Climate and sea level change: observations, projections and implications*, pp. 229–245. Cambridge University Press, Cambridge.

Forster P, Hegerl G, Knutti R *et al.* (2007). Assessing uncertainty in climate simulations. *Nature Reports Climate Change*, **4**, pp. 63–64. doi:10.1038/climate/2007.46a.

Francis R and Read DJ (1984). Direct transfer of carbon between plants connected by vesicular–arbuscular mycorrhizal mycelium. *Nature*, **307**, 53–56.

Frazier JG (1993). Dry coastal ecosystems of Kenya and Tanzania. In E van der Maarel, ed., *Ecosystems of the world 2B. Dry coastal ecosystems: Africa, America, Asia and Oceania*, pp. 129–150. Elsevier, Amsterdam.

Fuller GD (1912). Germination and growth of the cottonwood upon the sand dunes of Lake Michigan near Chicago. *Transactions of the Illinois Academy of Science*, **5**, 137–143.

Gaël' AG (1975). Effect of aeol sandy deposits on *Pinus* stem form, roots growth and soil formation (in Russian). *Lesovedenie*, **4**, 12–20.

Gagné J-M and Houle G (2001). Facilitation of *Leymus mollis* by *Honckenya peploides* on coastal dunes in subarctic Quebec, Canada. *Canadian Journal of Botany*, **79**, 1327–1331.

Galvin CJ Jr (1968). Breaker type classification on three laboratory beaches. *Journal of Geophysical Research*, **73**, 3651–3659.

Galvin CJ (1972). Wave breaking in shallow water. In RE Meyer, ed., *Waves on beaches*, pp. 413–456. Academic Press, New York.

García D, Bañuelos MJ and Houle G (2002). Differential effects of acorn burial and litter cover on *Quercus rubra* recruitment at the limit of its range in eastern North America. *Canadian Journal of Botany*, **80**, 1115–1120.

García-Mora MR, Gallego-Fernández JB and García-Novo F (1999). Plant functional types in coastal foredunes in relation to environmental stress and disturbance. *Journal of Vegetation Science*, **10**, 27–34.

Garcia Novo F (1979). The ecology of vegetation of the dunes of Doñana National Park (south-west Spain). In RL Jefferies and AJ Davy, eds, *Ecological processes in coastal environments*, pp. 571–592. Blackwell, Oxford.

Gedge KE and Maun MA (1992). Effects of simulated herbivory on growth and reproduction of two beach annuals, *Cakile edentula* and *Corispermum hyssopifolium*. *Canadian Journal of Botany*, **70**, 2467–2475.

Gedge KE and Maun MA (1994). Compensatory response of two dune annuals to simulated browsing and fruit predation. *Journal of Vegetation Science*, **5**, 99–108.

Gemma JN and Koske RE (1988). Seasonal variation in spore abundance and dormancy of *Gigaspora gigantea* and in mycorrhizal inoculum potential of a dune soil. *Mycologia*, **80**, 211–216.

Gemma JN and Koske RE (1997). Arbuscular mycorrhizae in sand dune plants of the North Atlantic coast of the US: field and greenhouse inoculation and presence of mycorrhizae in planting stock. *Journal of Environmental Management*, **50**, 251–264.

Gemma JN, Koske RE and Carreiro M (1989). Seasonal dynamics of selected species of V-A mycorrhizal fungi in a sand dune. *Mycological Research*, **92**, 317–321.

George E, Marschner H and Jakobsen I (1995). Role of arbuscular mycorrhizal fungi in uptake of phosphorus and nitrogen from soil. *Critical Reviews in Biotechnology*, **15**, 257–270.

Gillham ME (1956). Ecology of the Pembrokeshire islands: V. Manuring by the colonial seabirds and mammals, with a note on seed distribution by gulls. *Journal of Ecology*, **44**, 429–454.

Gillham ME (1963). Some interactions of plants, rabbits and sea-birds on South African islands. *Journal of Ecology*, **51**, 275–294.

Gillham ME (1970). Seed dispersal by birds. In F Perring, ed., *The flora of a changing Britain*, pp. 90–98. Pendragon Press, Cambridge.

Giovannetti M (1985). Seasonal variations of vesicular–arbuscular mycorrhizas and endogonaceous spores in a maritime sand dune. *Transactions of the British Mycological Society*, **84**, 679–684.

Gleason HA (1926). The individualistic concept of the plant association. *Bulletin Torrey Botanical Club*, **53**, 7–26.

Goodall DW (1953). Objective methods for the classification of vegetation. I. The use of positive interspecific correlation. *Australian Journal of Botany*, **1**, 39–63.

Gooding EGB (1947). Observations on the sand dunes of Barbados, British West Indies. *Journal of Ecology*, **34**, 111–125.

Gray AJ (1985). Adaptation in perennial coastal plants—with particular reference to heritable variation in *Puccinellia maritima* and *Ammophila arenaria*. *Vegetatio*, **61**, 179–188.

Greaver TL and Da Sternberg LLS (2006). Linking marine resources to ecotonal shifts of water uptake by terrestrial dune vegetation. *Ecology*, **87**, 2389–2396.

Green DR and King SD (2003). Progress in geographical information systems and coastal modeling: an overview. In VC Lakhan, ed., *Advances in coastal modeling*, pp. 553–580. Elsevier, Amsterdam.

Greipsson S and Davy AJ (1996). Sand accretion and salinity as constraints on the establishment of *Leymus arenarius* for land reclamation in Iceland. *Annals of Botany*, **78**, 611–618.

Greipsson S, Ahokas H and Vähämiko S (1997). A rapid adaptation to low salinity of inland-colonizing populations of the littoral grass *Leymus arenarius*. *International Journal of Plant Sciences*, **158**, 73–78.

Grime JP (1979). *Plant strategies and vegetation processes*. Wiley, Chichester.

Grootjans AP, Ernst WHO and Stuyfzand PJ (1998). European dune slacks: strong interactions of biology, pedogenesis and hydrology. *Trends in Ecology and Evolution*, **13**, 96–100.

Grootjans AP, Adema EB, Bekker RM and Lammerts EJ (2004). Why young coastal dune slacks sustain a high biodiversity. In ML Martínez and NP Psuty, eds, *Coastal dunes: ecology and conservation*, pp. 85–101. Springer-Verlag, Berlin.

Guire KE and Voss EG (1963). Distributions of distinctive shoreline plants in the Great Lakes region. *Michigan Botanist*, **2**, 99–114.

Guppy HB (1917). *Plants, seeds, and currents in the West Indies and Azores*. Williams and Norgate, London.

Gupta R and Mukerji KG (2000). The growth of VAM fungi under stress conditions. In KG Mukerji, BP Chamola and J Singh, eds, *Mycorrhizal biology*, pp. 57–66. Kluwer/Plenum, New York.

Gupta V, Satyanarayana T and Garg S (2000). General aspects of mycorrhiza. In KG Mukerji, BP Chamola and J Singh, eds, *Mycorrhizal biology*, pp. 27–39. Kluwer/Plenum, New York.

Hallam A (1989). The case for sea-level change as a dominant causal factor in mass extinction of marine invertebrates. *Philosophical Transactions of the Royal Society London*, **B325**, 437–455.

Halwagy R (1963). Studies on the succession of vegetation on some islands and sand banks in the Nile near Khartoum, Sudan. *Vegetatio*, **11**, 217–234.

Hancock G (2002). *Underworld: the mysterious origins of civilization*. Anchor Canada, Toronto.

Hansen J, Ruedy R, Sato M *et al.* (2001). A closer look at United States and global surface temperature change. *Journal of Geophysical Research*, **106**, 23947–23963.

Hansen JE (2007). Scientific reticence and sea level rise. *Environmental Research Letters*, **2**, 024002 (6pp.). http://www.iop.org/EJ/article/1748–9326/2/2/024002/er117_2_024002.

Haq B, Hardenbol J and Vail P (1987). Chronology of fluctuating sea levels since the Triassic. *Science*, **235**, 1156–1167.

Hardjosuwarno S and Hadisumarno S (1993). Dry coastal ecosystems of the southern coast of Java. In E van der Maarel, ed., *Ecosystems of the world 2B. Dry coastal ecosystems: Africa, America, Asia and Oceania*, pp. 189–196. Elsevier, Amsterdam.

Harper JL (1977). *Population biology of plants*. Academic Press, London.

Harris D and Davy AJ (1986a). Strandline colonization by *Elymus farctus* in relation to sand mobility and rabbit grazing. *Journal of Ecology*, **74**, 1045–1056.

Harris D and Davy AJ (1986b). Regenerative potential of *Elymus farctus* from rhizome fragments and seed. *Journal of Ecology*, **74**, 1057–1067.

Harris D and Davy AJ (1987). Seedling growth in *Elymus farctus* after episodes of burial with sand. *Annals of Botany*, **60**, 587–593.

Harris D and Davy AJ (1988). Carbon and nutrient allocation in *Elymus farctus* seedlings after burial with sand. *Annals of Botany*, **61**, 147–157.

Harty RL and McDonald TJ (1972). Germination behaviour in beach spinifex (*Spinifex hirsutus* Labill.). *Australian Journal of Botany*, **20**, 241–251.

Hassouna MG and Wareing PF (1964). Possible role of rhizosphere bacteria in the nitrogen nutrition of *Ammophila arenaria*. *Nature*, **202**, 467–469.

Hawke MA and Maun MA (1988). Some aspects of nitrogen, phosphorus, and potassium nutrition of three colonizing beach species. *Canadian Journal of Botany*, **66**, 1490–1496.

Hawke MA and Maun MA (1989). Intrapopulation variation in reproduction and seed mass of a beach annual; *Cakile edentula* var. *lacustris*. *Journal of Coastal Research*, **5**, 103–112.

Hayman DS and Mosse B (1972). Plant growth responses to vesicular-arbuscular mycorrhiza. III. Increased uptake of labile P from soil. *New Phytologist*, **71**, 41–47.

Hayman DS and Tavares M (1985). Plant growth responses to vesicular-arbuscular mycorrhiza. XV. Influence of soil pH on the symbiotic efficiency of different endophytes. *New Phytologist*, **100**, 367–377.

Hendrix SD (1979). Compensatory reproduction in a biennial herb following insect defloration. *Oecologia*, **42**, 107–118.

Hermesh R (1972). A study of the ecology of the Athabasca sand dunes with emphasis on the phytogenic aspects of dune formation. MSc thesis, University of Saskatchewan, Saskatoon, SK.

Hertling UM (1997). *Ammophila arenaria* (L.) Link (Marram grass) in South Africa and its potential invasiveness. PhD thesis, Rhodes University, Grahamstown.

Hesp P (2002). Foredunes and blowouts: initiation, geomorphology and dynamics. *Geomorphology*, **48**, 245–268.

Hesp PA (1981). The formation of shadow dunes. *Journal of Sedimentary Petrolology*, **51**, 101–112.

Hesp PA (1989). A review of biological and geomorphological processes involved in the initiation and development of incipient foredunes. In CH Gimingham, W Ritchie, BB Willetts and AJ Willis, eds, *Coastal sand dunes*, pp. 181–201. *Proceedings Royal Society of Edinburgh*, **96B**, Edinburgh.

Hesp PA (1991). Ecological processes and plant adaptations on coastal dunes. *Journal of Arid Environments*, **21**, 165–191.

Hesp PA (1999). The beach backshore and beyond. In AD Short, ed., *Handbook of beach and shoreface morphodynamics*, pp. 145–170. Wiley, London.

Hesp PA (2004). Coastal dunes in the tropics and temperate regions: location, formation, morphology and vegetation processes. In ML Martínez and NP Psuty, eds, *Coastal dunes: ecology and conservation*, pp. 29–49. Springer-Verlag, Berlin.

Hesp PA and Short AD (1999). Barrier morphodynamics. In AD Short, ed., *Handbook of beach and shoreface morphodynamics*, pp. 307–333. Wiley, London.

Hesp PA and Thom BG (1990). Geomorphology and evolution of active transgressive dunefields. In K Nordstrom, N Psuty and B Carter, eds, *Coastal dunes: form and process*, pp. 253–288. Wiley, Chichester.

Hetrick BAD, Wilson GWT and Todd TC (1990). Differential responses of C_3 and C_4 grasses to mycorrhizal symbiosis, phosphorus fertilization, and soil microorganisms. *Canadian Journal of Botany*, **68**, 461–467.

Hockey PAR, Siegfried WR, Crowe AA and Cooper J (1983). Ecological structure and energy requirements of the sandy beach avifauna of southern Africa. In A McLachlan and T Erasmus, eds, *Sandy beaches as ecosystems*, pp. 507–521. Junk, The Hague.

Hodgkin SE (1984). Scrub encroachment and its effects on soil fertility on Newborough Warren, Anglesey, Wales. *Biological Conservation*, **29**, 99–119.

Holgate SJ and Woodworth PL (2004). Evidence for enhanced coastal sea level rise during the 1990s. *Geophysical Research Letters*, **31**, L07305, doi:10.1029/2004GL019626.

Holton B Jr (1980). Some aspects of the nitrogen cycle in a northern California coastal dune-beach ecosystem, with emphasis on *Cakile maritima*. PhD thesis, University of California, Davis, CA.

Hope-Simpson JF and Jefferies RL (1966). Observations relating to vigour and debility in marram grass (*Ammophila arenaria* (L.) Link). *Journal of Ecology*, **54**, 271–274.

Houle G (1997). Interactions between resources and abiotic conditions control plant performance on subarctic coastal dunes. *American Journal of Botany*, **84**, 1729–1737.

Houle G (1998). Plant response to heterospecific neighbor removal and nutrient addition in a subarctic coastal dune system (northern Québec, Canada). *Ecoscience*, **5**, 526–533.

Howe HF (1986). Seed dispersal by fruit-eating birds and mammals. In DR Murray, ed., *Seed dispersal*, pp. 123–190. Academic Press, Orlando.

Huang Z, Dong M and Gutterman Y (2004). Caryopsis dormancy, germination and seedling emergence in sand, of *Leymus racemosus* (Poaceae), a perennial sand-dune grass inhabiting the Junggar Basin of Xinjiang, China. *Australian Journal of Botany*, **52**, 519–528.

Huiskes AHL (1977). The natural establishment of *Ammophila arenaria* from seed. *Oikos*, **29**, 133–136.

Huiskes AHL (1979). Biological flora of the British Isles: *Ammophila arenaria* (L.) Link. (*Psamma arenaria* (L.) Roem. et Schult.; *Calamagrostis arenaria* (L.) Roth). *Journal of Ecology*, **67**, 363–382.

Huiskes AHL and Harper JL (1979). The demography of leaves and tillers of *Ammophila arenaria* in a dune sere. *Oecologia Plantarum*, **14**, 435–446.

Hussey RS and Roncadori RW (1982). Vesicular–arbuscular mycorrhizae may limit nematode activity and improve plant growth. *Plant Disease*, **66**, 9–14.

Ignaciuk R and Lee JA (1980). The germination of four annual strand-line species. *New Phytologist*, **84**, 581–591.

Imbert E and Houle G (2000). Persistence of colonizing plant species along an inferred successional sequence on a subarctic coastal dune (Québec, Canada). *Ecoscience*, **7**, 370–378.

Inman DL (2002). Nearshore processes. Available at AccessScience@McGraw-Hill, http://www.accessscience.com, doi: 10.1036/1097-8542.446200.

Inman DL (2005). Littoral cells. In M Schwartz, ed., *Encyclopedia of coastal science*, pp. 594–599. Kluwer Academic, Dordrecht.

Inman DL and Masters PM (1991). Coastal sediment transport concepts and mechanisms. In *Coast of California storm and tidal waves study: state of the coast report*, San Diego Region, Chapter 5. US Army Engineer District, Los Angeles.

IPCC and McCarthy JJ, Canziani OF, Leary NA, Dokken DJ and White KS (eds) (2001). *Climate Change 2001: Impacts, adaptations and vulnerability. Contribution of Working Group I to the Third Assessment Report of the Intergovernmental Panel on Climate Change (IPCC)*. Cambridge University Press, Cambridge.

IPCC and Nicholls RJ, Wong PP, Burkett VR *et al.* (2007a). Coastal systems and low-lying areas. In ML Parry, OF Canziana, JP Palutikof, PJ van der Linden and CE Hanson, eds, *Climate change 2007: impacts, adaptation and vulnerablility. Contribution of Working Group II to the Fourth Assessment Report of the Intergovernmental Panel on Climate Change (IPCC)*, pp. 315–356. Cambridge University Press, Cambridge.

IPCC and Solomon DF, Qin D, Manning M *et al.* (2007b). *Climate change 2007: The physical science basis. Contribution of Working Group I to the Fourth Assessment Report of the Intergovernmental Panel on Climate Change (IPCC)*. Cambridge University Press, Cambridge.

Isayeva LL and Velicho AA (1998). Landscape components in the Quaternary (1.6 million years ago to the present day)—reconstruction: environment of the last glacial maximum (20,000–18,000 years BP). *Resources and environment world atlas*, p. 10. Ed Hölzel Vienna and Institute of Geology, Russian Academy of Sciences, Moscow.

Jagschitz JA and Bell RS (1966). American beachgrass (establishment–fertilization–seeding). *Agricultural Experimental Station*, Bulletin **383**, Contribution No. 1150. University of Rhode Island, Kingston, RI.

Jennings JN (1964). The question of coastal dunes in tropical humid climates. *Zeitschrift für Geomorphologie*, **8**, 150–154.

Jennings JN (1965). Further discussion of factors affecting coastal dune formation in the tropics. *Australian Journal of Science*, **28**, 166–167.

Johnson AF (1993a). Dry coastal ecosystems of California. In E van der Maarel, ed., *Ecosystems of the world 2B. Dry coastal ecosystems: Africa, America, Asia and Oceania*, pp. 359–364. Elsevier, Amsterdam.

Johnson AF (1993b). Dry coastal ecosystems of Northwestern Mexico. In E van der Maarel, ed., *Ecosystems of the world 2B. Dry coastal ecosystems: Africa, America, Asia and Oceania*, pp. 365–374. Elsevier, Amsterdam.

Johnson AF and Barbour MG (1990). Dunes and maritime forests. In RL Myers and JJ Ewel, eds, *Ecosystems of Florida*, pp. 429–480. University of Central Florida Press, Orlando.

Johnson PN (1993). Dry coastal ecosystems of New Zealand. In E van der Maarel, ed., *Ecosystems of the world 2B. Dry coastal ecosystems: Africa, America, Asia and Oceania*, pp. 197–221. Elsevier, Amsterdam.

Jones MD and Smith SE (2004). Exploring functional definitions of mycorrhizas: are mycorrhizas always mutualisms? *Canadian Journal of Botany*, **82**, 1089–1109.

Joughin I, Abdalati W and Fahnestock M (2004). Large fluctuations in speed on Greenland's Jakobshavn Isbrae glacier. *Nature*, **432**, 608–609.

Jungerius PD and Schoonderbeek D (1992). The use of Leica Quantimet 970 for scanning blowout development on sequential air photos of the 'Dunes de Slack', NW France. *Catena Supplement*, **23**, 59–73.

Kachi N and Hirose T (1983). Limiting nutrients for plant growth in coastal sand dune soils. *Journal of Ecology*, **71**, 937–944.

Kachi N and Hirose T (1985). Population dynamics of *Oenothera glazioviana* in a sand-dune system with special reference to the adaptive significance of size-dependent reproduction. *Journal of Ecology*, **73**, 887–901.

Kachi N and Hirose T (1990). Optimal time of seedling emergence in a dune-population of *Oenothera glazioviana*. *Ecological Research*, **5**, 143–152.

Karpan R and Karpan A (1998). *Northern sandscapes, exploring Saskatchewan's Athabasca sand dunes*. Parkland Publishing, Saskatoon.

Kattsov VM, Källén E, Cattle H *et al.* (2005). Future climate change: modeling and scenarios for the arctic. In C Symon, L Arris and B Heal, eds, *Arctic climatic impact assessment*, pp. 99–150. Cambridge University Press, Cambridge.

Kearney TH (1904). Are plants of sea beaches and dunes true halophytes? *Botanical Gazette*, **37**, 424–436.

Keddy PA (1980). Population ecology in an environmental mosaic: *Cakile edentula* on a gravel bar. *Canadian Journal of Botany*, **58**, 1095–1100.

Keddy PA (1981). Experimental demography of the sand-dune annual, *Cakile edentula*, growing along an environmental gradient in Nova Scotia. *Journal of Ecology*, **69**, 615–630.

Kemp PR (1989). Seed banks and vegetation processes in deserts. In MA Leck, VT Parker and RL Simpson, eds, *Ecology of soil seed banks*, pp. 257–281. Academic Press, New York.

Kendrick B (1992). *The fifth kingdom*, 2nd edn. Focus Information Group, Newburyport.

Kent M, Owen NW and Dale MP (2005). Photosynthetic responses of plant communities to sand burial on the Machair Dune Systems of the Outer Hebrides, Scotland. *Annals of Botany*, **95**, 869–877.

Killham K (1994). *Soil ecology*. Cambridge University Press, Cambridge.

Kirkpatrick JB (1993). Dry coastal ecosystems of southeastern Australia. In E van der Maarel, ed., *Ecosystems of the world 2B. Dry coastal ecosystems: Africa, America, Asia and Oceania*, pp. 273–288. Elsevier, Amsterdam.

Klimeš L, Klimešová J and Osbornová J (1993). Regeneration capacity and carbohydrate reserves in a clonal plant, *Rumex alpinus*: effect of burial. *Vegetatio*, **109**, 153–160.

Klinkhamer PGL, De Jong TJ and van der Meijden E (1988). Production, dispersal and predation of seeds in the biennial *Cirsium vulgare*. *Journal of Ecology*, **76**, 403–414.

Klironomos JN and Hart MM (2001). Animal nitrogen swap for plant carbon. *Nature*, **410**, 651–652.

Klironomos JN and Kendrick WB (1993). Research on mycorrhizas: trends in the past 40 years as expressed in the 'MYCOLIT' database. *New Phytologist*, **125**, 595–600.

Klironomos JN, Hart MM, Gurney JE and Moutoglis P (2001). Interspecific differences in the tolerance of arbuscular mycorrhizal fungi to freezing and drying. *Canadian Journal of Botany*, **79**, 1161–1166.

Klironomos JN, McCune J, Hart M and Neville J (2000). The influence of arbuscular mycorrhizae on the relationship between plant diversity and productivity. *Ecology Letters*, **3**, 137–141.

Klomp H (1970). The determination of clutch size in birds: a review. *Ardea*, **58**, 1–124.

Knelman F, Dombrowski N and Newitt DM (1954). Mechanism of the bursting of bubbles. *Nature*, **173**, 261.

Koehler H, Munderloh E and Hofmann S (1995). Soil microarthropods (Acari, Collembola) from beach and dune: characteristics and ecosystem context. *Journal of Coastal Conservation*, **1**, 77–86.

Koske RE (1975). *Endogone* spores in Australian sand dunes. *Canadian Journal of Botany*, **53**, 668–672.

Koske RE (1982). Evidence for a volatile attractant from plant roots affecting germ tubes of a VA mycorrhizal fungus. *Transactions of the British Mycological Society*, **79**, 305–310.

Koske RE (1987). Distribution of VA mycorrhizal fungi along a latitudinal temperature gradient. *Mycologia*, **79**, 55–68.

Koske RE and Gemma JN (1990). VA mycorrhizae in strand vegetation of Hawaii: evidence for long-distance codispersal of plants and fungi. *American Journal of Botany*, **77**, 466–474.

Koske RE and Gemma JN (1997). Mycorrhizae and succession in plantings of beachgrass in sand dunes. *American Journal of Botany*, **84**, 118–130.

Koske RE and Halvorson WL (1981). Ecological studies of vesicular–arbuscular mycorrhizae in a barrier sand dune. *Canadian Journal of Botany*, **59**, 1413–1422.

Koske RE and Polson WR (1984). Are VA mycorrhizae required for sand dune stabilization? *BioScience*, **34**, 420–424.

Koske RE and Tews LL (1987). Vesicular–arbuscular mycorrhizal fungi of Wisconsin sandy soils. *Mycologia*, **79**, 901–905.

Koske RE, Sutton JC and Sheppard BR (1975). Ecology of *Endogone* in Lake Huron sand dunes. *Canadian Journal of Botany*, **53**, 87–93.

Koske R, Bonin C, Kelly J and Martinez C (1996). Effects of sea water on spore germination of a sand dune-inhabiting arbuscular mycorrhizal fungus. *Mycologia*, **88**, 947–950.

Koske RE, Gemma JN, Corkidi L, Sigüenza C and Rincón E (2004). Arbuscular mycorrhizas in coastal dunes. In ML Martínez and NP Psuty, eds, *Coastal dunes: ecology and conservation*, pp. 173–187. Springer-Verlag, Berlin.

Kowalchuk GA, De Souza FA and Van veen JA (2002). Community analysis of arbuscular mycorrhizal fungi associated with *Ammophila arenaria* in Dutch coastal sand dunes. *Molecular Ecology*, **11**, 571–581.

Kozlowski TT (1984). Plant responses to flooding of soil. *Bioscience*, **34**, 162–167.

Krajnyk I and Maun MA (1981a). Vegetative reproduction in the juvenile phase of *Ammophila breviligulata*. *Canadian Journal of Botany*, **59**, 883–892.

Krajnyk IS and Maun MA (1981b). Incidence of ergot in populations of *Ammophila breviligulata*. *Canadian Plant Disease Survey*, **61**, 19–21.

Krajnyk I and Maun MA (1982). Reproductive biology of *Ammophila breviligulata*. *American Midland Naturalist*, **108**, 346–354.

Krebs CJ (1985). *Ecology: the experimental analysis of distribution and abundance*, 3rd edn. Harper and Row, New York.

Krebs CJ (1994). *Ecology: the experimental analysis of distribution and abundance*, 4th edn. Harper and Collins, New York.

Kurz H (1939). The reaction of magnolia, scrub live-oak, slash-pine, palmetto and other plants to dune activity on the western Florida coast. *Proceedings Florida Academy of Sciences*, **4**, 195–203.

Laing CC (1954). The ecological life history of the *Ammophila breviligulata* community on Lake Michigan dunes. PhD thesis, University of Chicago, Chicago, IL.

Laing CC (1958). Studies in the ecology of *Ammophila breviligulata*. I. Seedling survival and its relation to population increase and dispersal. *Botanical Gazette*, **119**, 208–216.

Laing CC (1967). The ecology of *Ammophila breviligulata*. II. Genetic change as a factor in population decline on stable dunes. *American Midland Naturalist*, **77**, 495–500.

Larson M and Kraus NC (1994). Temporal and spatial scales of beach profile change, Duck, North Carolina. *Marine Geology*, **117**, 75–94.

Larson M, Rosati JD and Kraus NC (2002). Overview of regional coastal sediment processes and control. In US Army Engineer Research Development Center. *Coastal and Hydraulic Engineering Technical Note (CHETN)*, **CHETN-XIV-4**. Vicksburg, MS.

Leck MA (1989). Wetland seed banks. In MA Leck, VT Parker and RL Simpson, eds, *Ecology of soil seed banks*, pp. 283–305. Academic Press, New York.

Leck MA, Parker VT and Simpson RL (eds) (1989). *Ecology of soil seed banks*. Academic Press, New York.

Lee JA (1993). Dry coastal ecosystems of West Africa. In E van der Maarel, ed., *Ecosystems of the world 2B. Dry coastal ecosystems: Africa, America, Asia and Oceania*, pp. 59–69. Elsevier, Amsterdam.

Lee JA and Ignaciuk R (1985). The physiological ecology of strandline plants. *Vegetatio*, **62**, 319–326.

Leont'yev IO (2003). Modeling the morphological response in a coastal zone for different temporal scales. In VC Lakhan, ed., *Advances in coastal modeling*, pp. 299–336. Elsevier, Amsterdam.

Leontiev IO (2006). Forecast of coastal changes based on morphodynamical modeling. *Oceanology*, **46**, 564–572.

Leuliette EW, Nerem RS and Mitchum GT (2004). Calibration of TOPEX/Poseidon and Jason altimeter data to construct a continuous record of mean sea level change. *Marine Geodesy*, **27**, 79–94.

Lichter J (1998). Primary succession and forest development on coastal Lake Michigan sand dunes. *Ecological Monographs*, **68**, 487–510.

Lichter J (2000). Colonization constraints during primary succession on coastal Lake Michigan sand dunes. *Journal of Ecology*, **88**, 825–839.

Little LR and Maun MA (1993). Salt tolerance of coastal and lacustrine varieties of *Lathyrus maritimus* with evolutionary implications. In E Özhan, ed., *Proceedings, First International Conference on the Mediterranean Coastal Environment*, **1**, pp. 89–101. Medcoast 93, Antalya.

Little LR and Maun MA (1996). The 'Ammophila problem' revisited: a role for mycorrhizal fungi. *Journal of Ecology*, **84**, 1–7. (JS Rowe Award).

Little LR and Maun MA (1997). Relationships among plant-parasitic nematodes, mycorrhizal fungi and the dominant vegetation of a sand dune system. *Ecoscience*, **4**, 67–74.

Liu PL-F, Lynett P, Fernando H *et al.* (2005). Observations by the international tsunami survey team in Sri Lanka. *Science*, **308**, 1595.

Lloyd DG (1980). Sexual strategies in plants. I. An hypothesis of serial adjustment of maternal investment during one reproductive session. *New Phytologist*, **86**, 69–79.

Lonard RI and Judd FW (2008). The biological flora of coastal dunes and wetlands. *Croton punctatus* N. von Jacquin. *Journal of Coastal Research*, preprint, doi.org/10.2112%2F07–0933.

Looney PB and Gibson DJ (1995). The relationship between the soil seed bank and above-ground vegetation of a coastal barrier island. *Journal of Vegetation Science*, **6**, 825–836.

Louda SM (1989). Predation in the dynamics of seed regeneration. In MA Leck, VT Parker and RL Simpson, eds, *Ecology of soil seed banks*, pp. 25–51. Academic Press, New York.

Lovett-Doust J and Eaton GW (1982). Demographic aspects of flower and fruit production in bean plants, *Phaseolus vulgaris* L. *American Journal of Botany*, **69**, 1156–1164.

Lubke RA (2004). Vegetation dynamics and succession on sand dunes of the eastern coasts of Africa. In ML Martínez and NP Psuty, eds, *Coastal dunes: ecology and conservation*, pp. 67–84. Springer-Verlag, Berlin.

Lugina KM, Groisman PYa, Vinnikov KYa *et al.* (2005). Monthly surface air temperature time series area—averaged over the 30-degree latitudinal belts of the globe, 1881–2004. In US Department of Energy, Carbon Dioxide Information Center, Oak Ridge National Laboratory. *Trends: a compendium of data on global change.* Oak Ridge, TN, http://cdiac.esd.ornl.gov/trends/temp/lugina.html.

Lugo MA, González Maza ME and Cabello MN (2003). Arbuscular mycorrhizal fungi in a mountain grassland. II: Seasonal variation of colonization studied, along with its relation to grazing and metabolic host type. *Mycologia*, **95**, 407–415.

Mack RN (1976). Survivorship of *Cerastium atrovirens* at Aberffraw, Anglesey. *Journal of Ecology*, **64**, 309–312.

Marin P and Filion L (1992). Recent dynamics of subarctic dunes as determined by tree-ring analysis of white spruce, Hudson Bay, Québec. *Quaternary Research*, **38**, 316–330.

Marshall DL, Levin DA and Fowler NL (1985). Plasticity in yield components in response to fruit predation and date of fruit initiation in three species of *Sesbania* (Leguminosae). *Journal of Ecology*, **73**, 71–81.

Marshall JK (1965). *Corynephorus canescens* (L.) P. Beauv. as a model for the *Ammophila* problem. *Journal of Ecology*, **53**, 447–463.

Marshall JK (1967). Biological flora of the British Isles: *Corynephorus canescens* (L.) Beauv. *Journal of Ecology*, **55**, 207–220.

Martin WE (1959). The vegetation of Island Beach State Park, New Jersey. *Ecological Monographs*, **29**, 1–46.

Martínez ML and Maun MA (1999). Response of dune mosses to experimental burial by sand under natural and greenhouse conditions. *Plant Ecology*, **145**, 209–219.

Martínez ML and Moreno-Casasola P (1996). Effects of burial by sand on seedling growth and survival in six tropical sand dune species from the Gulf of Mexico. *Journal of Coastal Research*, **12**, 406–419.

Martínez ML and Psuty NP (eds) (2004). *Coastal dunes: ecology and conservation*. Springer-Verlag, Berlin.

Martínez ML and Vásquez G (2001). Spatial and temporal dynamics during primary succession on tropical coastal sand dunes. *Journal of Vegetation Science*, **12**, 361–372.

Martínez ML, Psuty NP and Lubke RA (2004). A perspective on coastal sand dunes. In ML Martínez and NP Psuty, eds, *Coastal dunes: ecology and conservation*, pp. 3–10. Springer-Verlag, Berlin.

Martínez ML, Vázquez G, Sánchez CS and Colón S (2001). Spatial and temporal variability during primary succession on tropical coastal sand dunes. *Journal of Vegetation Science*, **12**, 361–372.

Martini IP (1975). Sedimentology of a lacustrine barrier system at Wasaga Beach, Ontario, Canada. *Sedimentary Geology*, **14**, 169–190.

Maun MA (1981). Seed germination and seedling establishment of *Calamovilfa longifolia* on Lake Huron sand dunes. *Canadian Journal of Botany*, **59**, 460–469.

Maun MA (1984). Colonizing ability of *Ammophila breviligulata* through vegetative regeneration. *Journal of Ecology*, **72**, 565–574.

Maun MA (1985). Population biology of *Ammophila breviligulata* and *Calamovilfa longifolia* on Lake Huron sand dunes. I. Habitat, growth form, reproduction, and establishment. *Canadian Journal of Botany*, **63**, 113–124.

Maun MA (1989). Population biology of *Ammophila breviligulata* and *Calamovilfa longifolia* on Lake Huron sand dunes. III. Dynamic changes in plant community structure. *Canadian Journal of Botany*, **67**, 1267–1270.

Maun MA (1993). Dry coastal ecosystems along the Great Lakes. In E van der Maarel, ed., *Ecosystems of the world 2B. Dry coastal ecosystems: Africa, America, Asia and Oceania*, pp. 299–316. Elsevier, Amsterdam.

Maun MA (1994). Adaptations enhancing survival and establishment of seedlings on coastal dune systems. *Vegetatio*, **111**, 59–70.

Maun MA (1996). The effects of burial by sand on survival and growth of *Calamovilfa longifolia*. *Ecoscience*, **3**, 93–100.

Maun MA (1998). Adaptations of plants to burial in coastal sand dunes. *Canadian Journal of Botany*, **76**, 713–738. (1997 George Lawson Medal Review).

Maun MA (2004). Burial of plants as a selective force in sand dunes. In ML Martínez and NP Psuty, eds, *Coastal dunes: ecology and conservation*, pp. 119–135. Springer-Verlag, Berlin.

Maun MA and Baye PR (1989). The ecology of *Ammophila breviligulata* Fern. on coastal dune systems. *CRC Critical Reviews in Aquatic Sciences*, **1**, 661–681.

Maun MA and Cavers PB (1971). Seed production and dormancy in *Rumex crispus*. I. The effects of removal of cauline leaves at anthesis. *Canadian Journal of Botany*, **49**, 1123–1130.

Maun MA and Crabe T (1995). Deer browsing of sand dune vegetation in southwestern Ontario, Canada. In AHPM Salman, H Berends and M Bonazountas, eds, *Coastal management and habitat conservation*, pp. 487–501. Proceedings IV European Dune Congress EUCC, Leiden.

Maun MA and Krajnyk I (1989). Stabilization of Great Lakes sand dunes: effect of planting time, mulches and fertilizer on seedling establishment. *Journal of Coastal Research*, **5**, 791–800.

Maun MA and Lapierre J (1984). The effects of burial by sand on *Ammophila breviligulata*. *Journal of Ecology*, **72**, 827–839.

Maun MA and Lapierre J (1986). Effects of burial by sand on seed germination and seedling emergence of four dune species. *American Journal of Botany*, **73**, 450–455.

Maun MA and Payne AM (1989). Fruit and seed polymorphism and its relation to seedling growth in the genus *Cakile*. *Canadian Journal of Botany*, **67**, 2743–2750.

Maun MA and Perumal J (1999). Zonation of vegetation on lacustrine coastal dunes: effects of burial by sand. *Ecology Letters*, **2**, 14–18.

Maun MA and Riach S (1981). Morphology of caryopses, seedlings and seedling emergence of the grass *Calamovilfa longifolia* from various depths in sand. *Oecologia (Berl.)*, **49**, 137–142.

Maun MA and Sun D (2002). Nitrogen and phosphorous budgets in a lacustrine sand dune ecosystem. *Ecoscience*, **9**, 364–374.

Maun MA, Elberling H and D'Ulisse A (1996). The effects of burial by sand on survival and growth of Pitcher's thistle (*Cirsium pitcheri*) along Lake Huron. *Journal of Coastal Conservation*, **2**, 3–12.

Mayr E (1977). *Populations, species, and evolution*. Harvard University Press, Cambridge, MA.

Maze KM and Whalley RDB (1992a). Effects of salt spray and sand burial on *Spinifex sericeus* R. Br. *Australian Journal of Ecology*, **17**, 9–19.

Maze KM and Whalley RDB (1992b). Germination, seedling occurrence and seedling survival of *Spinifex sericeus* R. Br. (Poaceae). *Australian Journal of Ecology*, **17**, 189–194.

Mazer SJ (1990). Seed mass of Indiana dune genera and families: taxonomic and ecological correlates. *Evolutionary Ecology*, **4**, 326–357.

McIntosh RP (1980). The relationship between succession and the recovery process in ecosystems. In J Cairns Jr, ed., *The recovery process in damaged ecosystems*, pp. 11–62. Ann Arbour Science, Michigan.

McLachlan A (1983). Sandy beach ecology-a review. In A McLachlan and T Erasmus, eds, *Sandy beaches as ecosystems*, pp. 321–380. Junk, The Hague.

McLachlan A (1990). Dissipative beaches and macrofauna communities on exposed intertidal sands. *Journal of Coastal Research*, **6**, 57–71.

McLachlan A, Wooldridge T and Dye AH (1981). The ecology of sandy beaches in southern Africa. *South African Journal of Zoology*, **16**, 219–231.

McLeod KW and Murphy PG (1977). Establishment of *Ptelea trifoliata* on Lake Michigan sand dunes. *American Midland Naturalist*, **97**, 350–362.

McLeod KW and Murphy PG (1983). Factors affecting growth of *Ptelea trifoliata* seedlings. *Canadian Journal of Botany*, **61**, 2410–2415.

Miller SL and Allen EB (1992). Mycorrhizae, nutrient translocation and interactions between plants. In MF Allen, ed., *Mycorrhizal functioning: an integrative plant–fungal process*, pp. 301–325. Chapman and Hall, London.

Milliman JD and Emery KO (1968). Sea levels during the past 35,000 years. *Science*, **162**, 1121–1123.

Milliman JD, Broadus JM and Gable F (1989). Environmental and economic implications of rising sea level and subsiding deltas: the Nile and Bengal examples. *Ambio*, **18**, 340–345.

Mitrovica JX (2003). Recent controversies in predicting post-glacial sea-level change. *Quaternary Science Reviews*, **22**, 127–133.

Miyanishi K and Johnson EA (2007). Coastal dune succession and the reality of dune processes. In EA Johnson and K Miyanishi, eds, *Plant disturbance ecology: the process and the response*, pp. 249–282. Academic Press, San Diego.

Miyawaki A and Suzuki K (1993). Dry coastal ecosystems of Japan. In E van der Maarel, ed., *Ecosystems of the world 2B. Dry coastal ecosystems: Africa, America, Asia and Oceania*, pp. 165–188. Elsevier, Amsterdam.

Moreno-Casasola P (1986). Sand movement as a factor in the distribution of plant communities in a coastal dune system. *Vegetatio*, **65**, 67–76.

Moreno-Casasola P (1993). Dry coastal ecosystems of the Atlantic coasts of Mexico and central America. In E van der Maarel, ed., *Ecosystems of the world 2B. Dry coastal ecosystems: Africa, America, Asia and Oceania*, pp. 389–405. Elsevier, Amsterdam.

Moreno-Casasola P and Vázquez G (1999). Succession in tropical dune slacks after disturbance by water-table dynamics. *Journal of Vegetation Science*, **10**, 515–524.

Morrison RG (1973). Primary succession on sand dunes at Grand Bend, Ontario. PhD thesis, University of Toronto, Toronto, ON.

Morrison RG and Yarranton GA (1973). Diversity, richness, and evenness during a primary sand dune succession at Grand Bend, Ontario. *Canadian Journal of Botany*, **51**, 2401–2411.

Morrison RG and Yarranton GA (1974). Vegetational heterogeneity during a primary sand dune succession. *Canadian Journal of Botany*, **52**, 397–410.

Morton JB and Benny GL (1990). Revised classification of arbuscular mycorrhizal fungi (Zygomycetes): a new order, Glomales, two new suborders, Glomineae and Gigasporineae, and two new families, Acaulosporaceae and Gigasporaceae, with an emendation of Glomaceae. *Mycotaxon*, **37**, 471–491.

Morton JK (1957). Sand-dune formation on a tropical shore. *Journal of Ecology*, **45**, 495–497.

Morton JK and Venn JM (1984). *The flora of Manitoulin Island and the adjacent islands of Lake Huron, Georgian Bay and the North Channel*, 2nd edn. University of Waterloo Biology Series, **28**, pp. 1–181. Waterloo, ON.

Mueller IM (1941). An experimental study of rhizomes of certain prairie plants. *Ecological Monographs*, **11**, 165–188.

Mukerji KG, Chamola BP and Singh J (eds) (2000). *Mycorrhizal biology*. Kluwer/Plenum, New York.

Murray DR (ed.) (1986a). *Seed dispersal*. Academic Press, Sidney.

Murray DR (1986b). Seed dispersal by water. In DR Murray, ed., *Seed dispersal*, pp. 49–86. Academic Press, Sidney.

Murray-Wallace CV (2007). Eustatic sea-level changes since the last glaciation. In SA Elias, ed., *Encyclopedia of quaternary science*, pp. 3034–3043. Elsevier, Amsterdam.

Nakanishi H (1988). Mechanical seed dispersal of herbaceous plants in southwestern Japan. *Hikobia*, **10**, 129–133.

Newman EI (1995). Phosphorus inputs to terrestrial ecosystems. *Journal of Ecology*, **83**, 713–726.

Newsham KK, Fitter AH and Watkinson AR (1994). Root pathogenic and arbuscular mycorrhizal fungi determine fecundity of asymptomatic plants in the field. *Journal of Ecology*, **82**, 805–814.

Newsham KK, Fitter AH and Watkinson AR (1995). Arbuscular mycorrhiza protect an annual grass from root pathogenic fungi in the field. *Journal of Ecology*, **83**, 991–1000.

Nicolson TH (1959). Mycorrhiza in the Gramineae. I. Vesicular–arbuscular endophytes, with special reference to the external phase. *Transactions British Mycological Society*, **42**, 421–438.

Nicolson TH (1960). Mycorrhiza in the Gramineae. II. Development in different habitats, particularly sand dunes. *Transactions British Mycological Society*, **43**, 132–145.

Nicolson TH and Johnston C (1979). Mycorrhiza in the Gramineae. III. *Glomus fasciculatus* as the endophyte of pioneer grasses in a maritime sand dune. *Transactions British Mycological Society*, **72**, 261–268.

Noble JC (1982). Biological flora of the British Isles: *Carex arenaria* L. *Journal of Ecology*, **70**, 867–886.

Noble JC, Bell AD and Harper JL (1979). The population biology of plants with clonal growth. I. The morphology and structural demography of *Carex arenaria*. *Journal of Ecology*, **67**, 983–1008.

Nobuhara H (1973). Sand-dune formation by coastal plants and their mode of reproduction. *Journal of Narashino High School*, **1**, 39–120.

Noest V (1991). Simulated impact of sea level rise on phreatic level and vegetation of dune slacks in the Voorne dune area (The Netherlands). *Landscape Ecology*, **6**, 89–87.

Nott J (2006). Tropical cyclones and the evolution of the sedimentary coast of northern Australia. *Journal of Coastal Research*, **22**, 49–62.

Olff H, Huisman J and van Tooren BF (1993). Species dynamics and nutrient accumulation during early primary succession in coastal sand dunes. *Journal of Ecology*, **81**, 693–706.

Olmsted CE (1937). Vegetation of certain sand plains of Connecticut. *Botanical Gazette*, **99**, 209–300.

Olson JS (1958a). Rates of succession and soil changes on southern Lake Michigan sand dunes. *Botanical Gazette*, **119**, 125–170.

Olson JS (1958b). Lake Michigan dune development. 1. Wind-velocity profiles. *Journal of Geology*, **66**, 254–263.

Olson JS (1958c). Lake Michigan dune development. 2. Plants as agents and tools in geomorphology. *Journal of Geology*, **66**, 345–351.

Olsson-Seffer P (1909). Relation of soil and vegetation on sandy sea shores. *Botanical Gazette*, **47**, 85–126.

Oosting HJ (1945). Tolerance to salt spray of plants of coastal dunes. *Ecology*, **26**, 85–89.

Oosting HJ (1954). Ecological processes and vegetation of the maritime strand in the southeastern United States. *Botanical Review*, **20**, 226–262.

Oosting HJ and Billings WD (1942). Factors effecting vegetational zonation on coastal dunes. *Ecology*, **23**, 131–142.

Open University (1989). *Waves, tides and shallow-water processes*. Pergamon Press, Oxford.

Owen NW, Kent M and Dale MP (2004). Plant species and community responses to sand burial on the machair of the Outer Hebrides, Scotland. *Journal of Vegetation Science*, **15**, 669–678.

Packham JR and Willis AJ (1997). *Ecology of dunes, salt marsh and shingle*. Chapman and Hall, London.

Parker VT and Kelly VR (1989). Seed banks in California chaparral and other Mediterranean climate shrublands. In MA Leck, VT Parker and RL Simpson, eds, *Ecology of soil seed banks*, pp. 231–255. Academic Press, New York.

Partridge TR and Wilson JB (1988). Vegetation patterns in salt marshes of Otago, New Zealand. *New Zealand Journal of Botany*, **26**, 497–510.

Pavlik BM (1983). Nutrient and productivity relations of the dune grasses *Ammophila arenaria* and *Elymus mollis*. III. Spatial aspects of clonal expansion with reference to rhizome growth and the dispersal of buds. *Bulletin of the Torrey Botanical Club*, **110**, 271–279.

Payne AM (1980). The ecology and population dynamics of *Cakile edentula* var. *lacustris* on Lake Huron beaches. MSc thesis, University of Western Ontario, London, ON.

Payne AM and Maun MA (1981). Dispersal and floating ability of dimorphic fruit segments of *Cakile edentula* var. *lacustris*. *Canadian Journal of Botany*, **59**, 2595–2602.

Payne AM and Maun MA (1984). Reproduction and survivorship of *Cakile edentula* var. *lacustris* along the Lake Huron shoreline. *American Midland Naturalist*, **111**, 86–95.

Peltier WR (2002). On eustatic sea level history: last glacial maximum to Holocene. *Quaternary Science Reviews*, **21**, 377–396.

Pemadasa MA and Lovell PH (1974). The mineral nutrition of some dune annuals. *Journal of Ecology*, **62**, 647–657.

Pemadasa MA and Lovell PH (1975). Factors controlling germination of some dune annuals. *Journal of Ecology*, **63**, 41–59.

Penrith M-L (1993). Dry coastal ecosystems of Namibia. In E van der Maarel, ed., *Ecosystems of the world 2B. Dry coastal ecosystems: Africa, America, Asia and Oceania*, pp. 71–74. Elsevier, Amsterdam.

Perumal JV (1994). Sand accretion and its effects on the distribution and ecophysiology of dune plants. PhD thesis, University of Western Ontario, London, ON.

Perumal JV and Maun MA (1999). The role of mycorrhizal fungi in growth enhancement of dune plants following burial in sand. *Functional Ecology*, **13**, 560–566.

Perumal VJ and Maun MA (2006). Ecophysiological response of dune species to experimental burial under field and controlled conditions. *Plant Ecology*, **184**, 89–104.

Peterson RL and Massicotte HB (2004). Exploring structural definitions of mycorrhizas, with emphasis on nutrient-exchange interfaces. *Canadian Journal of Botany*, **82**, 1074–1088.

Peterson RL, Massicotte HB and Melville LH (2004). *Mycorrhizas: anatomy and cell biology*. NRC Research Press, Ottawa, ON.

Pethick J (1984). *An introduction to coastal geomorphology*. Edward Arnold, London.

Pfadenhauer J (1993). Dry coastal ecosystems of temperate Atlantic South America. In E van der Maarel, ed., *Ecosystems of the world 2B. Dry coastal ecosystems: Africa, America, Asia and Oceania*, pp. 495–500. Elsevier, Amsterdam.

Pfeffer WT, Harper JT and O'Neel S (2008). Kinematic constraints on glacier contributions to 21st-century sea-level rise. *Science*, **321**, 1340–1343. doi:10:1126/science 1159099.

Phillips T and Maun MA (1996). Population ecology of *Cirsium pitcheri* on Lake Huron sand dunes. I. Impact of white-tailed deer. *Canadian Journal of Botany*, **74**, 1439–1444.

Pickett STA and McDonnell MJ (1989). Seed bank dynamics in temperate deciduous forest. In MA Leck, VT Parker and RL Simpson, eds, *Ecology of soil seed banks*, pp. 123–147. Academic Press, New York.

Pignatti S, Moggi G and Raimondo FM (1993). Dry coastal ecosystems of Somalia. In E van der Maarel, ed., *Ecosystems of the world 2B. Dry coastal ecosystems: Africa, America, Asia and Oceania*, pp. 31–36. Elsevier, Amsterdam.

Piotrowski JS, Denich T, Klironomos JN, Graham JM and Rillig MC (2004). The effects of arbuscular mycorrhizas on soil aggregation depend on the interaction between plant and fungal species. *New Phytologist*, **164**, 365–373.

Pirazolli PA (1996). *Sea-level changes: the last 20000 years*. Wiley, Chichester.

Planisek SL and Pippen RW (1984). Do sand dunes have seed banks? *Michigan Botanist*, **23**, 169–177.

Pluis JLA and De Winder B (1990). Natural stabilization. In TWM Bakker, PD Jungerius and PA Klijn, eds, *Dunes of the European coasts: geomorphology–hydrology–soils. Catena Supplement*, **18**, 195–208.

Poulson TL (1999). Autogenic, allogenic, and individualistic mechanisms of dune succession at Miller, Indiana. *Natural Areas Journal*, **19**, 172–176.

Poulson TL and McClung C (1999). Anthropogenic effects on early dune succession at Miller, Indiana. *Natural Areas Journal*, **19**, 177–179.

Pringle A and Bever JD (2002). Divergent phenologies may facilitate the coexistence of arbuscular mycorrhizal fungi in a North Carolina grassland. *American Journal of Botany*, **89**, 1439–1446.

Protz R, Ross GJ, Martini RP and Tarasmae J (1984). Rate of podsolic soil formation near Hudson Bay, Ontario. *Canadian Journal of Soil Science*, **64**, 31–49.

Pugh DT (2004). *Changing sea levels: effects of tides, weather and climate*. Cambridge University Press, Cambridge.

Pugh GJF (1979). The distribution of fungi in coastal regions. In RL Jefferies and AJ Davy, eds, *Ecological processes in coastal environments*, pp. 415–427. Blackwell, London.

Purer EA (1942). Anatomy and ecology of *Ammophila arenaria* Link. *Madroño*, **6**, 167–171.

Pye K (1983). Formation and history of Queensland coastal dunes. *Zeitschrift für Geomorphologie*, **45**, 175–204.

Pye K and Tsoar H (1990). *Aeolian sand and sand dunes*. Unwin Hyman, London.

Pyykkö M (1977). Effects of salt spray on growth and development of *Pinus sylvestris* L. *Annales Botanici Fennici*, **14**, 49–61.

Rabinowitz D and Rapp JK (1980). Seed rain in a North American tall grass prairie. *Journal of Applied Ecology*, **17**, 793–802.

Ralph RD (1978). Dinitrogen fixation by *Azotobacter* in the rhizosphere of *Ammophila breviligulata*. PhD thesis, University of Delaware, Newark, DE.

Rampino MR (1979). Holocene submergence of southern Long Island, New York. *Nature*, **280**, 132–134.

Ramsey DSL and Wilson JC (1997). The impact of grazing by macropods on coastal foredune vegetation in southeast Queensland. *Australian Journal of Ecology*, **22**, 288–297.

Randall R (1970). Salt measurement on the coast of Barbados, West Indies. *Oikos*, **21**, 65–70.

Ranwell D (1958). Movement of vegetated sand dunes at Newborough Warren, Anglesey. *Journal of Ecology*, **46**, 83–100.

Ranwell D (1960). Newborough Warren, Anglesey. II. Plant associes and succession cycles of the sand dune and dune slack vegetation. *Journal of Ecology*, **48**, 117–141.

Ranwell DS (1972). *Ecology of salt marshes and sand dunes*. Chapman and Hall, London.

Rao TA and Meher-Homji VM (1993). Dry coastal ecosystems of the Indian sub-continent and islands. In E van der Maarel, ed., *Ecosystems of the world 2B. Dry coastal ecosystems: Africa, America, Asia and Oceania*, pp. 151–164. Elsevier, Amsterdam.

Raper SCB (1993). Observational data on the relationships between climatic change and the frequency and magnitude of severe tropical storms. In RA Warrick, EM Barrow and TML Wigley, eds, *Climate and sea level change: observations, projections and implications*, pp. 192–212. Cambridge University Press, Cambridge.

Rauch EM and Bar-Yam Y (2005). Estimating the total genetic diversity of a spatial field population from a sample and implications of its dependence on habitat area. *Proceedings of the National Academy of Sciences*, **102**, 9826–9829.

Ray GC and McCormick-Ray J (2004). *Coastal–marine conservation: science and policy*. Blackwell, Oxford.

Read DJ (1989). Mycorrhizas and nutrient cycling in sand dune ecosystems. In CH Gimingham, W Ritchie, BB Willetts and AJ Willis, eds, *Coastal sand dunes; Proceedings Royal Society of Edinburgh*, **96B**, pp. 89–110. Royal Society of Edinburgh, Edinburgh.

Read DJ, Francis R and Finley RD (1985). Mycorrhizal mycelia and nutrient cycling in plant communities. In AH Fitter, D Atkinson, DJ Read and MB Usher, eds, *Ecological interactions in soil: plants, microbes and animals*, pp. 193–217. Blackwell, London.

Remy W, Taylor TN, Hass H and Kerp H (1994). Four hundred-million-year-old vesicular arbuscular mycorrhizae. *Proceedings National Academy of Science*, **91**, 11841–11843.

Ridley HN (1930). *The dispersal of plants throughout the world*. L Reeve and Co., Kent.

Rignot E and Kanagaratnam P (2006). Changes in the velocity structure of the Greenland ice sheet. *Science*, **311**, 986–990.

Rihan JR (1986). Origin, status and ecology of the hybrid marram grass in Britain. PhD thesis, University of Southampton, Southampton.

Rillig MC and Allen MF (1999). What is the role of arbuscular mycorrhizal fungi in plant-to-ecosystem responses to elevated atmospheric CO_2? *Mycorrhiza*, **9**, 1–8.

Ripley BS and Pammenter NW (2004). Physiological characteristics of coastal dune pioneer species from the Eastern Cape, South Africa, in relation to stress and disturbance. In ML Martínez and NP Psuty, eds, *Coastal dunes: ecology and conservation*, pp. 137–154. Springer-Verlag, Berlin.

Robinson AF (1994). Movement of five nematode species through sand subjected to natural temperature gradient fluctuations. *Journal of Nematology*, **26**, 46–58.

Rodman JE (1974). Systematics and evolution of the genus *Cakile* (Crucifereae). *Contributions of the Gray Herbarium, Harvard University*, **205**, 3–146.

Rodríguez-Echeverría S and Freitas H (2006). Diversity of AMF associated with *Ammophila arenaria* ssp. *arundinacea* in Portuguese sand dunes. *Mycorrhiza*, **16**, 543–552.

Rodríguez-Echeverría S, Crisóstomo J and Freitas H (2004). Arbuscular mycorrhizal fungi associated with *Ammophila arenaria* L. in European coastal sand dunes. In M Arianoutsou and VP Papanastasis, eds, *Proceedings 10th MEDECOS Conference*, Rhodes Island, Greece. Millpress, Rotterdam.

Roman CT and Nordstrom KF (1988). The effect of erosion rate on vegetation patterns of an east coast barrier island. *Estuarine, Coastal, and Shelf Science*, **26**, 233–242.

Ross CA and Ross JRP (1987). Timing and depositional history of eustatic sequences: constraints on seismic stratigraphy. *Cushman Foundation for Foraminiferal Research Special Publication*, **24**, 137–149.

Ross CA and Ross JRP (1988). Late Paleozoic transgressive-regressive deposition. In CK Wilgus, BJ Hastings, H Posamentier *et al.*, eds, *Sea-level change: an integrated approach*, **42**, 71–108. Society of Economic Paleontologists and Mineralogists, Tulsa.

Rowe JS and Abouguendia ZM (1982). The Lake Athabasca sand dunes of Saskatchewan: a unique area. *Musk-Ox*, **30**, 1–22.

Rowland J and Maun MA (2001). Restoration ecology of an endangered plant species: establishment of new populations of *Cirsium pitcheri*. *Restoration Ecology*, **9**, 60–70.

Rozema J, van Manen Y, Vugts HF and Leusink A (1983). Airborne and soilborne salinity and the distribution of coastal and inland species of the genus *Elytrigia*. *Acta Botanica Neerlandica*, **32**, 447–456.

Salinas LM, DeLaune RD and Patrick WH (1986). Changes occurring along a rapidly submerging coastal area: Louisiana, USA. *Journal of Coastal Research*, **2**, 269–284.

Salisbury E (1952). *Downs and dunes: their plant life and its environment*. G Bell and Sons, London.

Salisbury E (1974). Seed size and mass in relation to environment. *Proceedings of the Royal Society of London Series B*, **186**, 83–88.

Salisbury EJ (1934). On the day temperatures of sand dunes in relation to the vegetation at Blakeney Point, Norfolk. *Transactions of Norfolk and Norwich Naturalist Society*, **13**, 333–355.

Salisbury EJ (1942). *The reproductive capacity of plants: studies in quantitative biology*. G Bell and Sons, London.

Salisbury RA (1805). XI. An account of a storm of salt, which fell in January, 1803. *Linnean Society of London Transactions*, **8**, 286–290.

Saye SE and Pye K (2007). Implications of sea level rise for coastal dune habitat conservation in Wales, UK. *Journal of Coastal Conservation*, **11**, 31–52. ISSN 1400-0350 (print), 1874–7841 (online). doi:10.1007/S11852-007-0004-5.

Schincariol RA, Maun MA, Steinbachs JN, Wiklund JA and Crowe AC (2004). Response of an aquatic ecosystem to human activity: hydro-ecology of a river channel in a dune watershed. *Journal of Freshwater Ecology*, **19**, 123–139.

Schlesinger ME (1993). Model projections of CO_2-induced equilibrium climate change. In RA Warrick, EM Barrow and TML Wigley, eds, *Climate and sea level change: observations, projections and implications*, pp. 169–191. Cambridge University Press, Cambridge.

Schnack EJ (1993). The vulnerability of the east coast of South America to sea level rise and possible adjustment strategies. In RA Warrick, EM Barrow and TML Wigley, eds, *Climate and sea level change: observations, projections and implications*, pp. 336–348. Cambridge University Press, Cambridge.

Schwartz ML (1971). The multiple causality of barrier islands. *Journal of Geology*, **79**, 91–94.

Scott GAM (1963). *Mertensia maritima* (L.) S. F. Gray. *Journal of Ecology*, **51**, 733–742.

Scott GAM and Randall RE (1976). Biological flora of the British Isles: *Crambe maritima* L. *Journal of Ecology*, **64**, 1077–1091.

Seliskar DM (1994). The effect of accelerated sand accretion on growth, carbohydrate reserves, and ethylene production in *Ammophila breviligulata* (Poaceae). *American Journal of Botany*, **81**, 536–541.

Seneca ED (1969). Germination response to temperature and salinity of four dune grasses from the outer banks of North Carolina. *Ecology*, **50**, 45–53.

Seneca ED (1972). Seedling response to salinity in four dune grasses from the outer banks of North Carolina. *Ecology*, **53**, 465–471.

Seneca ED and Cooper AW (1971). Germination and seedling response to temperature, daylength, and salinity by *Ammophila breviligulata* from Michigan and North Carolina. *Botanical Gazette*, **132**, 203–215.

Seneca ED, Woodhouse WW Jr and Broome SW (1976). Dune stabilization with *Panicum amarum* along the North Carolina coast. *Miscellaneous report. Coastal Engineering Research Center*, **76–3**, 1–42.

Serreze MC, Holland MM and Stroeve J (2007). Perspectives on the arctic's shrinking sea-ice cover. *Science*, **315**, 1533–1536.

Shaw J (1989). Drumlins, subglacial meltwater floods and ocean responses. *Geology*, **17**, 853–856.

Shaw J and Forbes DL (1990). Relative sea level change and coastal response, northeast Newfoundland. *Journal of Coastal Research*, **6**, 641–660.

Shaw J, Taylor RB, Forbes DL, Ruz M-H and Solomon S (1998). Sensitivity of the coasts of Canada to sea-level rise. *Geological Survey of Canada Bulletin*, **505**. Ottawa, ON.

Shelford VE (1977). *Animal communities in temperate America as illustrated in the Chicago region*. Arno Press, New York.

Shennan I (2007). Sea level studies. In SA Elias, ed., *Encyclopedia of quaternary science*, pp. 2967–2974. Elsevier, London.

Shepard FP (1973). *Submarine geology*, 3rd edn. Harper and Row, New York.

Shepherd A and Wingham D (2007). Recent sea-level contributions of the Antarctic and Greenland ice sheets. *Science*, **315**, 1529–1532.

Shi L, Zhang ZJ, Zhang CY and Zhang JZ (2004). Effects of sand burial on survival, growth, gas exchange and biomass allocation of *Ulmus pumila* seedlings in the Hunshandak Sandland, China. *Annals of Botany*, **94**, 553–560.

Siever R (1988). *Sand*. WH Freeman, New York.

Sigüenza C, Espejel I and Allen EB (1996). Seasonality of mycorrhizae in coastal sand dunes of Baja California. *Mycorrhiza*, **6**, 151–157.

Simon L, Bousquet J, Lévesque RC and Lalonde M (1993). Origin and diversification of endomycorrhizal fungi and coincidence with vascular land plants. *Nature*, **363**, 67–69.

Sjögren E (1993). Dry coastal ecosystems of Madeira and the Azores. In E van der Maarel, ed., *Ecosystems of the world 2B. Dry coastal ecosystems: Africa, America, Asia and Oceania*, pp. 37–49. Elsevier, Amsterdam.

Smith SE and Read DJ (1997). *Mycorrhizal symbiosis*, 2nd edn. Academic Press, New York.

Smith TM and Reynolds RW (2005). A global merged land and sea surface temperature reconstruction based on historical observation (1880–1997). *Journal of Climatology*, **18**, 2021–2036.

Sparling JH (1965). The sand dunes of the Grand Bend region of Lake Huron. *The Ontario Naturalist*, **March**, 16–23.

Sprent JI (1993). The role of nitrogen fixation in primary succession on land. In J Miles and DWH Walton, eds, *Primary succession on land*, pp. 209–219. Blackwell, Oxford.

Sridhar KR and Beena KR (2001). Arbuscular mycorrhizal research in coastal sand dunes: a review. *Proceedings of the National Academy of Sciences, India*, **71**, 179–205.

Staddon PL and Fitter AH (1998). Does elevated atmospheric carbon dioxide affect arbuscular mycorrhizas? *Trends in Ecology and Evolution*, **13**, 455–458.

Stairs AF (1986). Life history variation in *Artemisia campestris* on a Lake Huron sand dune system. PhD thesis, University of Western Ontario, London, ON.

Stalter R (1993). Dry coastal ecosystems of the Gulf coast of the United States of America. In E van der Maarel, ed., *Ecosystems of the world 2B. Dry coastal ecosystems: Africa, America, Asia and Oceania*, pp. 375–387. Elsevier, Amsterdam.

Stanley SM (1985). Rates of evolution. *Paleobiology*, **11**, 13–26.

Stephenson AG (1992). The regulation of maternal investment in plants. In C Marshall and J Grace, eds, *Fruit and seed production: aspects of development, environmental physiology and ecology*, pp. 151–171. Cambridge University Press, Cambridge.

Stürmer SL and Bellei MM (1994). Composition and seasonal variation of spore populations of arbuscular mycorrhizal fungi in dune soils on the island of Santa Catarina, Brazil. *Canadian Journal of Botany*, **72**, 359–363.

Sun D (2000). Some aspects of nitrogen and phosphorus status in a lacustrine sand dune ecosystem. MSc thesis, University of Western Ontario, London, ON.

Sutton JC and Sheppard BR (1976). Aggregation of sand-dune soil by endomycorrhizal fungi. *Canadian Journal of Botany*, **54**, 326–333.

Sutton OG (1953). *Micrometeorology: a study of physical processes in the lowest layers of the earth's atmosphere*. McGraw-Hill, New York.

Swan B (1979). Sand dunes in the humid tropics: Sri Lanka. *Zeitschrift für Geomorphologie*, **23**, 152–171.

Swift DJP (1976). Coastal sedimentation. In DJ Stanley and DJP Swift, eds, *Marine sediment transport and environmental management*, pp. 255–310. Wiley, New York.

Sykes MT and Wilson JB (1988). An experimental investigation into the response of some New Zealand sand dune species to salt spray. *Annals of Botany*, **62**, 159–166.

Sykes MT and Wilson JB (1989). The effect of salinity on the growth of some New Zealand sand dune species. *Acta Botanica Neerlandica*, **38**, 173–182.

Sykes MT and Wilson JB (1990a). An experimental investigation into the response of New Zealand sand dune species to different depths of burial by sand. *Acta Botanica Neerlandica*, **39**, 171–181.

Sykes MT and Wilson JB (1990b). Dark tolerance in plants of dunes. *Functional Ecology*, **4**, 799–805.

Sylvia DM (1986). Spatial and temporal distribution of vesicular–arbuscular mycorrhizal fungi associated with *Uniola paniculata* in Florida foredunes. *Mycologia*, **78**, 728–734.

Szafer W (1966). *The vegetation of Poland*. Pergamon Press, Oxford.

Tansley AG (1953). *The British islands and their vegetation*. Cambridge University Press, Cambridge.

Taylor HC and Boucher C (1993). Dry coastal ecosystems of the South African south coast. In

E van der Maarel, ed., *Ecosystems of the world 2B. Dry coastal ecosystems: Africa, America, Asia and Oceania*, pp. 89–107. Elsevier, Amsterdam.

Thannheiser D (1984). The coastal vegetation of Eastern Canada. *Occasional Papers in Biology*, **No 8**. Memorial University of Newfoundland, NFL.

Thompson CH (1981). Podzol chronosequences on coastal dunes of eastern Australia. *Nature*, **291**, 59–61.

Thompson K and Grime JP (1979). Seasonal variation in the seed banks of herbaceous species in ten contrasting habitats. *Journal of Ecology*, **67**, 893–921.

Thompson K, Bakker JP and Bekker RM (1997). *The soil seed banks of north west Europe: methodology, density and longevity.* Cambridge University Press, Cambridge.

Thompson K, Band SR and Hodgson JG (1993). Seed size and shape predict persistence in soil. *Functional Ecology*, **7**, 236–241.

Thompson K, Bakker JP, Bekker RM and Hodgson JG (1998). Ecological correlates of seed persistence in soil in the north-west European flora. *Journal of Ecology*, **86**, 163–169.

Thompson PA (1970). Characterization of the germination response to temperature of species and ecotypes. *Nature*, **225**, 827–831.

Thorarinsson S (1966). *Surtsey: the new island in the North Atlantic.* Viking Press, New York.

Tilman D (1985). The resource–ratio hypothesis of plant succession. *American Naturalist*, **125**, 827–852.

Tilman D (1986). Nitrogen-limited growth in plants from different successional stages. *Ecology*, **67**, 555–563.

Tilman D (1988). *Plant strategies and the dynamics and structure of plant communities.* Princeton University Press, Princeton.

Tilman D (1993). Species richness of experimental productivity gradients: how important is colonization limitation? *Ecology*, **74**, 2179–2191.

Titus JG (1989). Sea level rise. In US EPA, Office of Policy, Planning and Evaluation, Report to Congress. *The potential effects of global climate change on the United States*, pp. 118–143. US Environmental Protection Agency, Washington.

Titus JG (1993). Regional effects of sea level rise. In RA Warrick, EM Barrow and TML Wigley, eds, *Climate and sea level change: observations, projections and implications*, pp. 395–400. Cambridge University Press, Cambridge.

Trembanis AC, Pilkey OH and Valverde HR (1999). Comparison of beach nourishment along the US Atlantic, Great Lakes, Gulf of Mexico, and New England shorelines. *Coastal Management*, **27**, 329–340.

Trenhaile AS (2003). Modeling shore platforms: present status and future developments. In VC Lakhan, ed., *Advances in coastal modeling*, pp. 393–409. Elsevier, Amsterdam.

Tsang A and Maun MA (1999). Mycorrhizal fungi increase salt tolerance of *Strophostyles helvola* in coastal foredunes. *Plant Ecology*, **144**, 159–166.

Ungar IA (1978). Halophyte seed germination. *Botanical Review*, **44**, 233–264.

US Environmental Protection Agency (2007). In US Sea Level Trends, *Monthly and annual mean sea level station. Permanent service for mean sea level (PSMSL).* Proudman Oceanographic Laboratory, Liverpool. http://www.epa.gov/climatechange/science/recentslc.html.

Van Asdall W and Olmsted CE (1963). *Corispermum hyssopifolium* on the Lake Michigan dunes its community and physiological ecology. *Botanical Gazette*, **124**, 155–172.

Van Breemen AMM (1984). Comparative germination ecology of three short-lived monocarpic Boraginaceae. *Acta Botanica Neerlandica*, **33**, 283–305.

Van Breemen AMM and van Leeuwen BH (1983). The seed bank of three short-lived monocarpic species, *Cirsium vulgare* (Compositae), *Echium vulgare* and *Cynoglossum officinale* (Boraginaceae). *Acta Botanica Neerlandica*, **32**, 245–246.

Van Leeuwen S, Dodd N, Calvete D and Falqués A (2007). Linear evolution of shoreface nourishment. *Coastal Engineering*, **54**, 417–431.

VandenBygaart AJ (1993). Soil genesis and gamma radioactivity on a dune soil chronosequence, Pinery Provincial Park. MSc thesis, University of Guelph, Guelph, ON.

VandenBygaart AJ and Protz R (1995). Soil genesis on a chronosequence, Pinery Provincial Park, Ontario. *Canadian Journal of Soil Science*, **75**, 63–72.

Van der Heijden MGA, Klironomos JN, Ursic M *et al.* (1998). Mycorrhizal fungal diversity determines plant biodiversity, ecosystem variability and productivity. *Nature*, **396**, 69–72.

Van der Maarel E (ed.) (1993). *Ecosystems of the world Vol. 2A, 2B, 2C. Dry coastal ecosystems: Africa, America, Asia and Oceania.* Elsevier, Amsterdam.

Van der Meulen F, Witter JV and Arens SM (1991). The use of a GIS in assessing the impacts of sea level rise on nature conservation along the Dutch coast: 1990–2090. A landscape ecological study of the foredunes with help of a geographic information system. *Landscape Ecology*, **6**, 105–113.

Van der Pijl L (1982). *Principles of dispersal in higher plants*, 3rd edn. Springer-Verlag, Berlin.

Van der Putten WH (1990). Establishment of *Ammophila arenaria* (marram grass) from culms, seeds and rhizomes. *Journal of Applied Ecology*, **27**, 188–199.

Van der Putten WH, Van Dijk C and Troelstra SR (1988). Biotic soil factors affecting the growth and development of *Ammophila arenaria*. *Oecologia*, **76**, 313–320.

Van der Valk AG (1974). Environmental factors controlling the distribution of forbs on coastal foredunes in Cape Hatteras National Seashore. *Canadian Journal of Botany*, **52**, 1057–1073.

Van der Valk AG (1977). The macroclimate and microclimate of coastal foredune grasslands in Cape Hatteras National Seashore. *International Journal of Biometeorology*, **21**, 227–237.

Van der Valk AG and Bliss LC (1971). Hydrarch succession and net primary production of oxbow Lakes in central Alberta. *Canadian Journal of Botany*, **49**, 1177–1199.

Vasseur B and Hequette A (2000). Storm surges and erosion of coastal dunes between 1957 and 1988 near Dunkerque (France), southwestern North Sea. In K Pye and JRL Allen, eds, *Coastal and estuarine environments: sedimentology, geomorphology and geoarchaeology. Special Publications*, **175**, 99–107. Geological Society, London.

Vázquez G (2004). The role of algal mats on community succession in dunes and dune slacks. In ML Martínez and NP Psuty, eds, *Coastal dunes: ecology and conservation*, pp. 189–203. Springer-Verlag, Berlin.

Vestergaard P (1991). Morphology and vegetation of a dune system in SE Denmark in relation to climate change and sea-level rise. *Landscape Ecology*, **6**, 77–87.

Vestergaard P (1997). Possible impact of sea-level rise on some habitat types at the Baltic coast of Denmark. *Journal of Coastal Conservation*, **3**, 103–112.

Vestergaard P (2006). Temporal development of vegetation and geomorphology in a man-made beach-dune system by natural processes. *Nordic Journal of Botany*, **24**, 309–326.

Vestergaard P (2008). Danish attitudes and reactions to the threat of sea-level rise. *Journal of Coastal Research*, **24**, 394–402.

Voesenek LACJ and Blom CWPM (1989). Growth responses of *Rumex* species in relation to submergence and ethylene. *Plant, Cell and Environment*, **12**, 433–439.

Voesenek LACJ, van der Putten WH, Maun MA and Blom CWPM (1998). The role of ethylene and darkness in accelerated shoot elongation of *Ammophila breviligulata* upon sand burial. *Oecologia*, **115**, 359–365.

Wagner RH (1964). The ecology of *Uniola paniculata* L. in the dune-strand habitat of North Carolina. *Ecological Monographs*, **34**, 79–96.

Wahab AMA (1975). Nitrogen fixation by *Bacillus* strains isolated from the rhizosphere of *Ammophila arenaria*. *Plant and Soil*, **42**, 703–708.

Walker D (1970). Direction and rate in some British post-glacial hydroseres. In D Walker and RG West, eds, *Studies in the vegetational history of the British Isles*, pp. 117–139. Cambridge University Press, Cambridge.

Walker IJ and Barrie JV (2003). Geomorphology and sea-level rise on one of Canada's most 'sensitive' coasts: northeast Graham Island, British Columbia. *Journal of Coastal Research*, **SI 39**. ICS 2004 Proceedings, Brazil. ISSN 0749–0208.

Walker LR and del Moral R (2003). *Primary succession and ecosystem rehabilitation.* Cambridge University Press, Cambridge.

Walker LR, Zasada JC and Chapin FS III (1986). The role of life history processes in primary succession on an Alaskan floodplain. *Ecology*, **67**, 1243–1253.

Wallace HR (1960). Movement of eelworms. VI. The influences of soil type, moisture gradients and host plant roots on the migration of the potato-root eelworm *Heterodera rostochiensis* Wollenweber. *Annals of Applied Biology*, **48**, 107–120.

Wallén B (1980). Changes in structure and function of *Ammophila* during primary succession. *Oikos*, **34**, 227–238.

Walsh JE, Anisimov O, Hagen JOM *et al.* (2005). Cryosphere and hydrology. In C Symons, L Arris and B Heal, eds, *Arctic climate impact assessment*, pp. 183–242. Cambridge University Press, Cambridge.

Wang GM, Stribley DP, Tinker PB and Walker C (1985). Soil pH and vesicular–arbuscular mycorrhizas. In AH Fitter, D Atkinson, DJ Read and MB Usher, eds, *Ecological interactions in soil: plants, microbes and animals*, pp. 219–224. Blackwell, Oxford.

Wardle DA, Bardgett RD, Klironomos JN, Setälä H, Van der Putten WH and Wall DH (2004). Ecological linkages between above ground and below ground biota. *Science*, **304**, 1629–1633.

Warrick RA (1993). Climate and sea level change: a synthesis. In RA Warrick, EM Barrow and TML Wigley, eds, *Climate and sea level change: observations, projections and implications*, pp. 3–21. Cambridge University Press, Cambridge.

Waterman WG (1919). Development of root systems under dune conditions. *Botanical Gazette*, **68**, 22–53.

Watkinson AR (1978). The demography of a sand dune annual: *Vulpia fasciculata*. III. The dispersal of seeds, *Journal of Ecology*, **66**, 483–498.

Watkinson AR and Davy AJ (1985). Population biology of salt marsh and sand dune annuals. *Vegetatio*, **62**, 487–497.

Watkinson AR and Harper JL (1978). The demography of a sand dune annual: *Vulpia fasciculata*. I. The natural regulation of populations. *Journal of Ecology*, **66**, 15–33.

Watkinson AR, Huiskes AHL and Noble JC (1979). The demography of sand dune species with contrasting life cycles. In RL Jefferies and AJ Davy, eds, *Ecological processes in coastal environments*, pp. 95–112. Blackwell, London.

Weaver JE (1968). *Prairie plants and their environment: a fifty year study in the Midwest*. University of Nebraska Press, Nebraska.

Weisser PJ and Cooper KH. (1993). Dry coastal ecosystems of the South African east coast. In E van der Maarel, ed., *Ecosystems of the world 2B. Dry coastal ecosystems: Africa, America, Asia and Oceania*, pp. 109–128. Elsevier, Amsterdam.

Weller SG (1985). The life history of *Lithospermum caroliniense*, a long-lived herbaceous sand dune species. *Ecological Monographs*, **55**, 49–67.

Weller SG (1989). The effect of disturbance scale on sand dune colonization by *Lithospermum caroliniense*. *Ecology*, **70**, 1244–1251.

Wells BW and Shunk IV (1938). Salt spray: an important factor in coastal ecology. *Bulletin of the Torrey Botanical Club*, **65**, 485–492.

Westelaken IL and Maun MA (1985a). Reproductive capacity, germination and survivorship of *Lithospermum caroliniense* on Lake Huron sand dunes. *Oecologia* (Berlin), **66**, 238–245.

Westelaken IL and Maun MA (1985b). Spatial pattern and seed dispersal of *Lithospermum caroliniense* on Lake Huron sand dunes. *Canadian Journal of Botany*, **63**, 125–132.

Westgate JM (1904). *Reclamation of Cape Cod sand dunes*. Bureau of Plant Industry, United States Department of Agriculture, Bulletin 65. Beltsville, MD.

Westhoff V and van der Maarel E (1978). The Braun-Blanquet approach. In RH Whittaker, ed., *Classification of plant communities*, 2nd edn, pp. 287–399. Junk, The Hague.

Whittingham J and Read DJ (1982). Vesicular–arbuscular mycorrhiza in natural vegetation systems. III. Nutrient transfer between plants with mycorrhizal interconnections. *New Phytologist*, **90**, 277–284.

Wiedemann AM (1990). The coastal parabola dune system at Sand Lake, Tillamook County, Oregon, USA. In R Davidson-Arnott, ed., *Proceedings of the symposium on coastal sand dunes*, pp. 171–194. Institute for Mechanical Engineering National Research Council of Canada, Ottawa, ON.

Wiedemann AM (1993). Dry coastal ecosystems of northwestern North America. In E van der Maarel, ed., *Ecosystems of the world 2B. Dry coastal ecosystems: Africa, America, Asia and Oceania*, pp. 341–358. Elsevier, Amsterdam.

Willis AJ (1963). Braunton Burrows: the effects on the vegetation of the addition of mineral nutrients to the dune soils. *Journal of Ecology*, **51**, 353–374.

Willis AJ (1965). The influence of mineral nutrients on the growth of *Ammophila arenaria*. *Journal of Ecology*, **53**, 735–745.

Willis AJ and Yemm EW (1961). Braunton Burrows: mineral nutrient status of the dune soils. *Journal of Ecology*, **49**, 377–390.

Willis AJ, Folkes BF, Hope-Simpson JF and Yemm EW (1959a). Braunton Burrows: the dune system and its vegetation. Part I. *Journal of Ecology*, **47**, 1–24.

Willis AJ, Folkes BF, Hope-Simpson JF and Yemm EW (1959b). Braunton Burrows: the dune system and its vegetation. Part II. *Journal of Ecology*, **47**, 249–288.

Wilson AT (1959). Surface of the ocean as a source of air-borne nitrogenous material and other plant nutrients. *Nature*, **184**, 99–101.

Wilson JB and Sykes MT (1999). Is zonation on coastal sand dunes determined primarily by sand burial or by salt spray? A test in New Zealand dunes. *Ecology Letters*, **2**, 233–236.

Wilson JM and Tommerup IC (1992). Interactions between fungal symbionts: VA mycorrhizae. In MF Allen, ed., *Mycorrhizal functioning: an integrative plant-fungal process*, pp. 199–248. Chapman and Hall, London.

Wind HG and Peerbolte EB (1993). Sea level rise: assessing the problems. In RA Warrick, EM Barrow and TML Wigley, eds, *Climate and sea level change: observations, projections and implications*, pp. 297–309. Cambridge University Press, Cambridge.

Wolfe BE, Husband BC and Klironomos JN (2005). Effects of a below ground mutualism on an above ground mutualism. *Ecology Letters*, **8**, 218–223.

Wolfe F (1932). Annual rings of *Thuja occidentalis* in relation to climatic conditions and movement of sand. *Botanical Gazette*, **93**, 328–335.

Woodcock AH, Kientzler CF, Arons AB and Blanchard DC (1953). Giant condensation nuclei from bursting bubbles. *Nature*, **172**, 1144–1145.

Woodell SRJ (1985). Salinity and seed germination patterns in coastal plants. *Vegetatio*, **61**, 223–229.

Woodhouse WW Jr (1982). Coastal sand dunes of the US. In RR Lewis III, ed., *Creation and restoration of coastal plant communities*, pp. 1–44. CRC Press, Boca Raton.

Woodhouse WW Jr, Seneca ED and Broome SW (1977). Effect of species on dune grass growth. *International Journal of Biometeorology*, **21**, 256–266.

Woodroffe CD (2002). *Coasts: form, process and evolution*. Cambridge University Press, Cambridge.

Woodruff DS (2001). Declines of biomes and biotas and the future of evolution. *Proceedings of the National Academy of Sciences*, **98**, 5471–5476.

Woodworth PL (1993). Sea level changes. In RA Warrick, EM Barrow and TML Wigley, eds, *Climate and sea level change: observations, projections and implications*, pp. 379–391. Cambridge University Press, Cambridge.

Wright SF and Upadhyaya A (1998). A survey of soils for aggregate stability and glomalin, a glycoprotein produced by hyphae of arbuscular mycorrhizal fungi. *Plant and Soil*, **198**, 97–107.

Yanful M (1988). Intrapopulation variation and the effects of sand burial on emergence, growth and establishment of seedlings of *Strophostyles helvola* (L.) Ell. MSc thesis, University of Western Ontario, London, ON.

Yanful M and Maun MA (1996a). Spatial distribution and seed mass variation of *Strophostyles helvola* along Lake Erie. *Canadian Journal of Botany*, **74**, 1313–1321.

Yanful M and Maun MA (1996b). Effects of burial of seeds and seedlings from different seed sizes on the emergence and growth of *Strophostyles helvola*. *Canadian Journal of Botany*, **74**, 1322–1330.

Yarranton GA and Morrison RG (1974). Spatial dynamics of a primary succession: nucleation. *Journal of Ecology*, **62**, 417–428.

Yim WWS (1993). Future sea level rise in Hong Kong and possible environmental effects. In RA Warrick, EM Barrow and TML Wigley, eds, *Climate and sea level change: observations, projections and implications*, pp. 349–376. Cambridge University Press, Cambridge.

Yu F, Dong M and Krüsi B (2004). Clonal integration helps *Psammochloa villosa* survive sand burial in an inland dune. *New Phytologist*, **162**, 697–704.

Yuan T, Maun MA and Hopkins WG (1993). Effects of sand accretion on photosynthesis, leaf-water potential and morphology of two dune grasses. *Functional Ecology*, **7**, 676–682.

Yun KW and Maun MA (1997). Allelopathic potential of *Artemisia campestris* ssp. *caudata* on Lake Huron sand dunes. *Canadian Journal of Botany*, **75**, 1903–1912.

Zahran MA (1993). Dry coastal ecosystems of the Asian Red Sea coast. In E van der Maarel, ed., *Ecosystems of the world 2B. Dry coastal ecosystems: Africa, America, Asia and Oceania*, pp. 17–30. Elsevier, Amsterdam.

Zaremba RE and Leatherman SP (1984). *Overwash processes and foredune ecology, Nauset Spit, Massachusetts*. Miscellaneous Paper, EL-84–8. US Army Corps of Engineers, Washington, DC.

Zhang J and Maun MA (1989a). Seed dormancy of *Panicum virgatum* L. on the shoreline sand dunes of Lake Erie. *American Midland Naturalist*, **122**, 77–87.

Zhang J and Maun MA (1989b). Effect of partial removal of endosperm on seedling sizes of *Panicum virgatum* and *Agropyron psammophilum*. *Oikos*, **56**, 250–255.

Zhang J and Maun MA (1990a). Effects of sand burial on seed germination, seedling emergence, survival, and growth of *Agropyron psammophilum*. *Canadian Journal of Botany*, **68**, 304–310.

Zhang J and Maun MA (1990b). Seed size variation and its effects on seedling growth in *Agropyron psammophilum*. *Botanical Gazette*, **151**, 106–113.

Zhang J and Maun MA (1990c). Sand burial effects on seed germination, seedling emergence and establishment of *Panicum virgatum*. *Holarctic Ecology*, **13**, 56–61.

Zhang J and Maun MA (1990d). Seed banks in lacustrine dune systems on Lakes Huron and Erie. In R Davidson-Arnott, ed., *Proceedings of the symposium on coastal sand dunes*, pp. 311–320. Institute for Mechanical Engineering, National Research Council of Canada, Ottawa, ON.

Zhang J and Maun MA (1991a). Establishment and growth of *Panicum virgatum* L. seedlings on a Lake Erie sand dune. *Bulletin of the Torrey Botanical Club*, **118**, 141–153.

Zhang J and Maun MA (1991b). Effects of partial removal of seed reserves on some aspects of seedling ecology of seven dune species. *Canadian Journal of Botany*, **69**, 1457–1462.

Zhang J and Maun MA (1992). Effects of burial in sand on the growth and reproduction of *Cakile edentula*. *Ecography*, **15**, 296–302.

Zhang J and Maun MA (1993). Components of seed mass and their relationships to seedling size in *Calamovilfa longifolia*. *Canadian Journal of Botany*, **71**, 551–557.

Zhang J and Maun MA (1994). Potential for seed bank formation in seven Great Lakes sand dune species. *American Journal of Botany*, **81**, 387–394.

Zhang K, Douglas BC and Leatherman SP (2004). Global warming and coastal erosion. *Climatic Change*, **64**, 41–58.

Zheng Y, Xie Z, Yu Y, Jiang L, Shimizu H and Rimmington GM (2005). Effects of burial in sand and water supply regime on seedling emergence of six species. *Annals of Botany*, **95**, 1237–1245.

Zoladeski CA (1991). Vegetation zonation in dune slacks on the Łeba Bar, Polish Baltic Sea coast. *Journal of Vegetation Science*, **2**, 255–258.

Zong Y (2007). Tropics. In SA Elias, ed., *Encyclopedia of quaternary science*, pp. 3087–3095. Elsevier, Amsterdam.

Zuloaga G (1966). The amateur scientist: a study of the salty rain of Venezuela. *Scientific American*, **December**, 136–138.

Index

Note: page numbers in *italics* refer to Figures and Tables, whilst those in **bold** refer to Glossary entries.

Scopimera species 155
Scutellospora species *135*, 138, *139*, 140, 141
  S. calospora, effect of soil moisture 141
sea level rise 214
  effect on coastal dunes 205–6
  effects of human manipulation 211–12
  impact of storm surges 202–3
  natural erosion and building 203–4
  responses of dune vegetation 206–9
  risk of extinction 212–14
  sudden changes 209–11
sea levels, changes over time 197–9, *201*
  predicted increase 199–200, *202*
seasonal changes
  in AM colonization 142
  in salt spray storms 184
  in seedling emergence 72–4
  survival mechanisms 166
seawater
  composition *120–1*
  salinity 120
seawater dispersal 40, 41
  fruits and seeds 41–4
  plant fragments 44–6
secondary dune ridges **216**
sediment movement
  by waves 6–8
  by wind 8–10
  effects of vegetation 10–11
sediment supply 2, 20–1
  cliff and coastal erosion 2–3
  river discharge 3
  tides, hurricanes and tsunamis 3–4
Sedum acre 208
seed bank potential 59–61
seed banks 53, 63, 213
  above-ground 61–3
  classification 53–4
  dynamics *56–8*
  persistent 55–6
  of stabilized dunes 59
  transient 54–5
seed characteristics, relationship to
        persistence in soil 55
seed dispersal 40
  by wind 46
    burial of inflorescences *48*
    micro-environmental factors 47–*8*
    travel distances 46–7
  in water 41–4
seed dormancy 64–5, 167
seed germination 64–5, 68–9
  effects of salinity 125–7, 133
seedling emergence 69–71, 85
  patterns 71–2, 81
    emergence in spring, summer and autumn 74

    entire emergence in autumn 72, 74
    entire emergence in spring 72
seedling establishment 65–6, 74, 85–6
  biennials 76–7
  limiting factors 78
    density-dependent processes 82–3
    desiccation 80–2
    human disturbance 85
    nutrients 79–80
    salt spray and soil salinity 82
    sand erosion or accretion 83–5
    soil temperature 80
  perennials 77–8
  summer annuals 74–6
  winter annuals 76
seedlings
  burial by sand 89–91
  dispersal 45
  effects of salinity 127
  root system growth 81
seed mass 66–7
  evolution 71
  frequency distributions *66*
  relationship to seedling emergence 70–*1*
seed predation 56–7
  relationship to seed size 67
Seneca, ED 125, 127
Seneca, ED and Cooper, AW 131
Seneca, ED *et al.* 13
Senecio californicus *168*
Senecio crassiflorus *172*
Senecio elegans *169*
  Sesuvium portulacastrum 165, *168*, *170*
    dispersal of plant fragments 45, 51
    fruit buoyancy *43*
shadow dunes *12*
shear stress of waves (τ, tau) 7
shear velocity of wind 8
shelducks (*Tadorna tadorna*) 156–7
Shelford, VE 157
  Shepherdia canadensis
    seed banks 59
    seed dispersal *50*
    nitrogen fixation 35
Shi, L *et al.* 92
shingle, definition 4–5
shingle beaches, formation 8, 204
shoot development, as response to
        burial 101–3, *102*
shoreline, changes over time *198–9*
short-term persistent seed banks 53
silt, particle sizes 4
silt fraction, relationship to age *31*, 38
Simon, L. *et al.* 134
singing sand 6
sink capacity 115, 116